92 Springer Series in Solid-State Sciences

Edited by Peter Fulde

Springer Series in Solid-State Sciences

Editors: M. Cardona   P. Fulde   K. von Klitzing   H.-J. Queisser

Managing Editor: H. K. V. Lotsch   Volumes 1–89 are listed at the end of the book

Y. Ishikawa N. Miura (Eds.)

Physics and Engineering Applications of Magnetism

With 223 Figures

With Contributions by
K. Adachi, S. Chikazumi, E. Hirota,
N. Imamura, Y. Ishikawa, Y. Makino,
N. Miura, T. Mizoguchi, Y. Nakamura,
Y. Sugita, K. Tajima, M. Takahashi,
A. Tonomura, and T. Wakiyama

Springer-Verlag Berlin Heidelberg New York
London Paris Tokyo Hong Kong Barcelona Budapest

Professor Yoshikazu Ishikawa †
Professor Noboru Miura

Institute for Solid State Physics, University of Tokyo,
Roppongi, Minato-ku, Tokyo 106, Japan

Series Editors:

Professor Dr., Dres. h. c. Manuel Cardona
Professor Dr., Dr. h. c. Peter Fulde
Professor Dr., Dr. h. c. Klaus von Klitzing
Professor Dr. Hans-Joachim Queisser

Max-Planck-Institut für Festkörperforschung, Heisenbergstrasse 1
D-7000 Stuttgart 80, Fed. Rep. of Germany

Managing Editor:

Dr. Helmut K. V. Lotsch

Springer-Verlag, Tiergartenstrasse 17
D-6900 Heidelberg, Fed. Rep. of Germany

Title of the original Japanese edition: Jisei butsurigaku to sono ōyō

ISBN-13: 978-3-642-84160-6   e-ISBN-13: 978-3-642-84158-3
DOI: 10.1007/978-3-642-84158-3

Library of Congress Cataloging-in-Publication Data.
Jisei butsurigaku to sono ōyō. English.
Physics and engineering applications of magnetism / Y. Ishikawa, N. Miura (eds.). Springer series in solid-state sciences). Translation of: Jisei butsurigaku to sono ōyō.
ISBN-13:978-3-642-84160-6
1. Magnetism. 2. Magnetic materials. 3. Magnetism-Industrial applications. I. Ishikawa, Yoshikazu, 1930. II. Miura, N. (Noboru), 1941. III. Title.
QC753.2.J5713 1991   538-dc20   90-9689

© Springer-Verlag Berlin Heidelberg 1991
Softcover reprint of the hardcover 1st edition 1991

Typesetting: Macmillan India Ltd., Bangalore, India

54/3020-543210 – Printed on acid-free paper

Preface

This book was originally published in Japanese in honour of Professor S. Chikazumi on the occasion of his retirement from the University of Tokyo in March 1982. Physicists who had been supervised by him or had closely collaborated with him wrote articles on recent developments in magnetism and its engineering applications. In the preface of his excellent textbook *Physics of Magnetism* (Wiley, 1964), Professor Chikazumi says that recent research in magnetism deals with fundamental physical problems and, at the same time, with more secondary magnetic phenomena, as well as with engineering applications of magnetic materials to electromagnetic machines, permanent magnets and electronic computers, and that the purpose of his textbook is to give a general view of these magnetic phenomena, focusing its main interest at the center of such a broad field. Always keeping such a viewpoint in mind, Professor Chikazumi has contributed a great deal to both fundamental physics and applications of magnetism. This is described in Chap. 1 of this book. Many books have been published on both the physics and applications of magnetism. However, no single book has a viewpoint covering both of them. The recent development of high technology needs such a broad viewpoint for scientists and engineers since it is a product of both fundamental science and technology.

Research in magnetism is based on the response which materials show to the application of magnetic fields. Thus, a detailed study of the magnetization process is the basis of both fundamental research and applications. This book focuses on the magnetization process and related problems. Chapter 2 deals directly with this problem, and readers will learn how useful information is obtained by increasing the applied magnetic field. Neutron scattering, described in Chap. 3, is known to be a most powerful means for the microscopic investigation of magnetism. The technique is also used to measure the response of materials to the microscopic field produced by neutrons. Chapters 4 and 5 deal with the magnetization process of typical magnetic materials, compound magnetic substances and Invar alloys. These studies have given important information concerning magnetism at finite temperatures, which is one of the major problems in magnetism at present. Chapters 6–8 discuss the problems that lie between fundamental physics and applied physics. Chapter 6 deals with recent progress in magnetic anisotropy and magnetostriction which are closely related to the magnetization process and play an important role in applications. Chapter 8 is devoted to fundamental problems of new amorphous magnetic materials, whose applications are discussed in Chaps. 9 and 10. In Chap. 7 readers will find the philosophy of an author who has long been investigating both the fundamentals and applications of magnetism with a similar viewpoint to that of Professor Chikazumi.

Chapters 9–13 deal with recent progress in the applications of magnetism. Magnetic recording, which is one of the most active research fields in Japan, is discussed in Chaps. 10–12 by leading scientists in the industrial world. Chapter 13 deals with the observation of magnetic domains governing the magnetization process, by a completely new technique developed by the author.

In editing the English edition, we have added new results since the original publication to each chapter to keep the book as up to date as possible. The editors will be very happy if this book can provide new ideas and stimulation to all those interested in magnetism, both in fundamental physics and in engineering applications. It should also play a useful role as the most recent review in this field. For the convenience of readers, the appendix explains various technical terms found in the text. We apologize for some unavoidable inconsistencies in the use of units.

We would like to thank Mr. K. Endoh and Mr. S. Makiya of the Shokabo Publishing Company who published the original book in Japanese. We are also grateful to Dr. H. Lotsch who made this publication possible. Thanks are also due to Professor K. Tajima of the Keio University who kindly translated Professor Ishikawa's original Japanese manuscript for this English edition. He also added to the chapter some topics concerning new developments in the field.

Finally, we have to mention the very sorrowful news that, during the process of editing, one of the editors, Professor Y. Ishikawa, suddenly passed away on February 28, 1986, at the age of 55. It is a great loss to the world of physics, and we would like to bless his memory and dedicate this book to him.

Tokyo, Japan   *Noboru Miura*
December 1990

Contents

Contributors

Adachi, Kengo
  Professor, Department of Physics, Nagoya University
Chikazumi, Sōshin
  Honorary Professor, University of Tokyo, and Professor,
  Department of Physics, Edogawa University, Yokohama
Hirota, Eiichi
  Matsushita Research Institute Tokyo, Inc.
Imamura, Nobutake
  Advanced Materials Research Laboratory, Toyo Soda
  Manufacturing Co., Ltd.
Ishikawa, Yoshikazu
  Deceased. Formerly Professor, Department of Physics,
  Tohoku University, Sendai
Makino, Yoshimi
  Research Center, Sony Corporation, Yokohama
Miura, Nobura
  Professor, Institute for Solid State Physics, University of Tokyo
Mizoguchi, Tadashi
  Professor, Department of Physics, Gakushuin University, Tokyo
Nakamura, Yoji
  Honorary Professor, Kyoto University, and Nisshin Steel Co., Ltd.
Sugita, Yutaka
  Hitachi Research Laboratory, Hitachi Ltd., Ibaraki
Tajima, Keisuke
  Associate Professor, Department of Physics, Keio University, Yokohama
Takahashi, Minoru
  Honorary Professor, Tohoku University, Sendai, and Guest Professor,
  Department of Electronic Engineering, Tohoku, Institute of Technology
Tonomura, Akira
  Advanced Research Laboratory, Hitachi Ltd., Tokyo
Wakiyama, Tokuo
  Professor, Department of Electronic Engineering,
  Tohoku University, Sendai

1. Progress in the Physics of Magnetism in the Past Forty-five Years

Sōshin Chikazumi

In October of 1944, that is about 45 years ago, I started the study of magnetism as a student of Professor S. Kaya. Professor Kaya had moved from Hokkaido Imperial University to Tokyo Imperial University one year before, that is in 1943, and was setting up his new laboratory there.

The history of magnetic research in Japan was initiated by Sir Alfred Ewing who was invited to Tokyo University from England in 1878 as a Visiting Professor when he was a 23-year-old young scientist investigating earthquakes and magnetism. After this Professor H. Nagaoka began the investigation of magnetostriction and this tradition was carried on by a young scholar K. Honda who was later appointed to a professorship at the newly established Tohoku Imperial University. There he investigated the periodicity of the magnetic susceptibility of the elements, magnetocrystalline anisotropy, etc., and also invented the KS permanent magnet, Sendust, and so on. He also had many disciples, including the eminent Professor Kaya, who measured the magneto-crystalline anisotropy of iron, nickel and cobalt with Professor Honda. After moving to Hokkaido Imperial University, he investigated the magnetic properties of superlattice alloys such as Ni_3Fe or Ni_3Mn. Then he moved to Tokyo as mentioned above.

The magnetic materials which were investigated at that time were limited to metals and alloys. Kaya devoted himself to the investigation of the magnetic properties of Permalloy, that is 21.5% Fe–Ni alloy. This alloy was known to form a so-called superlattice, which influenced the magnetic properties for some unknown reason. This problem was known as the "Permalloy Problem", and was the subject of my undergraduate thesis as well as that of my graduate degree course for five years thereafter. The essential point of this problem was that when the alloy was cooled from high temperatures in a magnetic field, a magnetic anisotropy was induced along the axis of magnetization, or along the direction of spontaneous magnetization. After performing various experiments. I had the idea of interpreting this phenomena in terms of the "directional order" which means the directional arrangement of Fe–Fe pairs in the Fe–Ni solid solution [1.1]. I assumed that such directional order must be associated with the lattice strain which gives rise to the induced anisotropy combined with magnetostriction. This idea was modified later by Néel [1.2] and Taniguchi et al. [1.3] by ascribing different pseudo-dipolar interactions for different kinds of atomic pairs (Sect. 6.5.1). They explained the mechanism of the induced anisotropy beautifully.

Thereafter, in 1950, I became assistant professor at the newly established Gakushuin University, where I continued to pursue this problem and also to investigate the mechanism of "roll magnetic anisotropy" which means the anisotropy induced by rolling ferromagnetic alloys. Néel [1.2] and Taniguchi et al. [1.3] also interpreted this phenomenon in terms of the directional order, which was, they thought, formed by the mixing of two kinds of atoms during rolling, just like the process of thermal agitation of atoms during annealing. I and my assistant K. Suzuki explained this phenomenon in terms of "slip-induced directional order" on the basis of observations of slip bands and magnetic domain patterns in rolled single crystals of Fe–Ni alloys [1.4]. Namely, when a slip deformation occurs along some atomic plane, new atomic pairs should be created across the slipped plane, resulting in the directional order. We succeeded in explaining almost all the experimental facts in terms of this "slip-induced directional order". This time we won the game against Néel, Taniguchi et al. (Sect. 6.5.2).

As the reader may have realized, in the above-mentioned model, the generally prevailing interpretation of metal magnetism was the so-called "localized model" (Appendix A.20), in which we regarded atoms as rigid spheres, and also the atomic magnetic moment as well-localized in the individual atoms. At this time people believed that electric conductivity was due entirely to the $4s$ electrons, and the atomic magnetic moments were formed by the localized $3d$ electrons.

1.1 Magnetism of $3d$ Transition Metals and Alloys

Among the over one hundred elements, there are only three—Fe, Co and Ni (atomic numbers 26, 27 and 28)—which are simple metals exhibiting metallic ferromagnetism at room temperature. Figure 1.1 shows the atomic saturation moment, or the saturation magnetization at 0 K divided by the number of atoms in a unit volume, as a function of the average number of electrons per atom in the $3d$ transition metals and alloys. This curve is called the Slater–Pauling curve (Appendix A.27). Looking at this curve, we see that ferromagnetism appears only in a narrow range of electron concentration. Namely the saturation moment emerges at electron number 24, has a maximum value $2.4M_B$ per atom at 26.3 and disappears at 28.6.

The right-hand side of the curve with gradient -45 degrees can be explained in terms of the rigid band model (Appendix A.27). Namely at electron number 28.6 for Ni–Cu alloys, there are 0.6 $4s$ electrons, 18 electrons filling up the Ar shell and 10 $3d$ electrons, thus all the spin magnetic moments cancel each other. With a decrease in the number of electrons, a vacancy appears at the top of the $-$ spin band, resulting in a difference between the number of electrons in $+$ and $-$ spin bands (Here, the sign of the spin is the same as that of the spin

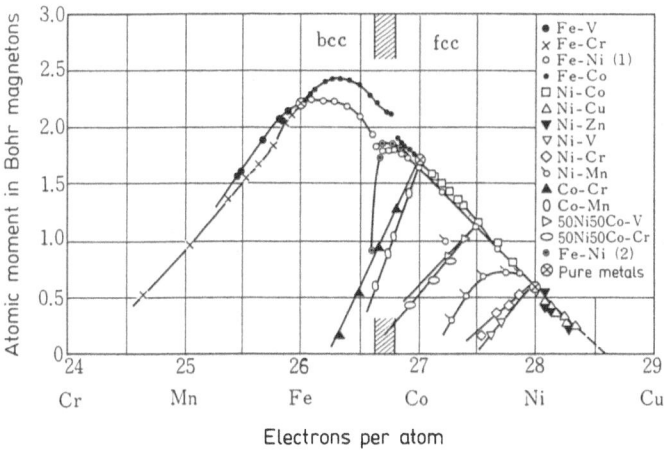

Fig. 1.1. Slater–Pauling Curve (After [1.8])

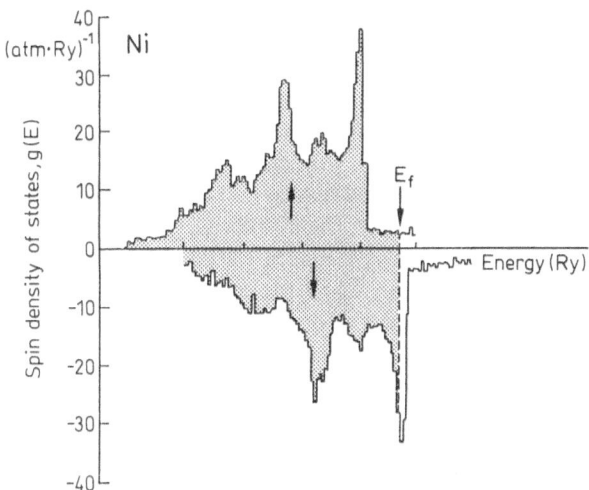

Fig. 1.2. Density-of-states curves of up and down spins for Ni (After [1.5])

magnetic moment). Figure 1.2 shows the density of states (DOS) as a function of energy for $+$ and $-$ spin bands as calculated by Connoly [1.5]. This curve is characterized by a steep peak at the top of the $3d$ band. When, therefore, the Fermi level is lowered (a shift towards the left in the figure), the void appears only in the $-$ spin band.

The left-hand side of the Slater–Pauling curve with a gradient of $+45$ degrees may be interpreted in terms of the DOS versus energy curve for the body-centered cubic lattice as calculated by Wakoh and Yamashita [1.6]

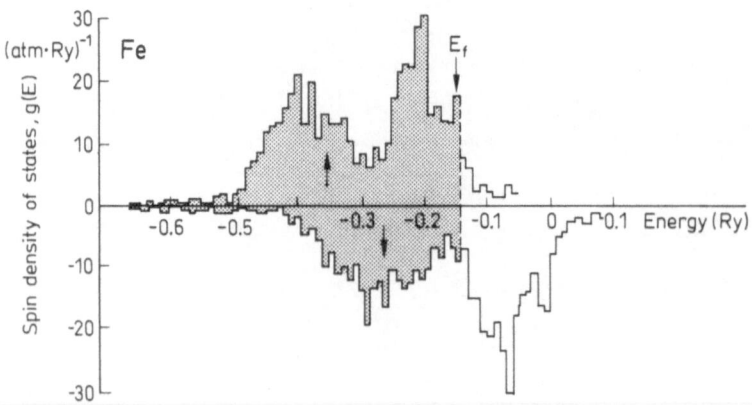

Fig. 1.3. Density-of-states curves of up and down spins for Fe (After S. Chikazumi: *Physics of Ferromagnetism* (Syokabo Publishing Co., Tokyo, 1978) (Original data is from [1.6])

(Fig. 1.3). The Fermi level for Fe coincides with a peak for the + spin band, and also with a valley for the − spin band, so that a decrease in the number of electrons causes a large decrease in + spins and rather a small decrease in − spins, resulting in a reduction of saturation moment.

The significance of such a peak may be understood by classifying the electrons into $d\varepsilon$ and $d\gamma$ character as calculated by Asano [1.7] (Fig. 1.4). When the d-electrons are placed in a cubic crystalline field, the energy levels are split into triply degenerate $d\varepsilon$ levels and doubly degenerate $d\gamma$ levels. The former wave functions are in the ⟨110⟩ directions, while the latter are in the ⟨100⟩ directions [1.8]. In Fig. 1.4, we see that the sharp peak at the top of the DOS curve for the body-centered cubic lattice (a) has $d\gamma$ character, while that for the face-centered cubic lattice (b) has $d\varepsilon$ character. This difference may be due to the difference in coordination of atoms between the two lattices. As seen in Fig. 1.1, almost all ferromagnetic alloys changes their crystal structure from face-centered cubic or hexagonal closed packed to body-centered cubic as the electron number decreases from a value of 26.7. This means that the crystal structure is also determined by the band structure. Contrary to the ferromagnetic metals, the 5d and 6d non-ferromagnetic alloys change their crystal structure at the electron number 24.5. In the unpolarized band, the Fermi level should be the same for + and − spin bands, so that the Fermi level comes to the same critical position as that of the − spin band of the polarized band at an electron number smaller than 28.7 by the difference in the number of electrons in + spin and − spin bands, or the Bohr magneton number of the saturation moment. This is actually the case as mentioned above.

Thus the saturation moment of the 3d transition metals and alloys can be understood in terms of their band structures. This, however, does not necessarily mean that the individual atoms in ferromagnetic alloys are uniformly magnetized. Neutron scattering experiments [1.9] on Fe–Ni alloys have revealed that

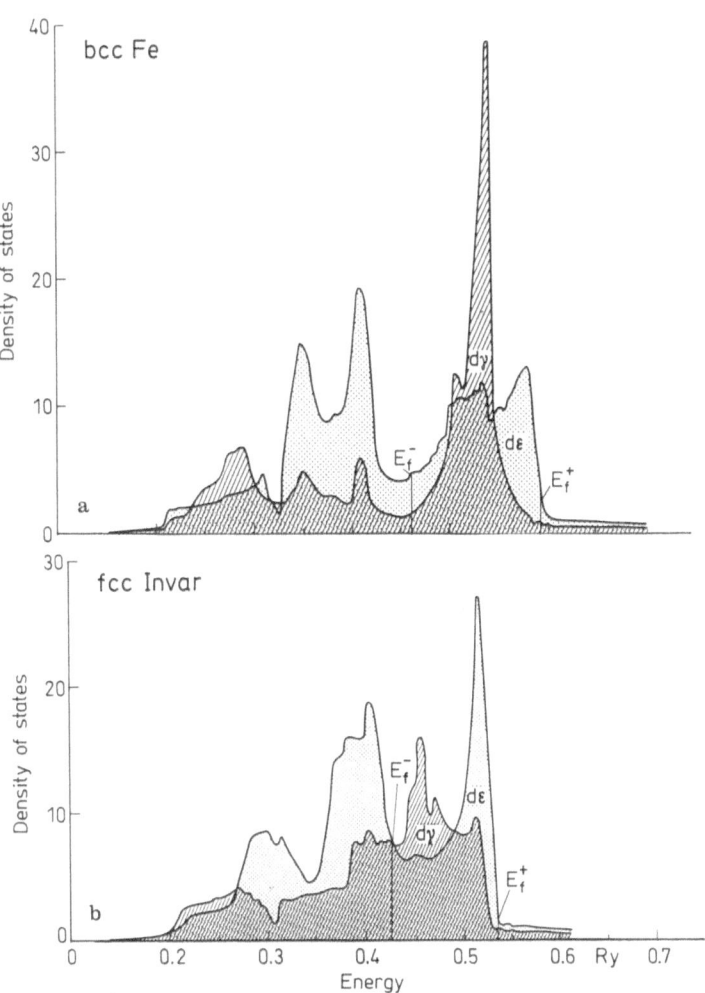

Fig. 1.4a, b. Density-of-states curves for $d\varepsilon$ and $d\gamma$ electrons. **a** for bcc Fe and **b** for fcc Invar (After [1.7])

different magnetic moments are to be associated with different kinds of atoms. Namely 2.6–$2.8 M_B$ are associated with Fe atoms, while 0.6–$0.8 M_B$ are associated with Ni atoms (Fig. 1.5). This fact was deduced by Hasegawa and Kanamori [1.10] from a band calculation using the coherent potential approximation (CPA) (Appendix A.5). According to this theory, $3d$ electrons are all itinerant, but the excess charge due to the difference in nuclear charge between Fe and Ni, ($\Delta Ze = 2e$), is shielded exclusively by the electrons in the $-$ spin band, so that a difference in atomic magnetic moment $2\mu_B$ results. The solid curves in Fig. 1.5 represent the results of calculations which agree well with experiment. Thus it

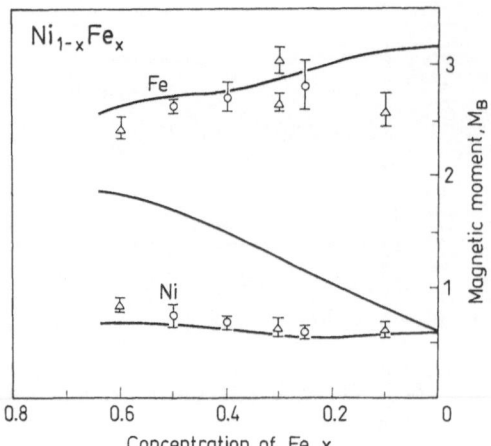

Fig. 1.5. Concentration dependence of the localized magnetic moment of Fe and Ni atoms in Fe–Ni alloys: circular and triangular points are due to neutron scattering experiments, while the solid curve is due to the CPA calculation. (After [1.10])

turns out that both $3d$ electrons and $4s$ electrons are itinerate and are both responsible for electric conductivity, but the atomic magnetic moments are well localized on individual atoms (Appendix A.20). The idea that the excess charge of the nucleus is shielded by the electrons in the $-$ spin band was first proposed by Friedel [1.11].

In Fe–Ni alloys, as the content of Fe is increased and reaches 35 at % Fe–Ni, which is the well-known low thermal expansion alloy Invar, the crystal structure becomes unstable. At this composition, the Fermi level is located at a valley between the peaks of the $d\varepsilon$ and $d\gamma$ states as seen in Fig. 1.4b. A further shift of the Fermi level towards the left (low energy side) results in a phase transition from the face-centered to the body-centered cubic lattice. The thermal expansion coefficient of Invar exhibits a complicated temperature dependence as shown in Fig. 1.6. As shown by the experimental curve (denoted by α_{exp}), the thermal expansion coefficient vanishes at room temperature [1.12]. As a result of this characteristic Invar has been utilized as a low thermal expansion alloy since its discovery by Guillaume in 1897 [1.14]. For comparison, the temperature dependence of the thermal expansion coefficient observed for pure Ni is shown in the insert in Fig. 1.6. The anomalous peak at the Curie point is caused by the disappearance of the negative volume magnetostriction (contraction) associated with the disappearance of spontaneous magnetization. In contrast, Invar exhibits a much larger positive volume magnetostriction (compare the shaded area with that in the insert), and its anomaly continues even above the Curie point. This anomaly is also related to the anomaly in elasticity, and is well explained phenomenologically in terms of the low-spin–high-spin transition of the Fe atoms [1.15]. This topic will be discussed in more detail in Chap. 5.

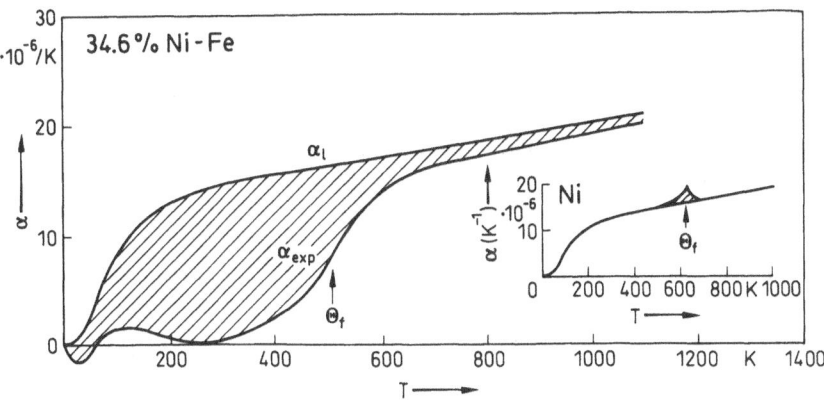

Fig. 1.6. Temperature dependence of the thermal expansion coefficient, α, of 34.6% Ni–Fe Invar alloy. (Insert is that of Ni) (After [1.13])

The saturation moment of body-centered cubic Fe, $2.2\mu_B$, is much smaller than that of Fe atoms in the face-centered cubic Fe–Ni alloys, as already mentioned. A lot of effort has been expended to try and increase this value by alloying body-centered cubic Fe with various other elements. The purpose of such effort is to increase the saturation magnetization of core materials, because the figure of merit for these materials is proportional to the square of the saturation magnetization. It was found that only three elements, Co, Pt and Pd, can increase the saturation magnetization. All of them, however, are too expensive to be used for practical purposes.

One thing to be noted here is the discovery by Kim and Takahashi [1.16] that thin films of iron evaporated in a poor vacuum showed an anomalously large saturation moment, that is 2.6 to $2.8\mu_B$ per Fe atom. They ascribed this effect to the appearance of the intermetallic compound $Fe_{16}N_2$ which has a large saturation magnetization. Another possibility is that the expansion and/or tetragonal distortion of the body-centered cubic lattice due to the interstitial insertion of N atoms might cause an increase of the saturation moment of Fe [1.17].

1.2 Magnetism of Rare Earth Metals and Alloys

Contrary to the magnetism of $3d$ transition metals and alloys with band character, the magnetism of compounds of $3d$ elements originates from localized atomic magnetic moments (Appendix A.20). Therefore it would be better to discuss the magnetism of rare earth metals and alloys which also have localized magnetic moments, before going into the magnetism of compounds associated with complicated crystal structures.

The term rare earth metal refers to any one of the 15 lanthanide elements, La, Ce, Pr, Nd, Pm, Sm, Eu, Gd, Tb, Dy, Ho, Er, Tm, Yb and Lu, plus Sc and Y. In contrast to the $3d$ transition metals, each of the rare earth metals has its own characteristic shape. A few years after I moved from Gakushuin University to the Institute for Solid State Physics, University of Tokyo in 1959, I became interested in the magnetism of rare earth metals and alloys. In 1962 we invited two eminent magneticians. Bozorth (Bell Labs.) and Graham (General Electric R & D Lab.), from the United States. One of them, Dr. Graham, brought a single crystal of gadolinium with him. I was much impressed by the gray block of rare earth metal which I saw for the first time. The electronic structure of a rare earth is composed of the inner shell, $5s^2 5p^6$, and three loosely bound electrons with the configuration $5d^1 6s^2$, which are easily lost with the formation of trivalent ions. The chemical properties of these trivalent ions are thus quite similar to each other, because their outer electronic structures are similar. As a result, the chemical separation of these rare earth elements has been quite difficult, until the ion exchange resin method, which was developed during the second world war, enabled them to be purified. Graham measured the magnetocrystalline aniso-tropy of Gd from room temperature down to 4.2 K [1.18], and Bozorth measured its magnetostriction constants from room temperature down to low temperatures with the assistance of Wakiyama [1.19] (Chap. 6).

Magnetism in the rare earths originates from the electrons in the inner $4f$ shell which are well protected by the $5s^2 5p^6$ shell. The number of electrons in the $4f$ shell changes from 0 to 14, as the atomic number increases from 57 (La) to 71 (Lu). The individual electron in the $4f$ shell possesses a spin and an orbital magnetic moment, which are combined to form an atomic magnetic moment by the so-called Hund's rules. According to the first rule, for up to 7 electrons, the spin angular momenta of all the electrons should be $+$, while for electrons in excess of 7 it should be $-$, thus forming the resultant spin S. Secondly, the resultant orbital angular momenta L should take a value as large as possible within the restriction of the first rule and the Pauli exclusion principle which allows only two electrons with $+$ and $-$ spins to enter the same orbit. Namely as the number of electrons, n, increases from 1 to 7, they should take the magnetic quantum number $m = 3, 2, 1, 0, -1, -2, -3$ successively, resulting in $L = 3, 5, 6, 6, 5, 3, 0$, respectively. The same sequence should be repeated from $n = 8$ to 14. The total angular momenta J is formed by combining S and J. The third rule asserts that for $n < 7$, $J = L - S$, while for $n > 7$, $J = L + S$.

In contrast to the $3d$ transition metals, the orbital magnetic moments of rare earth metals are well preserved, because the $4f$ shell is well protected by the outer cores. However, the orientation of L is strongly influenced by the crystal-line field, so that the total atomic magnetic moment produces a strong magneto-crystalline anisotropy through the spin-orbit interaction $\lambda L \cdot S$. For instance, in the case of Tb, the orbit with $L = 3$ is strongly fixed in the c-plane of the hexagonal crystal, so that the spontaneous magnetization is difficult to rotate out of the c-plane, even if we apply a high magnetic field parallel to the c-axis. We observed that the application of a field as strong as 400 kOe (= 32 MA/m) resulted in only a partial rotation of the magnetization, and also unexpectedly, it resulted in a partial destruction of the crystal [1.20].

We also measured the magnetocrystalline anisotropy of Gd-based dilute alloys which contained a few per cent of various rare earth metals [1.21]. Since Gd has no orbital moment, no magnetocrystalline anisotropy should be produced from the 4f electrons of Gd, but the impurity atoms, with their finite orbital moments, produced an extraordinarily large magnetic anisotropy. The results were well explained in terms of the anisotropic shape of the impurity atoms (Sect. 6.3.2 (1)). In this experiment the exchange interaction between Gd and the impurity atoms was used to rotate the magnetic moments of the anisotropic atoms instead of a strong external magnetic field.

1.3 Magnetism of Ferrimagnetic Oxides

Black iron oxide is frequently found in our daily life. For instance, the black surface of a frying pan is covered with iron oxide. This oxide is called magnetite and its chemical formula is given by Fe_3O_4 or $FeO \cdot Fe_2O_3$. The crystal structure of this oxide belongs to the spinel structure which has cubic symmetry. Generally speaking, the ionic radius of the O^{2-} ion is 1.32 Å which is much larger than that of metal ions such as Mn^{2+}, Fe^{2+}, Fe^{3+} etc., whose ionic radii range from 0.6 to 0.8 Å. Therefore the framework of the oxide crystal consists of close packed large O^{2-} ions and small metal ions which are inserted into narrow interstitial sites. In the case of the spinel structure, the large O^{2-} ions form a close packed face-centered cubic structure, in which the small metal ions occupy interstitial A sites surrounded by four O^{2-} ions tetrahedrally and interstitial B sites surrounded by six O^{2-} ions octahedrally.

In general, these iron oxides are known as ferrites and are expressed by the chemical formula $M^{2+}O \cdot Fe_2^{3+}O_3$, where M represents metal elements such as Mn, Fe, Co, Ni, Cu, Zn, Cd and Mg. From the standpoint of charge neutralization, we expect that A sites are occupied by M^{2+} ions, while B sites are occupied by Fe^{3+} ions, because then the electric charges of these metal ions are just compensated by the surrounding O^{2-} ions. For $M = Zn$, this is the case and such a crystal structure is called the normal spinel structure. In contrast, many magnetic ferrites have the inverse spinel structure, in which the A sites are occupied by some Fe^{3+} ions, and the B sites are occupied by the remaining Fe^{3+} and M^{2+} ions. The spins of the A site ions are coupled with those of the B site ions antiparallel to each other through a superexchange interaction (Appendix A.35) via the O^{2-} ions, thus exhibiting ferrimagnetism.

Magnetite is one of the magnetic ferrites in which $M = Fe$, and its crystal structure is the inverse spinel structure. In the B sites, there exist the same number of Fe^{2+} and Fe^{3+} ions. If an electron leaves an Fe^{2+}, this ion becomes Fe^{3+}, whereas the Fe^{3+} ion which accepts this electron becomes an Fe^{2+} ion. Such an exchange of electrons between Fe^{2+} and Fe^{3+} ions results in no change in the total energy, so that hopping of electrons is easily induced by thermal agitation. This is the reason why magnetite has a resistivity $\rho = 4 \times 10^{-3} \, \Omega \, cm$ which is extraordinary low for an oxide. By lowering the temperature, however, the electron hopping decreases, and finally stops at 125 K, where the resistivity is increased by a factor of 50–150. At the same time, the crystal deforms and transforms to a crystal structure with lower symmetry. This low temperature transition is accompanied by an anomalous specific heat and by a slight decrease in the saturation magnetization of 0.1% (see Fig. 1.7).

In 1947, Verwey [1.23] proposed an ordered arrangement of Fe^{2+} and Fe^{3+} on the B sites, in such a way that these different ions occupied different alternative c planes leading to orthorhombic symmetry. This symmetry was reported to be consistent with X-ray diffraction results and moreover the superlattice from such ionic ordering was also observed by neutron diffraction. Thus this problem seemed to be completely settled in 1960.

This conclusion, however, was completely overturned by subsequent experiments [1.22]. The previously observed superlattice was found to be due to multiple scattering and no superlattice with significant intensity could be identified [1.24]. Instead, a number of extra reflections caused by an oscillatory distortion of the lattice were found by electron [1.25] and neutron [1.26] diffraction. Moreover the crystal symmetry was reexamined and found to be monoclinic or triclinic [1.27]. Although the final answer has not yet been

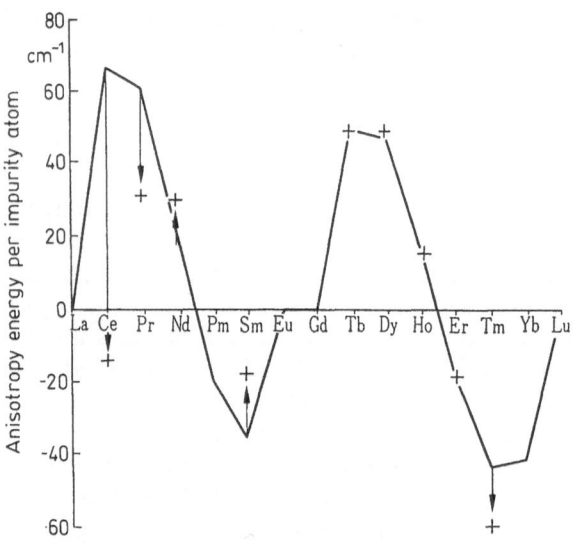

Fig. 1.7. Variation of various physical quantities at and below the Verwey temperature. (After [1.22])

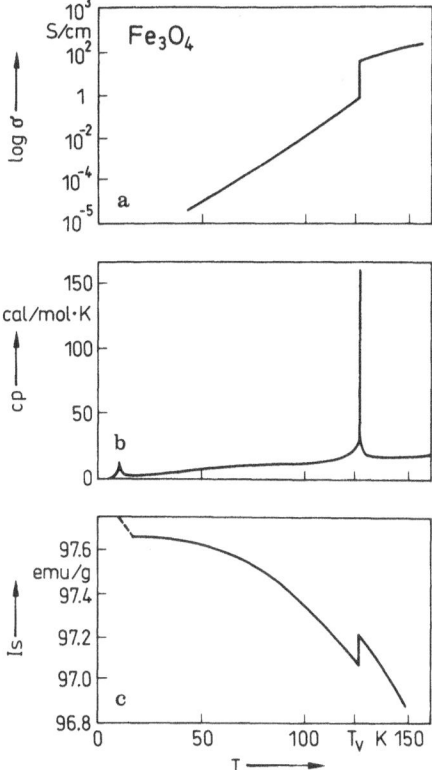

Fig. 1.8a–c.

obtained, one thing which must be noted is that all the distortions of the crystal are of the order of 1%. Generally speaking, the positions of the O^{2-} ions are slightly shifted from the A site position. This shift is described by a u-parameter, which is 3/8 or 0.375, when there is no distortion. In the case of magnetite, $u = 0.379$, so that the lattice distortion is given by $(0.379 - 0.375)/0.375 = 0.01$ or the order of 1%. All these anomalous distortions in the low temperature phase of magnetite may still originate from an instability of the lattice related to some sort of charge ordering. In contrast greigite, or Fe_3S_4, whose anion S^{2-} has an ionic radius 1.74 Å, much larger than that of O^{2-}, exhibits no low temperature transition [1.28], in spite of its ferromagnetic thermal behavior similar to magnetite. It was found that the u-parameter of geigite is just 0.375, the ideal value.

Magnetite exhibits a remarkable magnetoelectric effect [1.29, 30], or M–E effect, which means electric polarization induced by a change in magnetization. The voltage induced by this effect goes up to several volts across a 1 mm thick specimen at very low temperatures. This effect may also be related to the above-mentioned lattice instability, and is well worth further investigation.

1.4 Observation of Ferromagnetic Domains

Generally speaking, ferro- and ferrimagnetic materials can be in a demagnetized state in spite of the presence of spontaneous magnetization. This is due to the magnetic domain structure, in which the spontaneous magnetization takes different orientations in different domains. This idea was first proposed by Weiss [1.31] in his famous paper which deduced the appearance of ferromagnetic spontaneous magnetization theoretically. Thereafter the investigation of ferromagnetic domains required more than forty years before magnetic domains were imaged by Williams et al. [1.32]. Since I myself also investigated domains, I shall describe some details of domain studies in the following.

The first experimental verification of magnetic domains was made in 1919 by Barkhausen [1.33] who discovered the so-called Barkhausen noise. He was an authority on vacuum tubes and discovered that after amplifying the voltage induced in the secondary coil, the magnetization of ferromagnetic materials consisted of many discontinuous jumps which resulted in irregular noise. By dividing one flux jump by the saturation magnetization, the volume of a domain was found to be of the order of 10^{-8} cm^3.

The first attempt to observe an image of a ferromagnetic domain was made in 1932 by Bitter [1.34] and Hamos and Thiessen [1.35] independently. They observed magnetic domains by means of an optical microscope after placing a colloidal suspension onto the magnetic specimen. In a photograph shown in Bitter's paper, we can clearly see a straight domain wall, which the author simply regarded as a magnetic inhomogeneity, being presumably influenced by the false current concept of domains. The domain patterns observed on a mechanically polished surface are more regular and like a maze, a piece of which is of the order of 10^{-8} cm^3 in agreement with Barkhausen. On the basis of this accidental coincidence, people were convinced of the false current concept of magnetic domains.

When I came to work at Professor Kaya's laboratory, he said "the purpose of our investigation is to clarify the nature of magnetic domains". He had already investigated the maze domain on the mechanically polished {100} surface of an iron single crystal [1.36] and found that the pattern was much influenced by the method of polishing. At that time, I thought that if the domain structure was influenced by the internal stress, the real domain might be observed after annealing the specimen at high temperatures in a hydrogen atmosphere. If I dared to perform this experiment, I might be the first discoverer of the true magnetic domains. Although the correct concept of magnetic domains was theoretically forseen by Landau and Lifshitz in 1935 [1.37] and by Néel in 1944 [1.38], their ideas had not been accepted, because the concepts were so different from the false current concept. Finally a true image of magnetic domains was observed by Williams et al. in 1949 as already mentioned.

At that time I was a PhD student at Professor Kaya's laboratory, and I tried to reproduce their experiment. I borrowed from Prof. Kaya a single crystal disk of pure iron with the (100) surface which he had prepared previously. I electropolished the surface, and prepared a magnetic colloid with fine particles of Fe_3O_4, but the image of the true domains did not appear. According to the paper by Williams et al., the colloidal particles are attracted to the free poles at domain walls as a result of the perpendicular component of the spins inside the Bloch-type wall. I struggled with this experiment for several months.

It was realized later that such a spin configuration is unrealistic, and that, as Hubert [1.39] calculated, spins at the specimen surface always stay parallel to the surface, even inside the walls, thus resulting in no free poles at the surface. The real reason why the colloidal particles were attracted to the walls is due to the field gradient produced by the surface free poles at adjacent domains on both sides. Professor Kaya's crystal was cut parallel to the (100) plane so accurately, that there were no free poles on the domains. After repeated electropolishing, the image of the domains began to appear at the rounded edges.

After moving to Gakushuin University, I confronted the problem of maze domains: Why do maze domains look so vague and thick? Finally my assistant Suzuki discovered that the walls of the maze domains have a zigzag form! When many colloidal particles are attracted to the zigzag walls, they appear as if they were thick walls! The angle of zigzag was found to be smaller at the places where the internal stress produced by mechanical scratches is larger. The maximum stress in the vicinity of a strong scratch was calculated to be 90 kg/mm^2 [1.40].

Nowadays, magnetic domains can be observed by various means: Lorentz microscopy, scanning electron microscopy (SEM), colloid SEM, magneto-optical means, X-ray topography, etc. Among them, the most advanced method is electron interference microscopy [1.41] (Appendix A.18), by which not only domains but also the distribution of spontaneous magnetization can be observed. This technique is described in more detail in Chap. 13 of this book. Moreover, the growth of single crystals which is a prerequisite for domain observation is now quite a common technique, and is utilized in industry for such applications as bubble domain technology as described in Chap. 11. In bubble domains the detailed spin configuration inside the wall and the related motion of bubbles have been investigated.

Recently more advanced methods for domain observation have been developed. This is spin SEM [1.42], by which the direction of domain magnetization in the picture is distinguished by thickness. This method utilizes the Mott detector by which the direction of spins of the electrons emitted by the collision of scanning electrons from the d shell can be detected.

1.5 Experimental Techniques and Environments

Experimental techniques used for magnetic research are classified into two categories: i.e. macroscopic and microscopic means. The former is concerned with fundamental magnetic properties such as the measurement of magnetization, magnetic susceptibility, magnetic anisotropy, magnetostriction etc.

The latter consists of rather recently developed techniques such as neutron diffraction, electron spin resonance (ESR), nuclear magnetic resonance (NMR) etc., which are described in Chaps. 2 and 4. By means of these microscopic experiments, we can get various pieces of information concerning the interior of magnetic materials.

Experimental environments to which the specimen is exposed are also important factors. For instance, the temperature range that we can use has been remarkably enlarged during the past forty years, particularly towards the low temperature side, thanks to the use of liquid helium. Thus we can now probe various magnetic effects that take place only at very low temperatures. High magnetic fields are also important for magnetic research. Various types of magnetic behavior under high fields are described in Chap. 2. Generation of ultra-high fields, that is several hundreds T (tesla), is one of the projects which I am working on at the Institute for Solid State Physics. It is a dream for all magneticians to one day realize a field as strong as the exchange field acting in ferromagnetic materials. For instance, the exchange field to form spontaneous magnetization in metallic iron is about 1000 T, while at the moment we have only succeeded in producing 350 T. The ultra-high field project is still going on at the Institute and is described in Chap. 2.

1.6   Engineering Applications of Magnetic Materials

As mentioned above, magnetism is an important field in solid state physics, because it is closely related to the electronic structures of materials. Magnetism is also widely utilized in engineering, particularly recently, in electronics.

One of the traditional applications is the core materials used for transformers, motors, generators etc. Magnetic materials used for this purpose must be magnetically soft: in other words, they must have high permeabilities, low coercivity, low hysteresis loss etc. For large scale electric machines, the traditional material is grain-oriented silicon steel, in which the crystal grains are so oriented that the easy axis, or the axis along which the spontaneous magnetization is stabilized, is parallel to the rolling direction. Previously grain-oriented silicon steel (GOSS) was used exclusively, whereas now, HIB and RGH, which were developed in Japan, are being widely used in the world [1.22]. More recently a quite new material has appeared: that is, amorphous magnetic materials, which have low magnetic anisotropy because of their random atomic arrangement, resulting in high permeability. Chapters 8 and 9 are devoted to this topic.

Another traditional engineering application is permanent magnets. The figure of merit of permanent magnets is given by the maximum $B-H$ product or $(BH)_{max}$, which is a measure of the energy of the magnetic field produced by unit

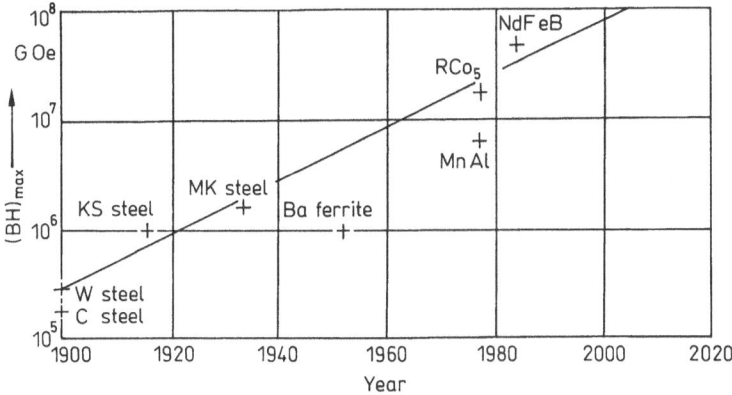

Fig. 1.9. Development of $(BH)_{max}$ of permanent magnet materials during this century

volume of the permanent magnet. Figure 1.9 shows the progress of $(BH)_{max}$ during this century on a logarithmic scale. Until 20 years ago, the permanent magnets commonly used were MK steel or Alnico and Ba ferrites. Both magnets obtain their high coercivities from a single-domain mechanism. However, recent rare earth permanent magnets such as RCo_5 and NdFeB owe their high coercivities to the difficulty of nucleation for domain walls. The record value of $(BH)_{max}$ was attained by a NdFeB magnet, which was invented recently by Sagawa et al. [1.43]. The main constituent of this magnet is the intermetallic compound $Nd_2Fe_{14}B$, which has a high saturation magnetization in addition to a large magnetocrystalline anisotropy. It is interesting that the material composed of the single phase of this compound is magnetically quite soft, exhibiting a low coercivity. The high coercivity is attained only by adding excess boron and sintering at high temperature. The best performance is obtained at the composition $Nd_{15}Fe_{77}B_8$, whose coercivity is as high as 12 kOe.

After retiring from the University of Tokyo, I moved to the newly established Physics Department at Keio University, where I and H. Miyajima built up a new magnetism laboratory. Since we had at our disposal a 100 kOe superconducting magnet and a very sensitive vibrating sample magnetometer, we decided to investigate the magnetic properties of Nd–Fe–B magnets down to low temperatures using these facilities. We were able to measure a hysteresis loop as wide as 100 kOe at 4.2 K. Based on the fact that the hysteresis loop is asymmetrical with respect to the direction in which the magnetic field is applied to the specimen, we proposed the presence of "magnetic seeds", from which a new domain is nucleated [1.44]. It was experimentally verified that the nucleation is controlled by the presence of a boron-rich phase at grain boundaries [1.45].

Before various facilities at Keio University were available, we started investigations on magnetism with magnetic fluids using simple apparatus [1.46].

Magnetic fluids are composite magnetic materials composed of magnetic colloidal particles suspended in nonmagnetic solvent. The magnetic particles are protected with surfactant to prevent aggregation by magnetic forces, so the fluid behaves like a ferromagnetic fluid. This material was originally invented for magnetic sealing of spacesuits in the Apollo Program.

It was found that a thin film of magnetic fluid exhibits very large magneto-optical effects such as magnetic dichroism and magnetic birefrengence. Figure 1.10 shows a change in the transmissivity of light for magnetite-based magnetic fluids with various solvents as a function of a magnetic field applied parallel to the magnetic fluid film [1.47]. Utilizing such phenomena, a high speed shutter has been constructed. Peculiar behavior observed for some magnetic fluids was found to be caused by the formation of elongated clusters of colloidal particles. We recently succeeded in observing such clusters by optical microscopy [1.48]. It was recognized that the formation of clusters by a magnetic field in well-dispersed colloidal suspensions is quite similar to the spinodal decomposition in precipitation alloys.

One of the major fields in engineering applications of magnetic materials is magnetic memory and recording, e.g., magnetic tapes, magnetic disks (Chap. 13), and bubble domain devices (Chap. 12). Moreover, magnetooptical disks which we can not only read but also write in digital signals have been developed (Chap. 11). This kind of disk has such an extraordinarily large memory density that it is expected to replace books and documents in the future.

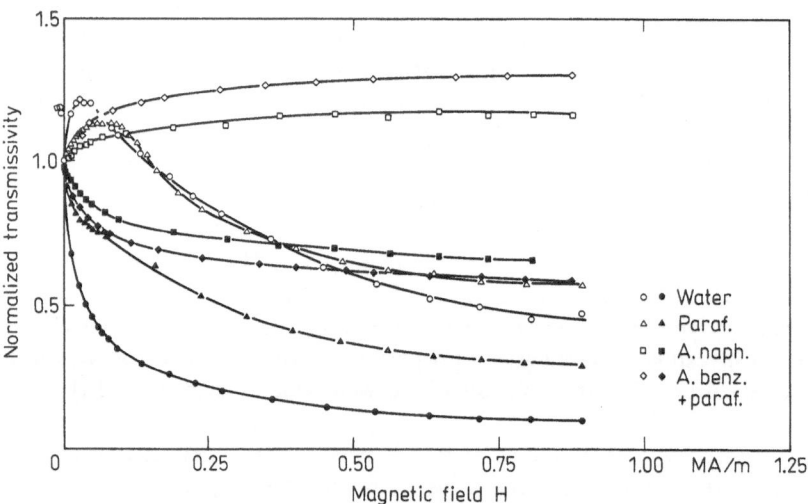

Fig. 1.10. Normalized transmissivity for ordinary (open symbols) and extraordinary (solid symbols) rays through 12 μm thick films of magnetite-based magnetic fluids with various slovents such as water, paraffine, alkylnaphthalene and a mixture of alkylbenzene and paraffine [1.47]

1.7 Conclusions

It was the purpose of this chapter to introduce a history of the development of magnetic research in the past 45 years and to provide an introductory chapter for this book. I am not sure whether or not this purpose has been achieved, but trust in any case that the reader will enjoy the following chapters. The authors of all the following chapters are either intimate friends of mine or my colleagues who are working actively in the field of magnetism. Each chapter has a different flavour, reflecting the difference in each author's personality. I hope that the book will serve as a record of the present status of experimental investigations on magnetism, as well as being enjoyable to read. I would also like to take this opportunity to express my sincere thanks to those who planned this book to commemerate my retirement from the University of Tokyo, to the authors who were willing to contribute to the book, and finally to the readers who are interested in the history and the present status of the physics and engineering applications of magnetism, in relation to my career in magnetism research over the past 45 years.

References

1.1. S. Chikazumi: J. Phys. Soc. Jpn. **5**, 327, 333 (1950)
1.2. L. Néel: J. Phys. Radium **15**, 225 (1954)
1.3. S. Taniguchi and M. Yamamoto: Sci. Rep. Res. Inst. Tohoku Univ. **A6**, 330 (1954)
1.4. S. Chikazumi, K. Suzuki, H. Iwata: J. Phys. Soc. Jpn. **12**, 1259 (1957)
1.5. J.W.D. Connoly: Phys. Rev. **159**, 415 (1967)
1.6. S. Wakoh, J. Yamashita: J. Phys. Soc. **21**, 1712 (1966)
1.7. S. Asano: Private Communication
1.8. For instance, S. Chikazumi: *Physics of Magnetism* (Wiley, New York, 1 964) p. 153
1.9. C.G. Shull, M.K. Wilkinson: Phys. Rev. **97**, 304 (1955)
1.10. H. Hasegawa, J. Kanamori: J. Phys. Soc. Jpn. **31**, 382 (1971); **33**, 1599, 1607 (1972)
1.11. J. Friedel: Nuovo Cimento Suppl. **VII**, 287 (1958)
1.12. M. Matsui, S. Chikazumi: J. Phys. Soc. Jpn. **45**, 458 (1978)
1.13. S. Chikazumi: J. Mag. Mag. Mat. **15–18**, 1130 (1980)
1.14. C.E. Guillaume: Compt. Rend. Acad. Sci. **125**, 235 (1897)
1.15. S. Chikazumi: J. Mag. Mag. Mat. **10**, 113 (1979)
1.16. T.K. Kim, M. Takahashi: Appl. Phys. Lett. **20**, 492 (1972)
1.17. K. Mitsuoka, H. Miyajima, S. Chikazumi, H. Ino: J. Phys. Soc. Jpn. **53**, 2381 (1984)
1.18. C.D. Graham, Jr.: J. Phys. Soc. Jpn. **17**, 1310 (1962)
1.19. R.M. Bozorth, T. Wakiyama: J. Phys. Soc. Jpn. **17**, 1669 (1962)
1.20. S. Chikazumi, S. Tanuma, I. Oguro, F. Ono, K. Tajima: Trans. Mag. IEEE MAG-**5**, 265 (1969)
1.21. S. Chikazumi, K. Tajima, K. Toyama: J. Phys. (France) c1, 179 (1971)

1.22. S. Chikazumi: J. Appl. Phys. **53**, 7631 (1982)
1.23. E.J.W. Verwey, P.W. Haayman, F.C. Romeijn: J. Chem. Phys. **15**, 181 (1947)
1.24. G. Shirane, S. Chikazumi, J. Akimitsu, K. Chiba, M. Matsui, Y. Fujii: J. Phys. Soc. Jpn. **39**, 947 (1975)
1.25. T. Yamada, K. Suzuki, S. Chikazumi: Appl. Phys. Lett. **13**, 172 (1968)
1.26. E. J. Samuelsen, E.J. Bleeker, L. Dobrzynski, K. Riste: J. Appl. Phys. **39**, 1114 (1968)
1.27. S. Chikazumi, K. Chiba, K. Suzuki, T. Yamada: *Ferrites* (Proc. Intn. Conf. Ferrites) (Univ. of Tokyo Press, Tokyo, 1971) p. 595
1.28. H. Nozaki: J. Appl. Phys. **51**, 486 (1980)
1.29. Y. Miyamoto, M. Ariga, A. Otuka, M. Morita, S. Chikazumi: J. Phys. Soc. Jpn. **46**, 1732 (1979)
1.30. K. Siratori, E. Kita, G. Kagi, A. Tasaki, S. Kimura, I. Shindo, K. Kohn: J. Phys. Soc. Jpn. **47**, 1779 (1979)
1.31. P. Weiss: J. Phys. **6**, 661 (1907)
1.32. H.J. Williams, R.M. Bozorth, W. Shockley: Phys. Rev. **75**, 155 (1949)
1.33. H. Barkhausen: J. Phys. **6**, 661 (1907)
1.34. F. Bitter: Phys. Rev. **38**, 1903 (1931); **41**, 507 (1932)
1.35. L.V. Hamos, P.A. Thiessen: Z. Phys. **71**, 442 (1932)
1.36. S. Kaya: Z. f. Phys. **89**, 796 (1934)
1.37. L. Landau, E. Lifshitz: Physik. Z. Sowjetunion **8**, 153 (1935); E. Lifshitz: J. Phys. USSR **8**, 337 (1944)
1.38. L. Neel: J. Phys. Radium **5**, 241, 265 (1944)
1.39. A. Hubert: Phys. Stat. Sol. **38**, 699 (1970)
1.40. S. Chikazumi, K. Suzuki: J. Phys. Soc. Jpn. **10**, 523 (1955); correction is given in S. Chikazumi, K. Suzuki: IEEE Trans. Mag. Mat. **15**, 1291 (1979)
1.41. A. Tonomura, T. Matsuda, H. Tanabe, N. Osakabe, I. Endo, A. Fukuhara, K. Shinagawa, H. Fujiwara: Phys. Rev. B**25**, 6799 (1982)
1.42. K. Koike, H. Matsuyama, K. Mitsuoka, K. Hayakawa: Jpn. J. Appl. Phys. **25**, L758 (1986)
1.43. M. Sagawa, S. Fujiwara, N. Togawa, H. Yamamoto, Y. Matsuura: J. Appl. Phys. **55**, 2083 (1984)
1.44. S. Chikazumi: J. Mag. Mag. Mat. **54–57**, 1151 (1986)
1.45. Y. Otani, H. Miyajima, S. Chikazumi: IEEE Trans. MAG-**25**, 3431 (1989)
1.46. H. Miyajima, Y. Kurihara, S. Chikazumi: IEEE Transl. J. Mag. Jpn. TJMJ -1, 806 (1985)
1.47. S. Chikazumi, S. Taketomi, M. Ukita, M. Mizukami, H. Miyajima, M. Setogawa, Y. Kurihara: J. Mag. Mag. Mat. **65**, 245 (1987)
1.48. N. Inaba, H. Miyajima, H. Takahashi, S. Taketomi, S. Chikazumi: IEEE Trans. MAG-**25**, 3866 (1989)

2. Generation of Megagauss Magnetic Fields and Their Application to Solid State Physics

Noboru Miura

"Nature does not favour high magnetic fields." This is the well-known dictum given by Professor S. Chikazumi who had been working on the generation of high magnetic fields for many years. The statement symbolizes well the difficulties involved in the generation of magagauss fields. Indeed, when we want to generate magnetic fields higher than some threshold value, all the physical laws are such that they seem to prevent it. There are so many difficult technical problems in the generation of ultra-high magnetic fields. However, once we overcome these difficulties and succeed in achieving convenient means for the generation of high fields, it will open up various new possibilities for application. For solid state physics, in particular, a magnetic field is one of the most important physical parameters that determine the existing form of matter. Therefore, it is a very powerful means for solid state physics to investigate phenomena which take place in solids when the applied magnetic field becomes extremely high.

Usually, when we investigate the magnetic properties of solids we apply magnetic fields to solids, measuring their susceptibility or magnetization. In such cases, the applied magnetic field is moderate, and we usually assume that it is a small perturbation which does not significantly modify the energy states in the solid. We obtain information about the energy states by simply observing the response to the perturbation given by the applied field. On the other hand, when the applied field becomes extremely high, the properties of the substance may undergo a large change. For instance, various magnetic field-induced phase transitions or non-linearities are brought about by ultra-high magnetic fields. In extremely high magnetic fields, electronic states are greatly influenced since they have spins and move around in crystals. The spin Zeeman energy or the cyclotron motion energy becomes enormous, and can exceed various characteristic energies in the solid. Therefore, when the applied field exceeds some threshold value, properties of solids which have been hidden in the background may show up. This is the most attractive aspect of the use of megagauss fields.

What are the technical difficulties in generating ultra-high magnetic fields? Magnetic fields are usually generated by supplying a current to coils. If we put in an iron core in the coil, higher fields are obtained due to the magnetization of the iron core. Iron cored electro-magnets are still conveniently utilized for laboratory experiments. In fields above 3 T, however, the use of iron cores is no longer useful because of the saturation of the iron magnetization, so that air-core solenoid coils are employed. As the field is proportional to the current in the solenoid, a large current is needed to generate high fields. One of the difficulties of high magnetic field generation is the large power consumption in the coil, namely the necessity of a large power supply and the large Joule heat produced by the large current, which gives rise to a temperature rise in the coil.

Superconducting magnets can generate high fields without the accompanying Joule heating. However, in superconducting materials there exists an upper critical field H_{c2} above which the superconductivity breaks down. At present, the highest field available by means of superconducting magnets is limited to about 20 T. Recently discovered high T_c superconductors possess a high H_{c2} which may enable higher field generation in the future, but it will take a few more years to develop these materials for their practical use. In several large facilities in the world, high magnetic fields are produced by using water cooled solenoid coils made from normal conductors (usually copper) [2.1]. However, for such facilities, a large power supply of the order of 10 MW and cooling water of the order of 400 ton/hour are needed. Hybrid magnets combining a superconducting magnet on the outside and a water cooled normal magnet mounted inside can generate a field above 30 T. Some facilities in the world now have new projects to produce a field of up to 40 T by employing hybrid magnets [2.1].

Much higher fields than steady fields can be readily generated in a pulsed form for a short duration. A large current can be more readily supplied to a solenoid from the point of view of power supply and Joule heating if the duration is short. The pulsed current can be obtained from a condenser bank. However, the current is limited by the electromagnetic force between the current and the field. Figure 2.1 shows a series of high speed photographs representing what will happen if we supply an excessive current to a pulsed magnet. The electromagnetic force is called Maxwell stress and is proportional to the square of the field. The direction of the force is radially outward and attractive between each winding. Figure 2.2 shows the Maxwell stress as a function of magnetic field. The stress increases rapidly with increasing field. At 100 T, it reaches 400 kg/mm^2 and exceeds the material strength from which the coil is constructed. As a result, the magnet is inevitably destroyed in such high magnetic fields. This problem of the destruction of the magnet is the second big problem for the production of high magnetic fields, and it is for this reason that especially developed techniques are required for producing ultra-high magnetic fields in the megagauss range (> 1 MG, or > 100 T).

2.1 Various Techniques for Generating Ultra-high Magnetic Fields

Figure 2.3 shows the maximum available field generated by various methods. As mentioned in the previous section, high magnetic fields above the range of the steady field can be generated as pulsed fields, whose duration is shorter as the field is increased.

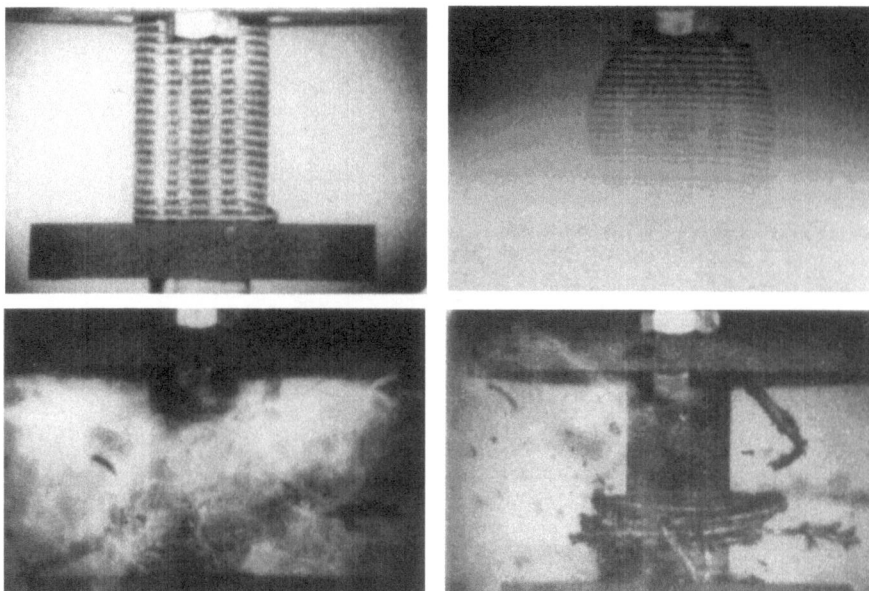

Fig. 2.1. High speed photographs showing the destruction of the solenoid coil when an excessive current is supplied. The time proceeds from top left to bottom right. The photographs were taken by a high speed movie camera with a speed of 10^4 frames/s

The prototype of pulsed magnetic fields can be seen in the pioneering work of Kapitza who first generated a pulsed high field in 1924 [2.2]. In addition to a condenser bank, Kapitza considered inductive storage in a coil, kinetic energy of a rotating fly wheel and a battery as the energy storage for obtaining a large current to a coil. By employing these methods of energy storage, he succeeded in producing a pulsed field up to 32 T. It is amazing that such high magnetic fields were produced and applied in solid state experiments in those early days when electronics had not been so much developed as in our present day. Today, condenser banks with a storage energy of about 100 kJ or more are usually employed for the energy source. Sometimes, a generator with a fly wheel is also used to obtain a large current with a long duration [2.3]. In Amsterdam, the pulsed current is taken directly from the city power line to generate a long pulsed field [2.4].

A multi-turn solenoid coil is made by winding a strong copper wire. The advantage of this type of coil is that the duration of the pulse is relatively long (well over 10 ms) due to the large coil inductance. The highest field obtainable by

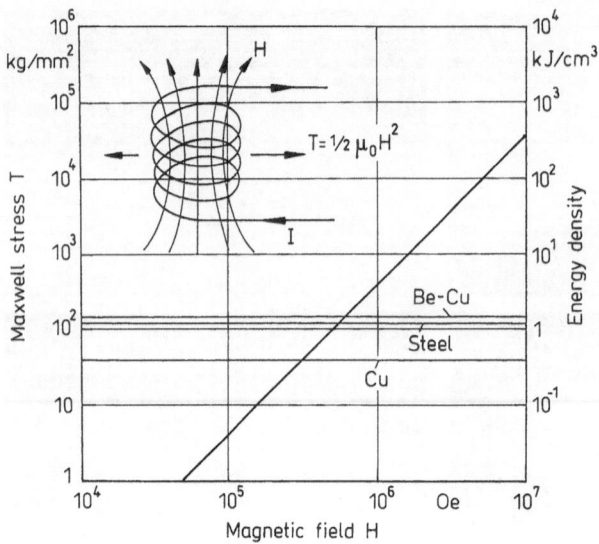

Fig. 2.2. The Maxwell stress and the energy density as a function of magnetic field. The horizontal lines indicate the yield strength of various materials

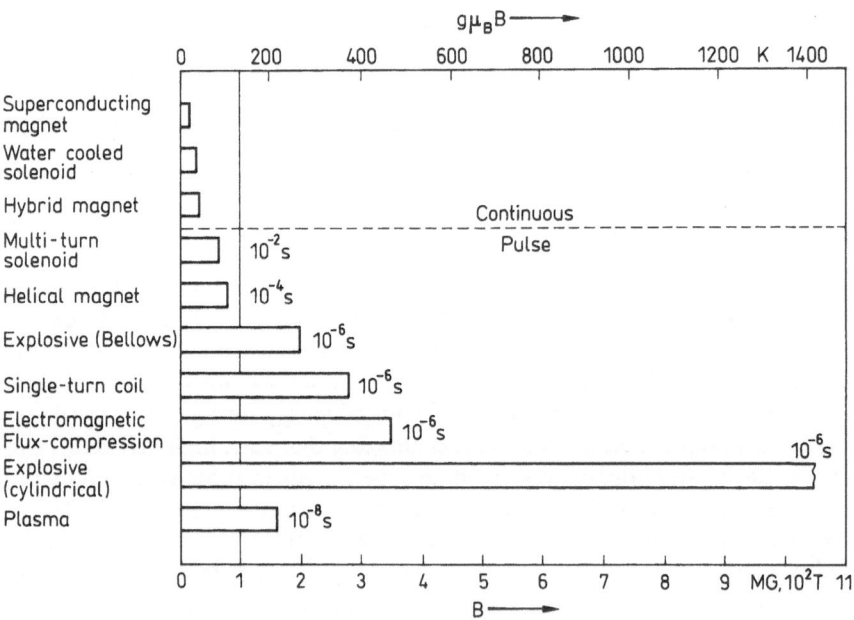

Fig. 2.3. The maximum field which can be produced by various methods and the pulse duration. On the upper scale, the Zeeman spin splitting energy $g\mu_B B$ of free electrons is shown in units of K

a copper wire wound coil is determined by the strength of the wire and the additional reinforcement from the outside. The copper wire is usually reinforced by adding Cr or Zr to the Cu. Recently, various other methods to reinforce the wire have also been developed. Wires, including fibers of Nb–Ti alloy embedded in a copper matrix commercially manufactured as superconducting wires, have more than twice the mechanical strength of normal strong copper wire, and higher fields above 50 T can be obtained by using these wires in the normal state at 77 K [2.5]. Foner has succeeded in producing a field up to 68 T by using a copper wire containing very thin Nb filaments [2.6]. The solenoid coils are usually impregnated with epoxy resin and strengthened further by mounting them in a stainless steel cylinder.

A helical coil was first developed by Foner and Kolm [2.7]. It has a shape like a screw, and is machined by using a lath from a rod of some strong metal. The Maxwell stress is sustained by the material strength itself, so that by using a strong material such as Be-doped Cu or Maraging steel, fields of up to about 80 T can be generated. However, very high precision machining is needed to make the coil, and the pulse duration is generally short due to small coil inductance.

As mentioned in the previous section, the Maxwell stress exceeds the material strength above 70–100 T. Therefore, 100 T or 1 MG is a sort of landmark above which non-destructive generation is extremely difficult. This is the reason why ultra-high magnetic fields are often referred to as "megagauss fields". In Osaka, Date and coworkers are trying to generate megagauss fields non-destructively by using a multilayer structure of helical coils which allows the Maxwell stress to be balanced [2.8]. Apart from this, megagauss fields have only been produced by destructive means.

Figure 2.4 illustrates various methods for generating megagauss fields. The explosive-driven flux compression utilizes the huge energy of chemical explosives to obtain a high field. Among these, is (a), the cylindrical compression method, where chemical explosives are set around a metal cylinder called a liner and the seed field is injected into the liner. By triggering the detonators, the explosives drive the liner rapidly towards the center. An eddy current is induced in the liner which keeps the magnetic flux constant, so that when the cross-section enclosed by the liner becomes sufficiently small, the magnetic flux density increases inversely proportional to it, producing ultra-high magnetic fields. Fowler et al. of the Los Alamos National Laboratory has reported a field of 1400 T by this method [2.9].

Although very high fields can be generated, the cylindrical compression method is very destructive, so that it is rather difficult to apply to various experiments. On the other hand, in (b), the bellows type generator is more convenient for application. The seed field is produced in a circuit enclosed by two flat metal plates and the coil. The explosives are set on one of the flat plates and the explosion is detonated at the edge of the plate. The explosion propagates from left to right pushing the plate onto the other plate, so that the reduction of the circuit area compresses the magnetic flux into the coil where a very high field

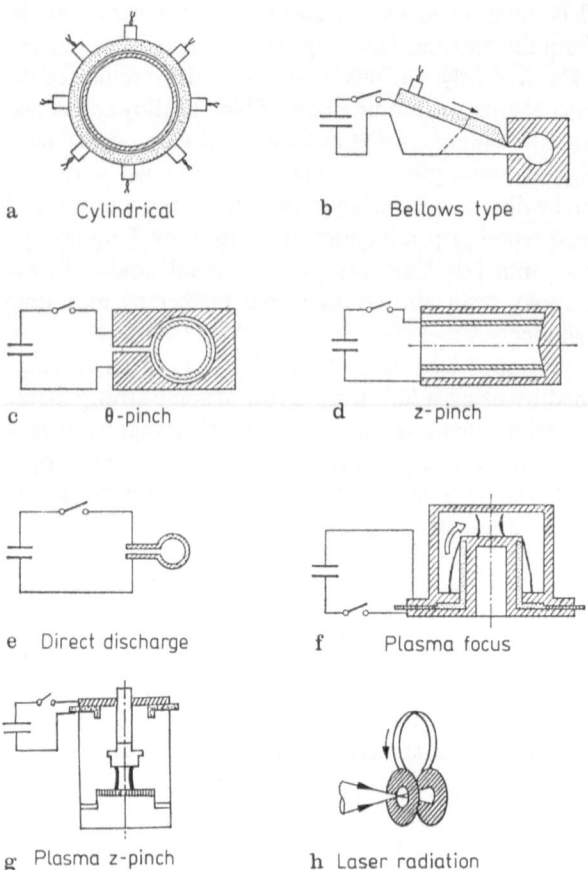

a   Cylindrical   b   Bellows type

c   θ-pinch   d   z-pinch

e   Direct discharge   f   Plasma focus

g Plasma z-pinch   h Laser radiation

Fig. 2.4a–h. Various techniques for generating ultra-high magnetic fields in the mega-gauss range

is generated. The advantage of this method is that the explosion occurs apart from the coil, which makes the setting up of experimental apparatus easier. Fowler et al. succeeded in generating a field of up to 200 T, and in measuring successfully magnetooptical spectra [2.10].

The method of electro-magnetic flux-compression utilizes the electro-magnetic force instead of explosives to compress the liner. Depending on the current direction, there are two kinds of technique, (c) θ-pinch were the current flows in the azimuthal direction and (d) z-pinch where the current flows in the axial direction. The θ-pinch was first developed by Cnare in 1966 [2.11], and is sometimes called the Cnare method. The details will be discussed in the next section. In the z-pinch configuration, the inner tube (liner) is squeezed by the circular field between two co-axial tubes. This method has the advantage that there is no feed gap in the coil as in the θ-pinch and that the compressing pressure becomes larger and larger as the liner diameter becomes smaller, in contrast to the θ-pinch. However, a faster condenser bank is required because of the smaller inductance and it is more difficult to access the center of the field for application in experiments in comparison to the θ-pinch. Alikhanov et al. have reported the generation of 300 T by this z-pinch method [2.12].

The first megagauss field which appeared in the literature was by (e) the single-turn coil technique. Furth et al. produced a field of up to 160 T by discharging a fast pulsed current to a massive single-turn coil [2.13]. Later, Herlach et al. developed the technique using light weight single-turn coils [2.14], and applied the fields to experiments in quantum electrodynamics [2.15] and cyclotron resonance [2.16]. The details of this technique will be discussed in Sect. 2.4.

Megagauss fields can also be generated by (f) a plasma focus which compresses the magnetic flux injected in advance. Hirano et al. succeeded in compressing the flux by a Fillipov-type plasma focus and were able to use this to carry out a Faraday rotation measurement [2.17]. Methods employing the plasma have the merit that the used gas can be more easily replaced after the shot in comparison to methods using the metal liners, although the duration of the field is very short, of the order of 10^{-8} s. Wessel et al. have reported a field of up to 160 T by (g) the plasma z-pinch [2.18]. Attempts have been made to produced megagauss fields by method (h) the use of high power laser radiation. Daido et al. have produced a field of up to 400 T by using a plasma produced by strong laser radiation [2.19].

2.2 Electromagnetic Flux-Compression

At ISSP (Institute for Solid State Physics) of the University of Tokyo, techniques have long been developed for generating pulsed high magnetic fields in the megagauss range by electromagnetic flux compression, and research in solid state physics has been carried out in these ultra-high magnetic fields up to several megagauss [2.20, 21].

Figure 2.5 shows a schematic diagram of the coil system currently used at ISSP [2.22, 23]. A large primary current is supplied from a condenser bank to a primary coil made from steel. Inside the primary coil, a metal ring (usually copper) called a liner is set coaxially. In the liner, the secondary current is induced in the opposite direction to the primary current to shield the field produced by the primary current. Then the repulsive force between the two currents squeezes the liner inwards. This is the θ-pinch in the terminology of plasma physics. During the process of the θ-pinch, if we inject a seed field in the liner in advance, the seed magnetic flux is compressed by the implosion of the liner, so that the total flux is nearly conserved. Consequently, when the liner diameter becomes sufficiently small, the flux density inside the liner becomes extremely high producing ultra-high magnetic fields. It is interesting to search the successive energy conversion in the process. First, the electric energy stored in the condenser bank is converted to the kinetic energy of the motion of the liner. Then the kinetic energy is transferred to the energy of the magnetic field.

At ISSP, a main condenser bank with a storage energy of 5 MJ (40 kV) has been installed to supply the primary current. The 5 MJ bank can be either utilized as an entire bank or divided into two banks of 4 MJ and 1 MJ [2.21, 23]. A sub-condenser bank of 1.5 MJ (10 kV) is installed to inject the seed field. The actual experimental setup near the coil is shown in Fig. 2.6 and Fig. 2.7 [2.23]. Between the primary coil and the liner, a phenolic pipe is inserted for insulation. The inside of the phenolic pipe is evacuated in order to avoid the generation of a shock wave at the rapid liner deformation. The currently used primary coil has dimensions of 210 mm in outer diameter, 160 mm in inner diameter and 60–80 mm in length. The primary coil is firmly clamped to the collector plates by a hydraulic press with a force of 100 tons. The primary coil is destroyed at every shot, so that a heavy steel protector with a wood lining is mounted around the primary coil to stop the burst. The sample is mounted in a small specially designed cryostat in which a large amount of liquid helium flows near to the sample so that the sample temperature can be cooled down to liquid helium temperature when necessary. At every shot of the field generation, the primary coil, the liner, the phenolic pipe, and other insulators are all destroyed.

When the energy of the condenser bank is small as in the old system at ISSP, which has a storage energy of 285 kJ [2.20], we can employ primary coils which allow multiple-use as shown in Fig. 2.8b–i. They are made of steel plates with a

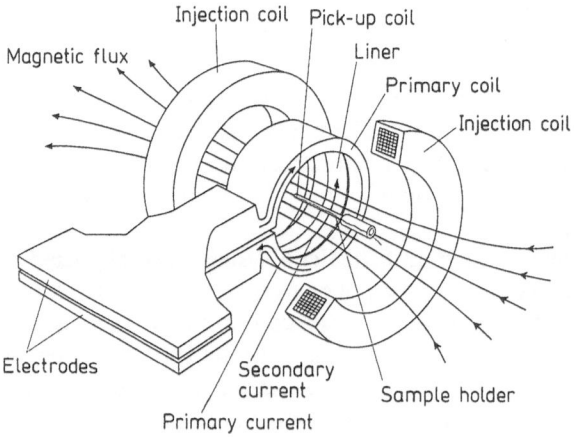

Fig. 2.5. Schematic diagram of the coil system in electromagnetic flux-compression showing the principle of the field generation

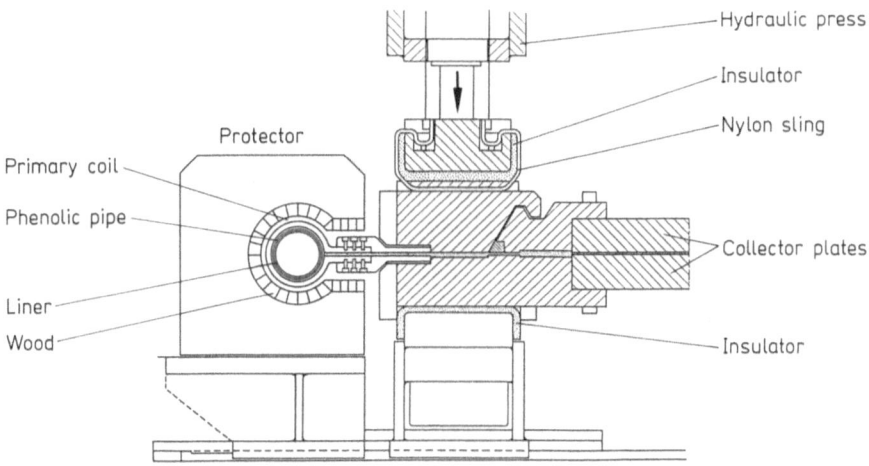

Fig. 2.6. Side view of the coil system for electromagnetic flux compression

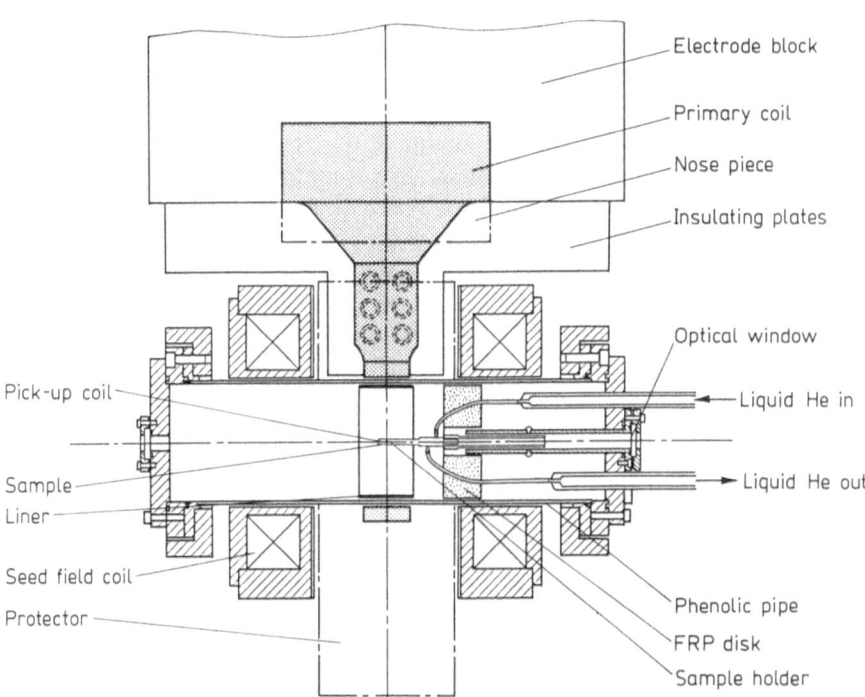

Fig. 2.7. Top view of the coil system for electromagnetic flux compression

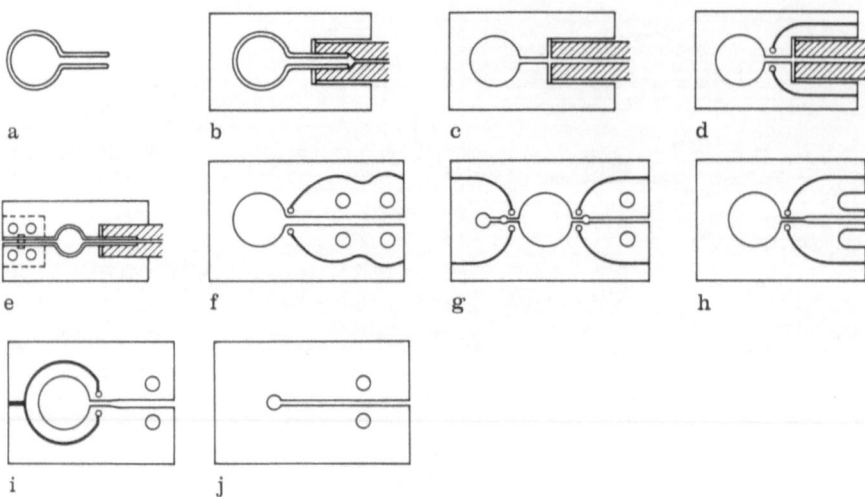

Fig. 2.8a–j. Different types of primary coils for multiple-use developed for a relatively small scale condenser bank of 285 kJ. **a** is the prototype of the primary coil, which can be used only once. **b–i** are the types for multiple use, being clamped at top and bottom by a steel plate. **b** and **e** have a copper lining in which the primary current flows, whereas in others the current flows in the steel plate itself. The grooves with insulation in them are cut for improving the symmetry of the field inside. **i** is a single-turn coil which produces high fields up to about 50 T directly by the current for preliminary measurements

hole defining the coil diameter. By clamping the plate at the top and bottom sides, the coil deformation is suppressed to a minimum, and can be used over several shots. For a condenser bank of 285 kJ storage energy, various types of primary coils have been developed, as shown in Fig. 2.8. To avoid the non-uniform deformation of the liner due to the feed gap, grooves are cut and insulators are immersed in them. These primary coils have long been successfully employed for megagauss field generation [2.20]. For a larger condenser bank energy as in the present system, the clamping of the primary coil becomes very difficult due to the extremely large electromagnetic force, so that we have to use one shot type primary coils as shown in Figs. 2.5–7, and destroy them in every shot.

To avoid any danger, the entire coil system is enclosed in a protecting chamber as shown in Fig. 2.9. The chamber is made of double walls of thick steel plates, which also serve as shields against the sound and electromagnetic noise of the shot. The chamber also has optical windows on both sides for optical measurements.

Figure 2.10 shows high speed frame photographs of the motion of the liner for the case of a primary coil with an inner diameter of 160 mm and a length of 60 mm, a liner with outer diameter of 150 mm, thickness of 1.5 mm, length of 70 mm, and a discharge from a condenser bank with 36.7 kV and 5 mF. They

Fig. 2.9. The protecting chamber enclosing the coil system for electromagnetic flux-compression. Front: for the 5 MJ bank. Back: for the 1 MJ bank

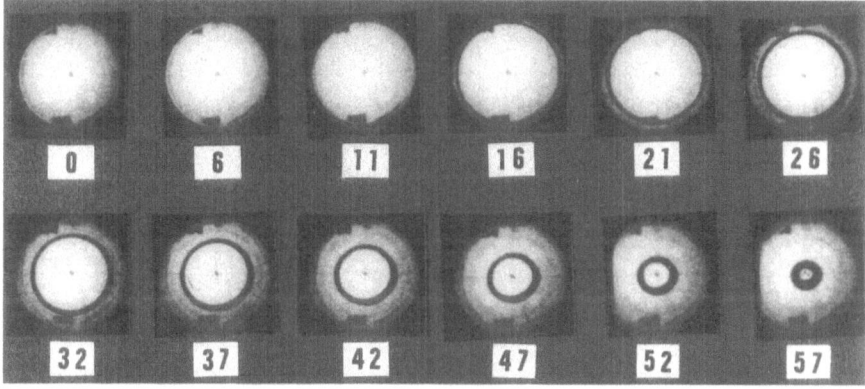

Fig. 2.10. High frame speed photographs of the linear motion. The time for each frame after the start of the primary current is indicated in μs

show how uniformly the liner is squeezed except for a slight protuberance due to the feed gap of the primary coil. From such photographs, the liner speed was estimated to be as high as 1.6 km/s. The magnetic field is measured by using a small pick-up coil wound around the sample and integrating the induced voltage [2.24]. The waveforms of the primary current and the magnetic field obtained in the same experimental condition as that of Fig. 2.10 are shown in

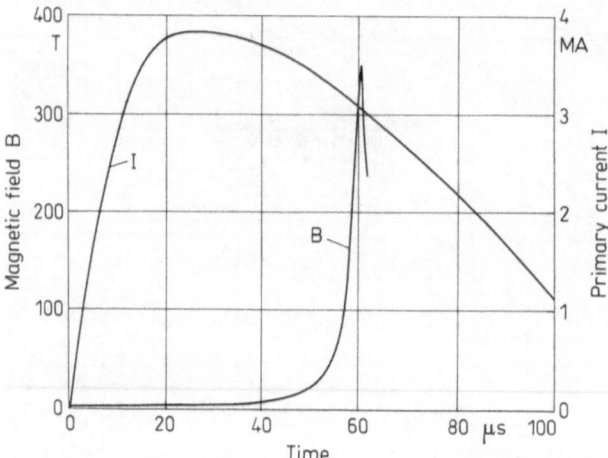

Fig. 2.11. Waveforms of the primary current and the magnetic field. The experimental conditions are the same as for Fig. 2.10

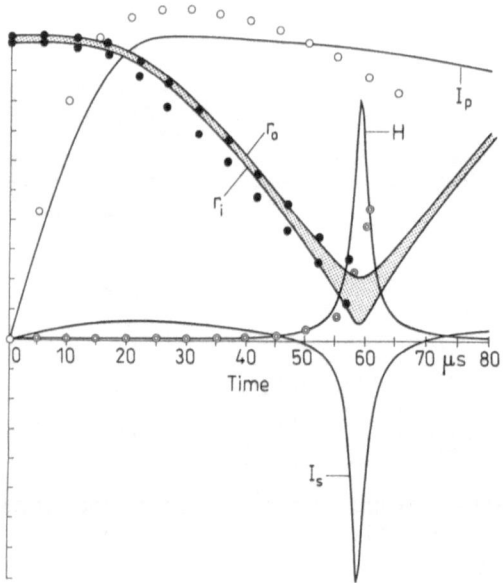

Fig. 2.12. An example of the results of computer simulation for electromagnetic flux compression. The calculated results are shown by solid lines for the primary current I_p, the secondary current in the liner I_s, the outer and inner diameter of the liner r_o and r_i and the magnetic field H. The experimental parameters corresponding to Fig. 2.10 and 2.11 were assumed. No adjustable parameters were employed. The experimental points are shown for comparison.

Fig. 2.11. The primary current deviates considerably from a sinusoidal form after about 20 µs because of the influence of the motion of the liner. The magnetic field reaches 350 T in this case.

The process of the magnetic flux compression is well represented by computer simulation, since it is a combination of the classical equations of motion of the liner and the electro-magnetic equations [2.25]. Because the current is nonuniformly distributed in the cross-section of the liner and the primary current, it is necessary to carry out a two-dimensional calculation dividing the cross-section into many small sections, and considering the current through each section [2.26]. Figure 2.12 shows an example of the final results. The experimental data are in good agreement with the theoretical calculation, despite the fact that no adjustable parameters were used in the calculation. Only for the field height is there a slight discrepancy between theory and experiment. This is due to the diffusion of the shock wave propagation out of the field [2.27] and the slight deviation from perfectly uniform compression.

2.3   Single-turn Coil Technique

Another convenient means for generating megagauss fields is the single-turn coil technique. This is based on the direct discharge of a large fast current to a small single-turn coil. When the current is of the order of mega-amperes and the coil size is less than 10 mm, magnetic fields of the order of 100 T can be generated. Of course, in such high fields the coil is destroyed. However, if the discharge is sufficiently fast, the magnetic field is generated before coil destruction occurs, because the coil has some inertia. Figures 2.13 and 2.14 illustrate the side and top views of the experimental set up around the coil employed at ISSP [2.28]. The coil is made by bending a copper plate with a thickness of 3 mm. It is firmly clamped by a hydraulic press to collector plates. A pulsed current up to 2.5 MA is supplied from a fast condenser bank with a storage energy of 100 kJ at 40 KV. The internal inductance L_b and resistance R_b of the bank are minimized to obtain the fastest possible current by reducing the length of the current path and using many cables (120). The bank at ISSP has $L_b = 18$ nH and $R_b = 2.9$ mΩ. A fragment catching box with a wood lining is placed around the coil to avoid the rebounding of exploding coil fragments which might damage the sample or sample holder. The sample is cooled to liquid helium temperatures by using a flow type cryostat as mentioned above.

The greatest advantage of the single-turn coil technique is that the coil explosion always occurs in the outward direction, so that the sample and the sample holder are not damaged at all by the shot. Thus we can repeat the experiment on the same sample and sample holder many times. This facilitates very accurate measurements which need comparisons between several shots in the same condition.

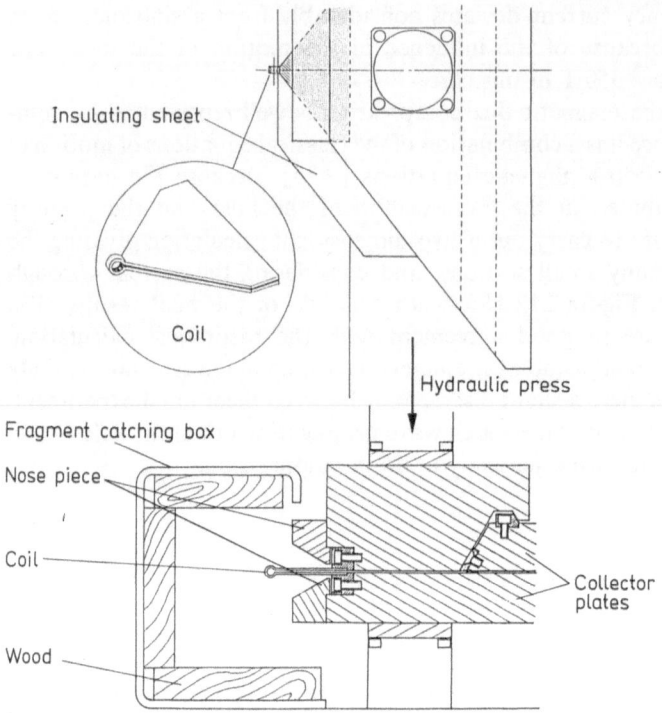

Fig. 2.13. Side view of the single-turn coil system

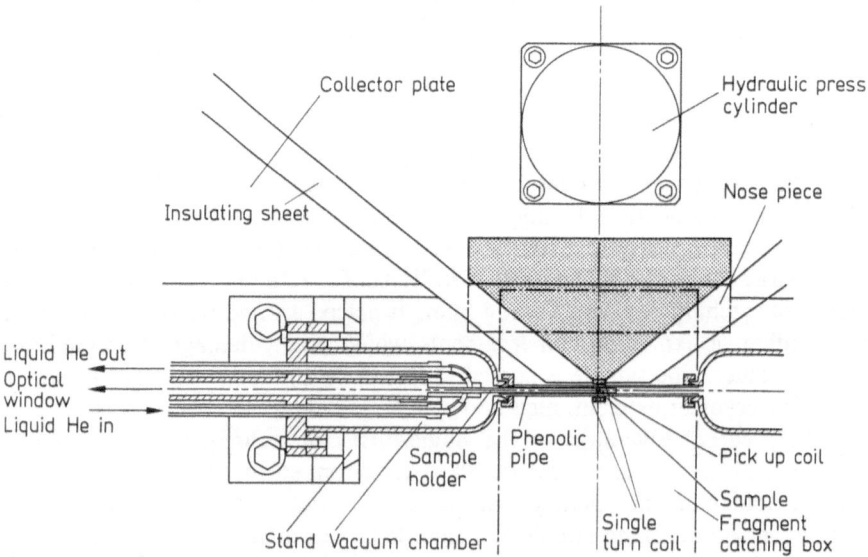

Fig. 2.14. Top view of the single-turn coil system

Moreover, because the coil and the energy consumed are much smaller in comparison to the flux-compression, experiments can be much more easily performed, and the megagauss field generation can be readily repeated about every half an hour.

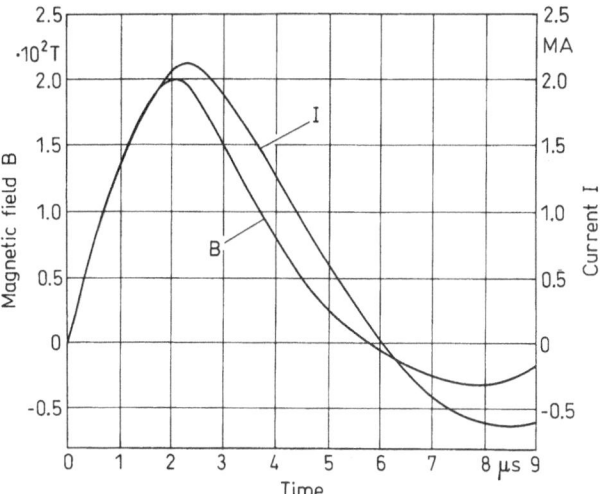

Fig. 2.15. Waveforms of current and magnetic field for the single-turn coil technique

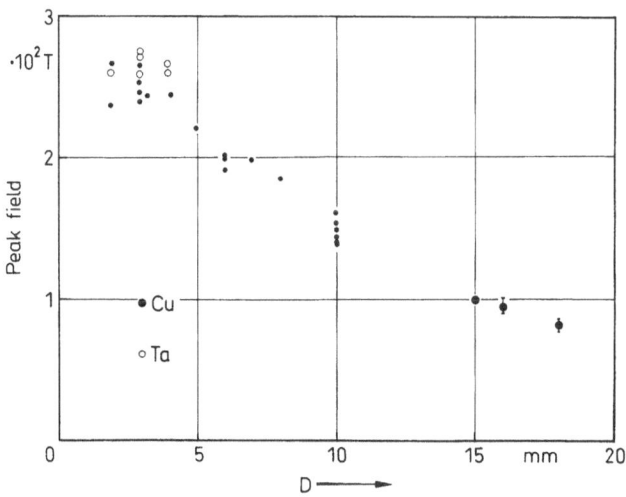

Fig. 2.16. Maximum fields produced by the single turn-coil technique as a function of the diameter of the coil D. The length of the coil l is nearly the same as D for each coil. The thickness of the coil is 3 mm. When Ta coil is used, a slightly higher field is obtained

Figure 2.15 shows the waveforms of the current and the magnetic field obtained using a coil with an inner diameter $D = 10$ mm and a length $l = 10$ mm. Because the coil is expanding, the maximum of the field is attained earlier than the current maximum. It is interesting to see that both of the curves are very smooth without any sign of the destruction of the coil, although the coil is destroyed at some point. This is because the current continues to flow in the plasma state of the vaporized coil even after the coil is broken.

The maximum available field H_m depends on the coil size, as shown in Fig. 2.16. With an inner diameter $D = 10$ mm and a length $l = 10$ mm, $H_m = 150$ T, with $D = 6$ mm and $l = 7$ mm, $H_m = 201$ T. When we use a Ta coil, a slightly higher field is obtained. The maximum field is $H_m = 276$ T for a Ta coil with $D = 3$ mm and $l = 3$ mm [2.28].

2.4 Magnetism Experiments in Megagauss Fields

2.4.1 Faraday Rotation and Magnetization

In megagauss fields, electronic states in solids are greatly altered. For example, the energy of the Zeeman splitting for free electron spins becomes as large as 136 K at 100 T and 1360 K at 1000 T as shown in Fig. 2.3. The external fields can become larger than various internal fields acting on the spins in solids. Thus megagauss fields make a variety of new interesting experiments possible which have never been considered in the lower field range. At ISSP, megagauss fields have been successfully applied to cyclotron resonance studies of semiconductors [2.29–31], magneto-optics of excitons [2.32, 33], phase transitions or a semi-metal-insulator transition of semimetals [2.34, 35], magnetic phase transitions in magnetic substances [2.36, 37] and so forth. For the application of megagauss fields in solid state physics in general, other review papers can be referred to [2.27, 38]. Here, we will mainly concentrate on discussing magnetism in megagauss fields.

Magnetization measurement in megagauss fields is by no means an easy task. When we measure the magnetization by an induction method, we put a sample in a coil, and measure the magnetic induction B in the coil when an external magnetic field H is applied. In megagauss fields, the contribution of the magnetic field H is much larger than the sample magnetization M. Therefore, in the total induction B the field part itself predominates over the magnetization M. Unless we measure the induction with enormous accuracy, it is very difficult to measure M.

If the sample is optically transparent, the Faraday rotation (Appendix A.9) is a convenient means of measuring the magnetization in high magnetic fields, because the Faraday rotation is generally proportional to or at least related to the magnetization [2.39]. Figure 2.17 shows the Faraday rotation in a magnetic semiconductor EuS [2.40]. We can see that the rate of increase of the Faraday

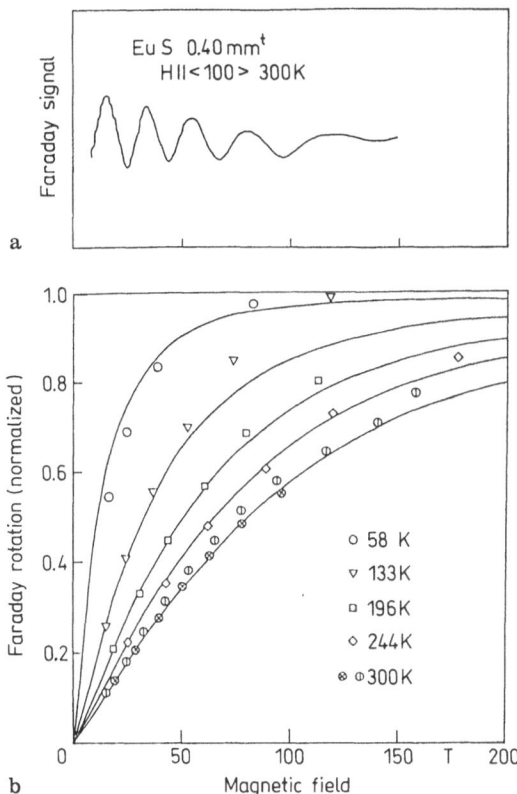

a

b

Fig. 2.17a, b. Data for Faraday rotation in EuS. The wavelength λ is 1.15 μm, and the temperature T is 300 K. **a** Signals from the Faraday rotation and the magnetic field. **b** Faraday rotation angle as a function of magnetic field at various temperatures. The solid lines are theoretical lines for the magnetization of Mn^{2+} ions [2.40]

rotation is reduced with increasing field. Figure 2.17b depicts a plot of the rotation angle as a function of magnetic field at various temperatures. The field dependence of the rotation is nearly proportional to the Brillouin function, and proportional to the magnetization of Eu^{2+} ions. This example clearly shows that the Faraday rotation is proportional to the magnetization. It is interesting to see that the nonlinear magnetization following the Brillouin function is observed even at room temperature due to the high magnetic field.

2.4.2   Magnetization in Dilute Magnetic Semiconductors

Another example of Faraday rotation is shown in Fig. 2.18 for dilute magnetic semiconductors (sometimes called semi-magnetic semiconductors)

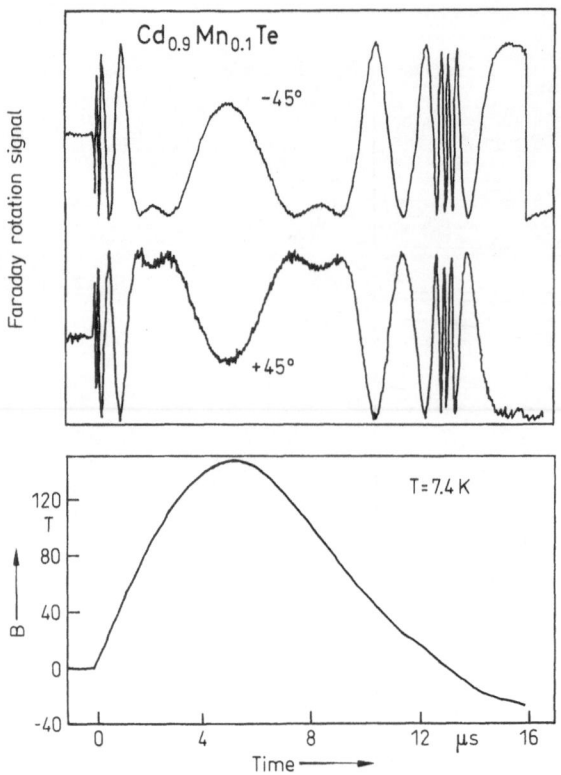

Fig. 2.18. Faraday rotation in $Cd_{0.9}Mn_{0.1}Te$ measured using the single-turn coil technique. $\lambda = 1.15$ μm. $T = 300$ K

$Cd_{1-x}Mn_xTe$ [2.41]. These experiments were done as a JSPS(Japan)–NSF(USA) joint research program in collaboration with the MIT group. In this substance, the Faraday rotation is enhanced by the magnetization of Mn^{2+} ions. When x is small, the sample shows almost a Brillouin function-like paramagnetism. With increasing x, the magnetization becomes more complicated because of the antiferromagnetic super-exchange interaction (Appendix A.35) between randomly distributed Mn ions. For $x > 0.1$, the magnetization does not show any saturation in the low field range below 40 T [2.42]. In the measurement in megagauss fields up to 150 T, Isaacs et al. found that the magnetization deduced from the Faraday rotation exhibits saturation for $x = 0.1$ and $x = 0.2$. For $x > 0.3$, however, even higher fields are needed. From such experiments, information can be obtained about the exchange interaction in the clusters of Mn ions in the crystal [2.41].

2.4.3 Spin-Flip Transitions

Megagauss fields can exceed the internal field due to exchange interactions in crystals and give rise to magnetic spin structure transitions. In ferrimagnetic crystals such as ferrites or iron garnets, there are two or three sublattices with opposite spin directions. Yttrium iron garnet (YIG) has two sublattices, the a-site (octahedral site) and the d-site (tetrahedral site) as shown in Fig. 2.19 [2.43]. The d-site has a larger sublattice moment M_d, and is thus directed towards the external magnetic field H_0. In sufficiently high external fields of the order of the exchange field between the two sublattice moments, the a-site moment M_a which is in the opposite direction becomes unstable. Above some critical field H_1, the a-site moment starts rotating, and this accompanies a d-site rotation due to the exchange interaction. Thus a spin cant phase is formed. The angle between the two sublattice moments decreases with increasing field H_0, and above a critical field H_2 the two moments become parallel. The process of the spin rotation is called a spin-flip transition.

Figure 2.20 shows data for the Faraday rotation in YIG indicating a spin-flip transition [2.44, 45]. At about 300 T, a kink is observed corresponding to the transition from the ferrimagnetic phase to the spin-cant phase. By measuring such transitions at various temperatures, we can construct a magnetic phase diagram such as the one in Fig. 2.21, from which the super-exchange interaction constants (Appendix A.35) can be directly determined.

In rare-earth iron garnets which possess another sublattice on the rare-earth ion site (c-site) in addition to the a- and d-sites, the three sublattices exhibit a more complicated magnetic phase diagram. When the exchange interactions of the c-site with the a- and d-sites $|J_{ca}|$ and $|J_{cd}|$ are much smaller than that between the a- and d-sites $|J_{ad}|$, the spin flip transition takes place in two stages. In the first step, the firm antiparallel coupling between the a- and d-sites is kept

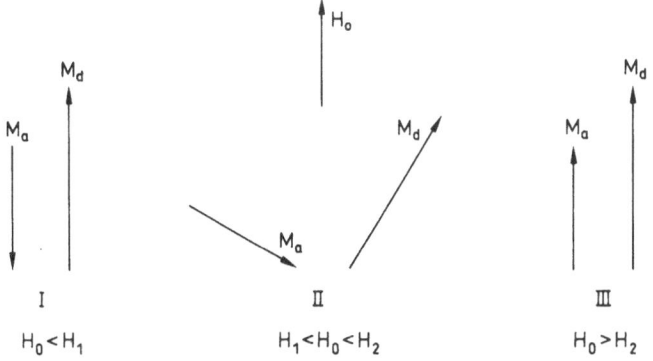

Fig. 2.19. Sublattice moments in YIG for various spin configurations

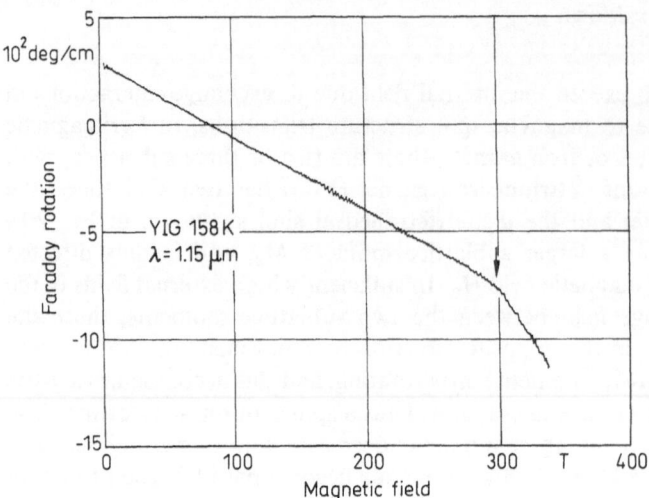

Fig. 2.20. Faraday rotation angle in YIG as a function of magnetic field. A kink is observed at about 300 T corresponding to the spin-flip transition [2.44, 45]

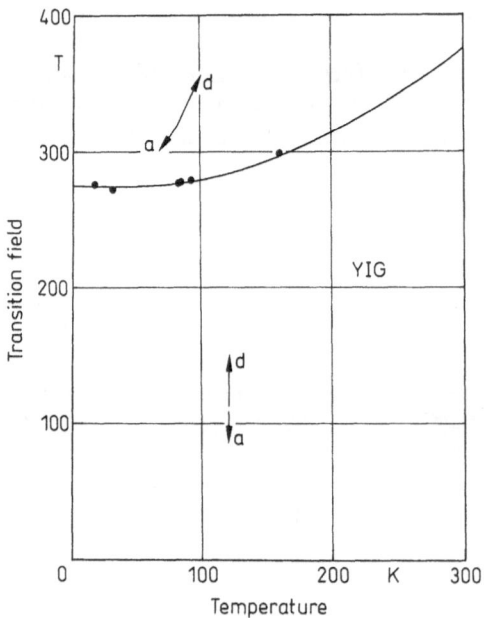

Fig. 2.21. The phase boundary between the ferrimagnetic phase and the spin cant phase in YIG

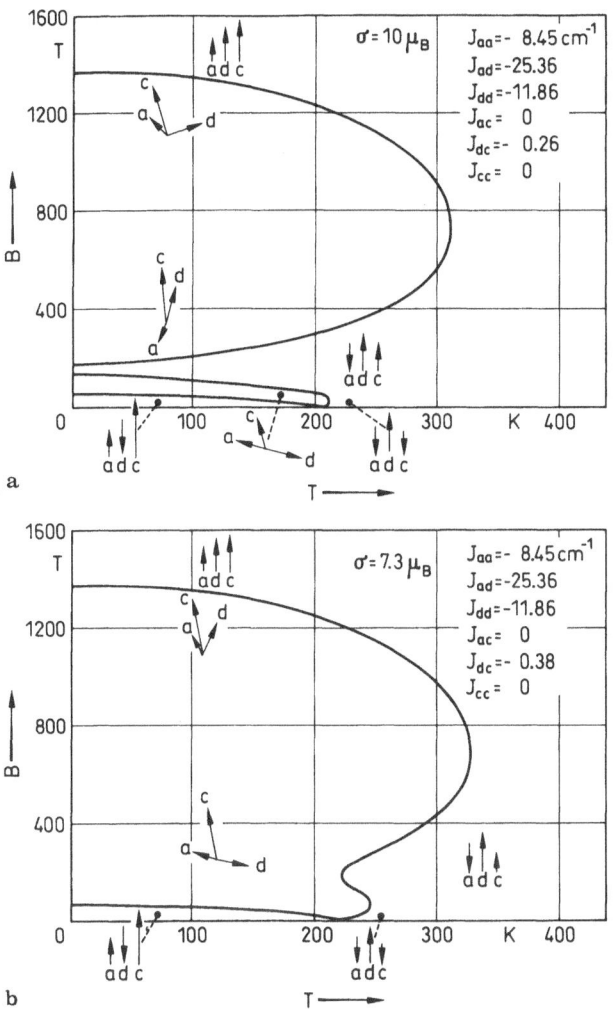

Fig. 2.22a, b. Magnetic phase diagram for three sublattice ferrimagnets. **a** the case $|J_{ca}|$, $|J_{cd}| \ll |J_{ad}|$, **b** the case $|J_{ca}|$, $|J_{cd}| \simeq |J_{ad}|$. In both cases, the exchange interaction constants close to those of DyIG are used

unchanged and the spin-flip occurs between the c-site and the assembly of the $a + d$ sites. Next in higher fields, the a- and the d-sites rotate together. Consequently, the spin-cant phase splits into two parts with a spin collinear phase in between as shown in Fig. 2.22a. On the other hand, if $|J_{ca}|$ or $|J_{cd}|$ is not so much different from $|J_{ad}|$, the two spin-cant phases mingle with each other and the phase diagram is like Fig. 2.22b [2.46]. It is interesting to see which configuration is realized in actual substances. The phase diagram determined by the

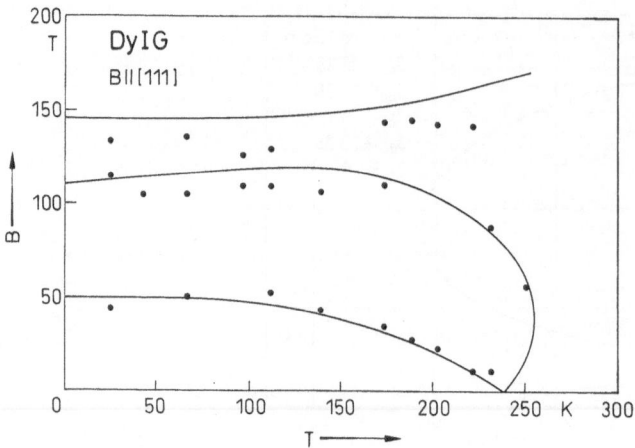

Fig. 2.23. Magnetic phase diagram of DyIG deduced from the Faraday rotation

Faraday rotation for DyIG is shown in Fig. 2.23 [2.47]. The diagram clearly shows that the latter is the case for DyIG.

2.4.4 Magnetization Measurements

Magnetization measurements by other than the Faraday rotation are needed for materials which are not transparent. A method to employ a high frequency field modulation combined with phase sensitive detection is useful to eliminate a large induction by the megagauss field [2.48]. The magnetization of a ferrite has been measured up to about 100 T by this method.

Recently, measurement by an ordinary induction method has become possible in a megagauss field produced by the single-turn coil technique, owing to the fact that this technique enables us to repeat the experiments on the same sample and measuring system. In the induction method, two pick-up coils which have nearly the same sensitivity but with opposite windings are connected in series and mounted in the field as shown in Fig. 2.24 [2.49]. In one of the windings, a sample is inserted. The other winding compensates the field part in the total induction. A slight miscompensation can be corrected electronically by mixing a signal from another compensation coil. By making an adjustment of the mixing and the position of the pick-up coil precisely in test shots at low fields, a compensation of the order of 10^{-6} can be achieved. This allows the measurement of the magnetization of the order of 1 emu in a megagauss field. Figure 2.25 shows an example of the magnetization data in a one dimensional Ising system $CsCoCl_3$ near 100 T. A step-wise metamagnetic transition due

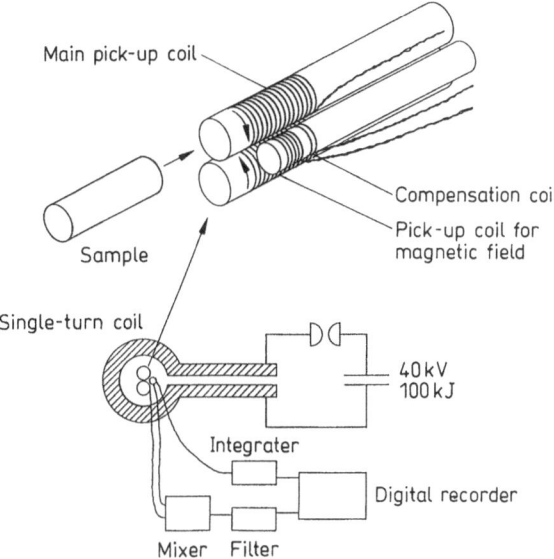

Fig. 2.24. A pick-up coil system for measuring the high field magnetization by an induction method using the single-turn coil technique

to spin level crossing is clearly observed in the magnetization at a field $H_c = 2J/g_{11}\mu_B$ [2.50].

2.4.5   Upper Critical Field of High T_c Superconductors

Another interesting example of the application of megagauss fields is the upper critical field measurement of high T_c superconductors. Recently discovered high T_c oxide superconductors possess not only a high critical temperature [2.51, 52], but also a high upper critical field H_{c2} (Appendix A.37). These are very important characteristics for the actual application of these materials. The extrapolation of the lower field data up to 40 T [2.53] by the conventional theory of dirty superconductors [2.54] predicts that H_{c2} is about 240 T for $B \perp c$ and 68 T for $B \| c$. The dc resistance measurement of the superconducting transition in megagauss fields is extremely difficult because of the large induced voltage. We have to employ the ac current technique [2.48] to measure the magnetoresistance in megagauss fields.

A more fundamental way of studying the superconducting transition is by magnetization measurements. It is well known that below the lower critical field H_{c1}, type II superconductors like the new high T_c materials show the Meissner effect. Above H_{c1}, magnetic flux starts penetrating into the crystals and the flux density increases until they finally undergo a transition to the normal state at the

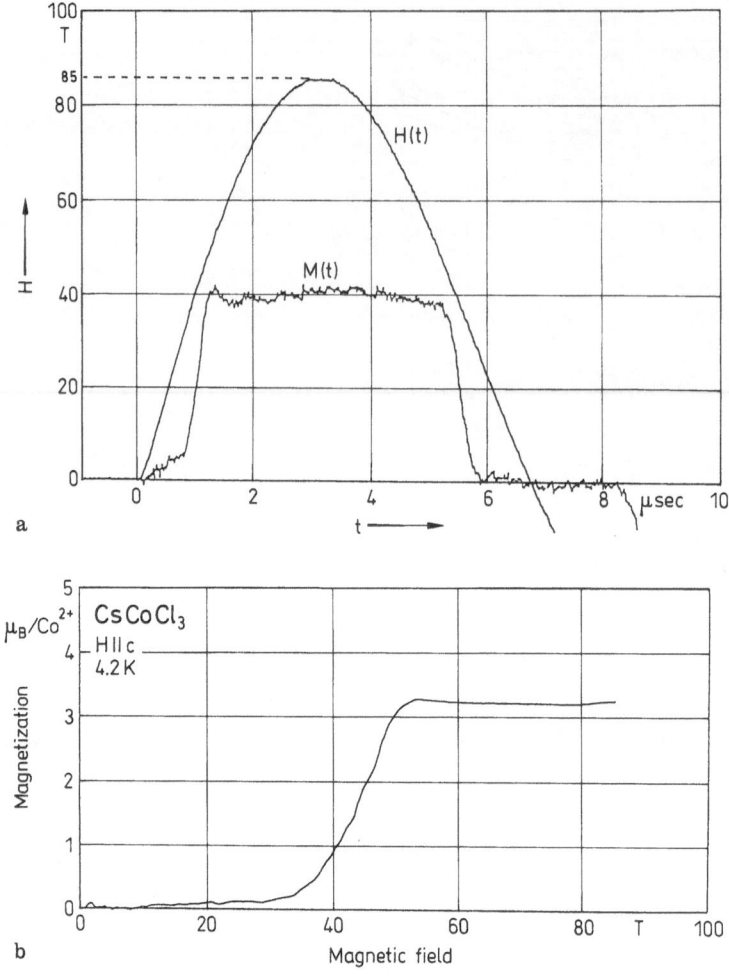

Fig. 2.25a, b. The magnetization curve for CsCoCl$_3$ up to 85 T measured with the single-turn coil technique. **a** Magnetization $M(t)$ and magnetic field $H(t)$ as a function of time. **b** Magnetization vs. magnetic field

upper critical field H_{c2}. The magnetization of the crystal generally exhibits a hysteresis due to flux pinning [2.55]. According to the critical state model [2.56], the magnitude of the hysteresis is proportional to the critical current of the sample. Therefore, the existence of a magnetization hysteresis is a good measure of the superconductivity of the sample. The magnetization shows a large jump when the direction of the field sweep changes the sign. Therefore, if we measure the time derivative of the magnetization in a pulsed magnetic field using a pick-up coil as shown in Fig. 2.24, a sharp pulsed signal would show up

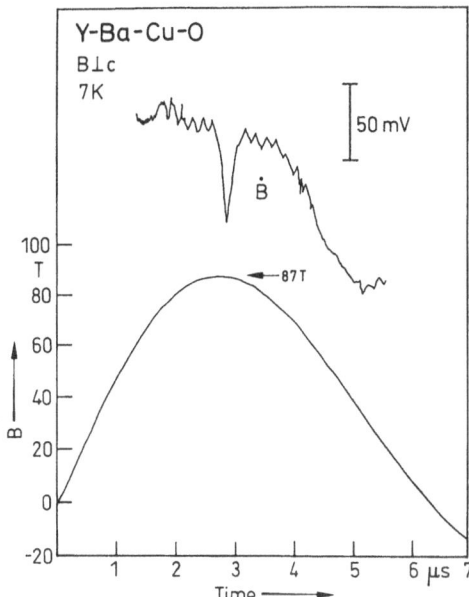

Fig. 2.26. An example of the experimental traces of the signal of $\dot{M}$ and the magnetic field for a single crystal of the high T_c superconductor YBaCuO. At the top of the field, a pulsed signal is observed in $\dot{M}$ corresponding to the magnetization hysteresis

[2.57]. Such a measurement of the pulse in M is easier than the measurement of the whole curve of M itself.

Figure 2.26 shows an example of the experimental traces of $\dot{M}$ for YBaCuO single crystals in pulsed high magnetic fields up to 87 T produced by the single-turn coil technique. A sharp pulse appeared in $\dot{M}$ at the top of the pulse field giving evidence for superconductivity at this field. The magnitude of the magnetization hysteresis ΔM thus measured is shown in Fig. 2.27 as a function of external fields for the field direction parallel and perpendicular to the c-axis. The starting temperature T_0 is 6 K, but the actual temperature at the top of the field is higher than this because of heating by the magnetic hysteresis. The temperature rise depends on the field, and it is estimated to be about 22 K at 100 T for $T_0 = 6$ K. In the low field range, ΔM is larger for $B \parallel c$ than for $B \perp c$ according to the larger critical current J_c for $B \parallel c$. However, at fields higher than 30 T, ΔM is larger for $B \perp c$ because of the larger H_{c2}. The field where ΔM crosses the zero corresponds to H_{c2}. By measuring H_{c2} in this manner at various temperatures, we can construct a phase diagram as shown in Fig. 2.28 [2.58]. We can see a large anisotropy of H_{c2}. It is interesting to see that H_{c2} is considerably lower than predicted by the conventional theory. Thus, we have obtained evidence of the superconductivity in megagauss fields.

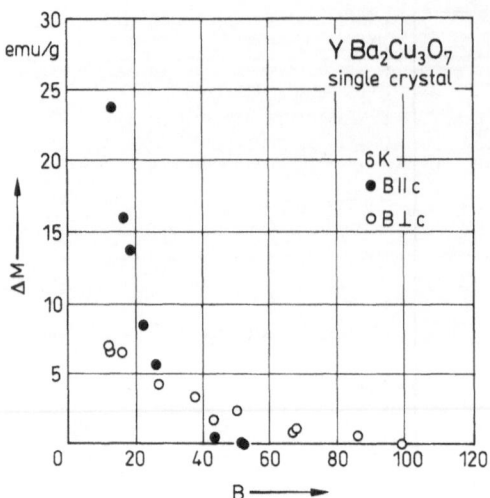

Fig. 2.27. The magnitude of the magnetic hysteresis ΔM as a function of magnetic field for a YBaCuO single crystal

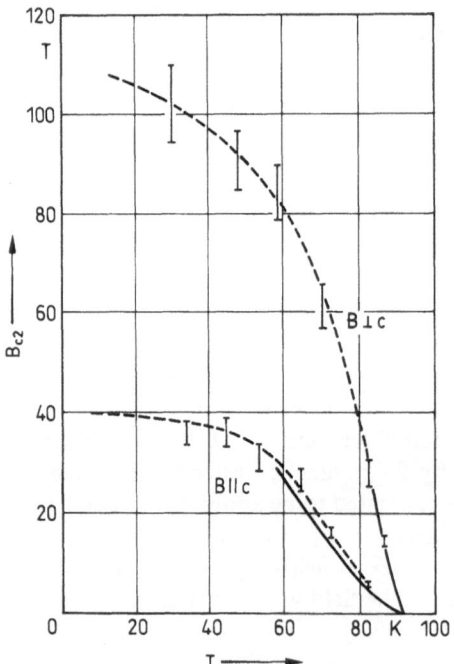

Fig. 2.28. The upper critical field H_{c2} of YBaCuO single crystals as a function of temperature. The solid lines show the data in the non-destructive fields [2.53]

Acknowledgment

The author would like to express his sincere gratitude to Professor S. Chikazumi for establishing a firm basis for the megagauss experiments at ISSP and for continual and valuable discussion and encouragement. Thanks are also due to his coworkers who contributed to the works discussed in this article, Drs. G. Kido, H. Miyajima, T. Goto, K. Nakao, S. Takeyama, T. Sakakibara, T. Haruyama, T. Kikuchi, M. Suekane, E.D. Isaacs, D. Heiman and K. Amaya.

References

2.1. For a recent review on the largest magnet facilities in the world see *Physica B* **155** (Proc. International Symposium on High Field Magnetism, Leuven, 1988) ed. F. Herlach (North Holland, Amsterdam 1989)
2.2. P. Kapitza: Proc. Roy. Soc. A**105**, 691 (1924); A**115**, 658 (1927)
2.3. H.C. Praddaude, S. Foner: Rev. Sci. Instrum. **10**, 1183 (1979)
2.4. R. Gersdorf, F.A. Muller, L.W. Roeland: Rev. Sci. Instrum. **36**, 1100 (1965)
2.5. N. Miura, T. Goto, K. Nakao, S. Takeyama, T. Sakakibara, T. Haruyama, S. Todo, K. Kikuchi: *Physica B* **155**, 106 (1989)
2.6. S. Foner: Appl. Phys. Lett. **49**, 982 (1986)
2.7. S. Foner, H.H. Kolm: Rev. Sci. Instrum. **27**, 547 (1956); **28**, 799 (1957)
2.8. M. Date: J. Phys. Soc. Jpn. **39**, 892 (1973)
2.9. C.M. Fowler, W.B. Garn, R.S. Caird: J. Appl. Phys. **31**, 588 (1960)
2.10. C.M. Fowler: Science **180**, 261 (1973)
2.11. E.C. Cnare: J. Appl. Phys. **37**, 3812 (1966)
2.12. S.G. Alikhanov, V.G. Belan, A.I. Ivanchenko, V.N. Karasjuk, G.N. Kichigin: J. Phys. E**1**, 543 (1968)
2.13. H.P. Furth, M.A. Levine, R.W. Waniek: Rev. Sci. Instrum. **28**, 949 (1957)
2.14. F. Herlach, R. McBroom: J. Phys. E**6**, 652 (1973)
2.15. F. Herlach, R. McBroom, T. Erber, J. Murray, R. Gearhart: IEEE Trans. **NS-18**, 809 (1971)
2.16. Herlach, J. Davis, R. Schmidt, H. Spector: Phys. Rev. B**10**, 682 (1974)
2.17. K. Hirano, K. Shimoda, M. Sato, H. Misaizumi, G. Kido, N. Miura, S. Chikazumi: J. Phys. Soc. Jpn. **52**, 3424 (1983)
2.18. F.J. Wessel, F.S. Felber, N.C. Wild, H.U. Rahma, A. Fisher, E. Ruden: Appl. Phys. Lett. **48**, 1119 (1986)
2.19. H. Daido, F. Miki, K. Mima, M. Fujita, K. Sawai, H. Fujita, Y. Kitagawa, S. Nakai, C. Yamanaka: Phys. Rev. Lett. **56**, 846 (1986); H. Daido: ILE Quarterly Progress Report **86-17**, 20 (1986)
2.20. N. Miura, G. Kido, M. Akihiro, S. Chikazumi: J. Mag. Mag. Mater. **11**, 275 (1979)
2.21. N. Miura, T. Goto, K. Nakao, S. Takeyama, T. Sakakibara, F. Herlach: J. Mag. Mag. Mater. **54–57**, 1409 (1986)
2.22. T. Goto, N. Miura, K. Nakao, S. Takeyama, T. Sakakibara: In *Megagauss Technology and Pulsed Power Applications*, ed. C.M. Fowler, R.S. Caird, D.J. Erickson (Plenum, New York 1987) p. 149

2.23. N. Miura, T. Goto, K. Nakao, S. Takeyama, T. Sakakibara, T. Haruyama, K. Kikuchi: *Physica B* **155**, 23 (1989)
2.24. G. Kido, N. Miura, K. Kawauchi, I. Oguro, S. Chikazumi: J. Phys. **E9**, 587 (1976)
2.25. N. Miura, S. Chikazumi: Jpn. J. Appl. Phys. **18**, 553 (1979)
2.26. N. Miura, K. Nakao: Jpn. J. Appl. Phys. **29**, 1580 (1990)
2.27. N. Miura, F. Herlach: In *Strong and Ultrastrong Magnetic Fields and Their Applications*, ed. F. Herlach Topics Appl. Phys. **57** (Springer, Berlin, Heidelberg 1985) p. 247
2.28. K. Nakao, F. Herlach, T. Goto, S. Takeyama, T. Sakakibara, N. Miura: J. Phys. **E18**, 1018 (1985)
2.29. N. Miura, G. Kido, S. Chikazumi: Solid State Commun. **18**, 885 (1976)
2.30. N. Miura, G. Kido, M. Suekane, S. Chikazumi: J. Phys. Soc. Jpn. **52**, 2838 (1983)
2.31. S.P. Najda, S. Takeyama, N. Miura, W. Zawadzki, P. Pfeffer: Phys. Rev. B**40**, 6189 (1989)
2.32. N. Miura, G. Kido, H. Katayama, S. Chikazumi: J. Phys. Soc. Jpn. **49**, Suppl. A 409 (1980)
2.33. N. Miura, S. Takeyama, Y. Iwasa: Proc. 18th Int. Conf. Phys. Semiconductors (World Scientific Pub. Co., 1987) p. 715
2.34. N. Miura, K. Hiruma, G. Kido, S. Chikazumi: Phys. Rev. Lett. **49**, 1339 (1982)
2.35. N. Miura, T. Osada, T. Goto: Proc. 17th Int. Conf. Phys. Semiconductors, ed. J.D. Chadi, W.A. Harrison (Springer, Berlin, Heidelberg 1985) p. 973
2.36. N. Miura, G. Kido, I. Oguro, K. Kawauchi, S. Chikazumi, J. F. Dillon. Jr., L.G. Van Uitert: Physica **86–88B**, 1219 (1977)
2.37. F.M. Yang, N. Miura, G. Kido, S. Chikazumi: J. Phys. Soc. Jpn. **48**, 71 (1980)
2.38. N. Miura: In *Infrared and Millimeter Waves*, ed. K.J. Button (Academic, New York 1984) p. 73
2.39. W.A. Crossley, R.W. Cooper, J.L. Page, R.P. van Stapele: Phys. Rev. **181**, 896 (1969)
2.40. M. Suekane, G. Kido, N. Miura, S. Chikazumi: J. Mag. Mag. Mater. **31–34**, 589 (1983)
2.41. E.D. Isaacs, D. Heiman, X. Wang, P. Becla, K. Nakao, S. Takeyama, N. Miura: To be published
2.42. D. Heiman, E.D. Isaacs, P. Becla, S. Foner: Phys. Rev. B**35**, 3307 (1987)
2.43. N. Miura, I. Oguro, S. Chikazumi: J. Phys. Soc. Jpn. **45**, 1534 (1978)
2.44. K. Nakao, T. Goto, N. Miura: J. de Phys. (Paris) Colloq. C8, Suppl. No. 12, 953 (1988)
2.45. T. Goto, K. Nakao, N. Miura: *Physica B* **155**, 285 (1989)
2.46. N. Miura, T. Goto, K. Nakao, S. Takeyama and T. Sakakibara: In *Megagauss Technology and Pulsed Power Applications*, ed. C.M. Fowler, R.S. Caird, D.J. Erickson (Plenum, New York 1987) p. 209
2.47. K. Nakao, N. Miura, T. Goto: In *Advances in Magneto-optics* ed. K. Tsushima and K. Shinagawa, J. Magnetics Society of Japan **11**, 11 Suppl. 153 (1987)
2.48. G. Kido, N. Miura, K. Nakamura, H. Miyajima, K. Nakao, S. Chikazumi: In *High Field Magnetism*, ed. M. Date (North Holland, Amsterdam 1983) p. 309
2.49. S. Takeyama, K. Amaya, T. Nakagawa, M. Ishizuka, K. Nakao, T. Sakakibara, T. Goto, N. Miura, Y. Ajiro, H. Kikuchi: J. Phys. E Sci. Instrum. **21**, 1025 (1988)
2.50. K. Amaya, S. Takeyama, T. Nakagawa, M. Ishizuka, K. Nakao, T. Sakakibara, T. Goto, N. Miura, Y. Ajiro, H. Kikuchi: *Physica B* **155**, 396 (1989)
2.51. J.G. Bednorz, K.A. Muller: Z. Phys. B**64**, 189 (1986)
2.52. S. Uchida, H. Takagi, K. Kitazawa, S. Tanaka: Jpn. J. Appl. Phys. **26**, L1 (1987)

2.53. T. Sakakibara, T. Goto, Y. Iye, N. Miura, H. Takeya, H. Takei: Jpn. J. Appl. Phys. **26**, L1892 (1987)
2.54. N.R. Werthamer, E. Helfand, P.C. Hohenberg: Phys. Rev. **147**, 295 (1966)
2.55. K. Nakao, K. Tatsuhara, N. Miura, S. Uchida, H. Takagi, T. Wada, S. Tanaka: J. Phys. Soc. Jpn. **57**, 2476 (1988)
2.56. C.P. Beans: Rev. Mod. Phys. **36**, 31 (1964)
2.57. K. Nakao, N. Miura, K. Tatsuhara, S. Uchida, H. Takagi, T. Wada, S. Tanaka: Nature **55**, 816 (1988)
2.58. K. Nakao, N. Miura, K. Tatsuhara, H. Takeya, H. Takei: Phys. Rev. Lett. **63**, 97 (1989)

3. Magnetism in Metals and Alloys Studied by Neutron Scattering

Yoshikazu Ishikawa and Keisuke Tajima

The study of the physics of magnetism started with the simple question: "why does a permanent magnet attract iron and nickel?" and, after a long journey, reached the question: "how can the magnetism of iron and nickel be understood?". This was one of the main topics of discussion at the International Conference on Magnetism held in Japan in 1982 [3.1]. However, a unified interpretation of the problem could not be obtained. One reason why this historical problem has not been resolved comes from the ambiguous character, localized or itinerant, (Appendix A.20) of d-electrons which are the origin of the magnetism in metals. In order to study the itinerant or localized character of d-electrons, the dynamical properties of the magnetic moment need to be examined by means of neutron scattering. In particular, for the study of Fe and Ni, a high flux of epithermal polarized neutrons is indispensable. However, it is very difficult to make such a neutron source with present nuclear reactors.

The contribution of neutron scattering to progress in the physics of magnetism is, of course, very large, especially its contribution to the understanding of magnetism in metals as mentioned above. In this chapter, we describe how neutron scattering studies have been used for the study of magnetism, and give a review of recent contributions of neutron scattering to the study of magnetism in metals. We begin with a brief survey of the present status of understanding of magnetism in metals.

The study of magnetism in metals so far has been the study of the degree of localization or itineracy of d-electrons [3.2]. At the beginning of the 1960's, measurements of the shapes of the Fermi surfaces and of the densities of states were performed and these results were explained very well by the band model based on the one electron approximation [3.3] which was developed with the development of the computer at that period. As a result, the concept of the itineracy of the d-electron in transition metals seemed to be well established. In particular, in the field of magnetism, the magnetic structure of the spin density waves (SDW) (Appendix A.30) in Cr metal which was found by means of neutron diffraction was explained by the singular structure of the Fermi surface. It was regarded as evidence of the itinerant character of d-electrons [3.4].

Subsequently, the magnetic moment of each kind of atom in disordered ferromagnetic alloys has been calculated by the band theory based on the coherent potential approximation (CPA) (Appendix A.5). The results were found to be in good agreement with the value obtained from neutron diffraction and other experiments [3.5]. Up to that time, many experimentalists had believed that the concept of localized character of the magnetic moments was established by the fact that each kind of atom in an alloy had a different magnetic moment. However, after these experimental and theoretical studies, they finally recognized the predominance of the band model for the description of the ground state of the magnetism in a metal. Around 1970, the spin wave dispersion relations of Fe and Ni were measured by neutron scattering techniques and the results were in good agreement with the band theory based on a generalized random phase approximation (RPA) [3.6], Appendix A.26. Thus, band theory can explain the magnetism in metals not only for the ground state but also for the excited states.

It became clear at the same time, however, that the band model based on simple approximations could not describe the magnetism at finite temperatures [3.2]. For example, Curie–Weiss behavior of the magnetic susceptibility $\chi(0) = C/(T - T_C)$ above the Curie temperature T_C *can not be derived* from simple band theory, yet in $ZrZn_2$, which is considered to be a typical weak itinerant electron ferromagnet, Appendix A.19, $\chi(0)$ above T_C exhibits Curie–Weiss behavior. Moreover, the temperature dependence of the magnetization in Fe is in good agreement with a spin wave theory based on the localized spin model (Appendix A.20) (Heisenberg model) and the value of the anomaly in the specific heat at T_C is also close to that expected from the localized spin model. On the other hand, band theory can explain this behavior only with rather unrealistic assumptions [3.7]. Therefore, many experimentalists apply the band theory to explain the ground state, while they analyse the magnetism at finite temperatures by the Heisenberg model.

In the 1970's, many theoretical attempts were made to resolve this contradiction and to understand the magnetism at finite temperatures. Among these, the most remarkable theory is the unified theory developed by Moriya and co-workers [3.2]. They took into account the effect of spin fluctuations on the band theory and the essential points of this theory will be introduced in the following section. For details, readers are referred to the excellent review by Moriya [3.2, 8]. We point out here that neutron scattering experiments have contributed a great deal to the development of this theory which can be applied to understand the magnetism of metals at finite temperatures. The main purpose of this chapter is to introduce the results of neutron scattering studies. In the next section, a review of the utility of neutron scattering for magnetism as well as its recent trends will be given. After that, we will demonstrate how a better understanding of the magnetism in metals can be obtained by neutron scattering studies, showing several recent experimental results.

3.1 Significance of Neutron Scattering for the Study of Magnetism

Neutron scattering experiments for research in material science use neutrons with a wavelength of 1–10 Å. The reason why neutron scattering is important for the study of magnetism is as follows [3.9]:

(1) The neutron is electrically neutral but has a magnetic moment. ($\mu_n =$ $- 1.913\mu_N$, μ_N is the nuclear Bohr magneton). Therefore, we can observe the magnetic moment of an atom through the magnetic scattering of neutrons by this magnetic moment.

(2) The relationship between the energy and the wavelength for neutrons is given as

$$E(k) = \frac{9.51 \times 10^2}{\lambda^2(\text{Å}^2)} \, . \tag{3.1}$$

The wavelength of the neutron for energy of about 300 K is 1.8 Å. This energy and wavelength is convenient for studying both the arrangement of the magnetic moments (space correlation) and their motions (time correlation). In particular, the correlation between the spin state of the i-th atom at $t = 0$, $S_i(0)$, and that of the j-th atom at $t = t$, $S_j(t)$, (the time space correlation function, $\langle S_i(0)S_j(t) \rangle$) can be studied by neutron scattering techniques. The spin wave is the propagating wave of the disturbance of the magnetic moments and is one of the typical examples of time-space correlations. In the case of electro- magnetic waves, X-rays of $\lambda \sim 1$ Å are used to study space correlations, but its energy is about 10^8 K which is unsuitable for studying time correlations. On the other hand, far infra red light ($E \sim 300$ K) is used to study the motion of atoms in various materials, but its wavelength is 50 µm and it is difficult to study space correlations by this technique. In contrast, neutron scattering can be used to study both space and time correlations, thus it has an advantage over electromagnetic scattering techniques.

(3) The magnetic behavior in a material is determined by its generalized susceptibility (Appendix A.14) $\chi(Q)$ or $\chi(Q, \omega)$ which expresses the response of the magnetic moment to the time-space dependent magnetic field $H(r, t)$ (or, its Fourier transform $H(Q, \omega)$). The neutron scattering technique is the only method which can measure this generalized susceptibility. In order to understand the physical meaning of $\chi(Q)$, it is illustrated schematically in Fig. 3.1b. $\chi(Q)$ is also compared in the figure with the static susceptibility, $\chi(0)$ which is the response to a uniform field.

The magnetic scattering of the neutron occurs through the interaction between the magnetic moment of the atom and the magnetic dipole field H_n which is produced by the magnetic moment of the neutron. Therefore, the scattering cross section is related to the magnetic susceptibility of the materials. As is shown in Fig. 3.1a, we assume that the unpolarized neutrons having energy E_i and wave vector k_i are scattered with the dipole interaction U between the neutron magnetic moment μ_n and the atomic moment M. The cross section $d^2\sigma/d\omega dE_f$ for the scattered neutron of energy E_f and wave vector k_f with the scattering vector Q can be calculated. The energy and momentum transfer are

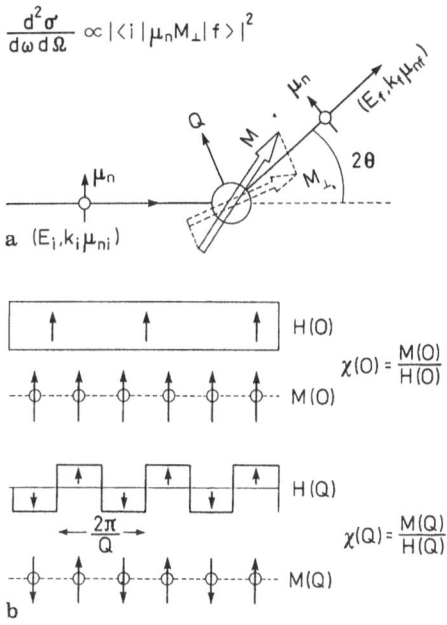

$$\frac{d^2\sigma'}{d\omega\,d\Omega} \propto |\langle i | \mu_n M_\perp | f \rangle|^2$$

Fig. 3.1. a The partial differential cross section for neutron magnetic scattering. **b** The generalized magnetic susceptibility $\chi(Q)$

defined as,

$$\hbar\omega = E_i - E_f \tag{3.2}$$

$$Q = k_i - k_f \tag{3.3}$$

where $\hbar\omega$ is the energy transfer in the scattering process. The cross section is proportional to Q and ω Fourier components of $|U|^2$. The Q Fourier component of U is given in terms of the scalar product of $M_\perp$ and μ_n, where $M_\perp$ is the component of M perpendicular to the scattering vector Q. Therefore, we can say that the neutron is scattered by the perpendicular component of M. The magnetic scattering cross section is given by

$$\frac{d^2\sigma}{d\omega\,d\Omega} = N \left(\frac{e^2\gamma}{mc^2}\right)^2 \frac{k_f}{k_i} |f(\boldsymbol{Q})|^2 \sum_\alpha (1 - \hat{e}_q^{\alpha^2}) S^{\alpha\alpha}(\boldsymbol{Q}, \omega) \tag{3.4}$$

where $f(\boldsymbol{Q})$ is the magnetic form factor, $\hat{e}_q^\alpha$ is the α-component ($\alpha = x, y, z$) of the unit vector in the direction of the scattering vector $\boldsymbol{Q}$ and the $(1 - \hat{e}_q^{\alpha^2})$ term represents the scattering caused by $M_\perp$. $S^{\alpha\alpha}(\boldsymbol{Q}, \omega)$ is the scattering function which is given by the Fourier transform of the time-space correlation function of the spin, if the magnetic moment of the atom is due only to the spin

$$S^{\alpha\alpha}(\boldsymbol{Q}, \omega) = \frac{1}{2\pi N} \sum_{ij} e^{i\boldsymbol{Q}\cdot(\boldsymbol{r}_i - \boldsymbol{r}_j)} \int_{-\infty}^{\infty} e^{-i\omega t} \langle S_i^\alpha(0) S_j^\alpha(t) \rangle \, dt \ . \tag{3.5}$$

$S(Q, \omega)$ is related to the imaginary part of the generalized susceptibility, $\mathrm{Im}\,\chi(Q, \omega)$, through Kubo's fluctuation-dissipation theorem (Appendix A.12),

$$S^{\alpha\alpha}(Q, \omega) = \frac{\hbar}{\pi} \frac{1}{(g\mu_B)^2} \frac{\mathrm{Im}\,\chi^{\alpha\alpha}(Q, \omega)}{1 - \exp(-\hbar\omega/kT)} + S_{st}^{\alpha\alpha}(Q)\delta(\omega) \tag{3.6}$$

where $S_{st}^{\alpha\alpha}(Q)$ is the time independent spin component which does not contribute to the susceptibility. This last term gives the Bragg scattering if there is magnetic order. At high temperatures ($\hbar\omega/kT \ll 1$), the integration of $S(Q, \omega)$ over energy gives,

$$\int_{-\infty}^{\infty} S^{\alpha\alpha}(Q, \omega)d\omega \propto \frac{kT}{\pi} \int \frac{\mathrm{Im}\,\chi^{\alpha\alpha}(Q, \omega)}{\omega} d\omega = kT\chi^{\alpha\alpha}(Q) . \tag{3.7}$$

The last term can be obtained from the Kramers–Kronig relation. Thus, the cross section without energy analysis $d\sigma/d\Omega$ is proportional to $\chi(Q)$,

$$\frac{d\alpha}{d\Omega} = \int \frac{d^2\sigma}{d\omega d\Omega} d\omega \propto kT \sum_\alpha (1 - \hat{e}_q^{\alpha^2})\chi^{\alpha\alpha}(Q) . \tag{3.8}$$

Neutron scattering is classified by its coherent or incoherent and elastic or inelastic character. Each type of scattering gives different information as follows:

1) coherent elastic scattering (Bragg scattering)
2) incoherent elastic scattering (impurity diffuse scattering)
3) coherent quasi-elastic scattering (critical and short range order scattering)
4) coherent inelastic scattering (spin wave scattering)
5) incoherent inelastic scattering (paramagnetic scattering).

In the following sections, information obtained from each type of scattering will be given in more detail. The difficulty of the scattering experiment increases with increasing number in the above list.

To end this section, we describe recent trends in the technical aspects of neutron scattering studies. Recent developments in neutron scattering experiments may be summarized as the remarkable extension of the measurable Q–ω range, which is due to the appearance of a new type of neutron source and new measuring techniques. The situation is demonstrated in Fig. 3.2. In this figure, a region of "thermal neutrons" indicates the region in which studies can be made with the thermal neutrons of an ordinary nuclear reactor. Magnetic structure and its disturbance have been studied extensively in this region. Here the triple axis spectrometer is the most powerful apparatus. In the 1970's, neutrons could be cooled down by installing a cold moderator in the reactor and low energy and low momentum transfer experiments were made possible, which is indicated as the "cold neutron source" region in Fig. 3.2. Small angle scattering instruments were constructed, parallel with the development of the cold neutron source, which made it possible to study semi-macroscopic structures with dimensions of 10–1000 Å, including long period structures and,

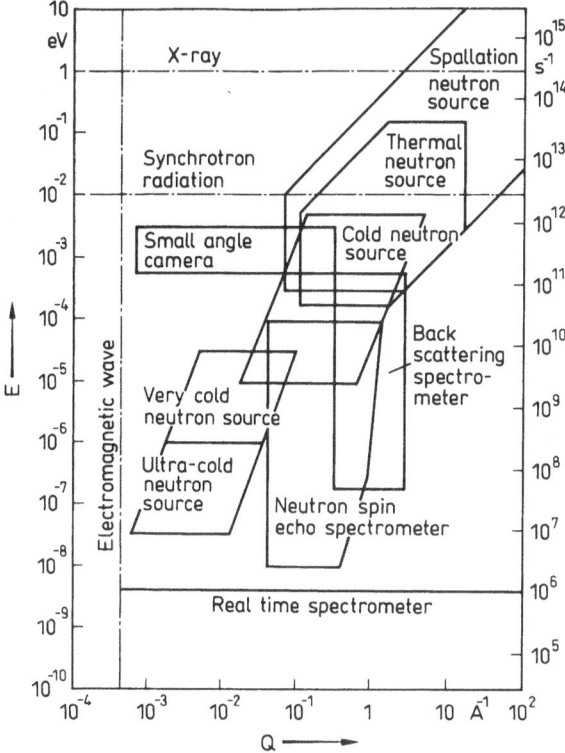

Fig. 3.2. The momentum–energy range that can be covered with various techniques [3.10]

especially, biomolecules. In the field of magnetism in metals, the long period of the spin structure of MnSi, which will be mentioned in the next section, and spin glass structures (Appendix A.32) have also been studied using small angle scattering.

On the other hand, back scattering and neutron spin echo spectrometers have considerably improved the energy (or time) resolution. The contribution of the spin echo spectrometer to research in magnetism is especially large. Slow spin fluctuations around the critical temperature of spin glasses have been measured up to 10^{-8}–10^{-7} seconds with this spectrometer. In the 1980's, the spallation neutron source which produces neutrons using high energy protons has appeared. The range in Q and ω space is considerably extended to high momentum and energy with this neutron source because the neutron intensity in the epithermal region (100 meV $< E <$ 10 eV) is very high in a spallation source. Various researches with this type of neutron source are now in progress. Therefore, the difficulty in the study of magnetism in metals, mentioned in the introduction to this chapter, is expected to be resolved in the near future. Thus, we may say that the most remarkable feature of recent neutron scattering studies is the extension of the accessible regions of Q–ω space.

3.2   Studies of Antiferromagnetic Metals with Elastic Scattering

The most important information obtained from neutron coherent elastic scattering is the magnetic structure and the magnetic form factor $f(Q)$. The latter describes the spacial distribution of electrons which carry the magnetic moment. Here, we introduce only the subjects related to the former, although the latter is very important in understanding the character of the magnetic electron and actually there are many studies on it using polarized neutron beams. Among the various magnetic structures determined by neutron scattering, the helimagnetic structure (Appendix A.15) is strongly related to magnetism in metals. This structure was proposed first by Yoshimori to interpret the magnetism of MnO_2 and was found later in Cr, $MnAu_2$ and many rare earth metals. The period of the spin arrangement in this structure is not commensurate with that of the crystal (incommensurate spin structure) and the spin S_j at the j-th lattice point is generally expressed as,

$$S_j = S_x \cos(q_1 R_j + \phi)\hat{u}_1 + S_y \sin(q_1 R_j + \phi)\hat{u}_2$$
$$+ S_z \cos(q_2 R_j + \psi)\hat{u}_3 \ . \tag{3.9}$$

Various kinds of helimagnets can be defined with values of S_x, S_y, S_z, q_1, etc., which are shown in Table 3.1. In these helimagnetic spin structures, Bragg scattering is observed at the position of $\pm q_i$ away from a reciprocal lattice point τ (satellite). The periodic length R of the spin arrangement is obtained as $R = 2\pi/|q_i|$. Usually, the satellite appears at both sides of the reciprocal lattice point. This is because the left and right handed screw structure usually has the same energy except in the special case of magnetic interactions.

Table 3.1. Various types of helimagnet, with examples

Structure		Examples
spiral	$S_z = 0$, $S_x = S_y$	Tb, Dy, Ho (high temperature phase), MnSi
cone spiral	$S_z = S_0$, $q_z = 0$	Ho (low temperature phase), MnSi ($H \geq 6$ kOe)
cycloid	$S_y = 0$, $q_1 = q_2$	Er
longitudinal-SDW	$S_x = S_y = 0$	Cr ($T < T_f = 121$ K)
transverse-SDW	$S_z = S_y = 0$	Cr ($T > T_f$)

SDW: spin density wave

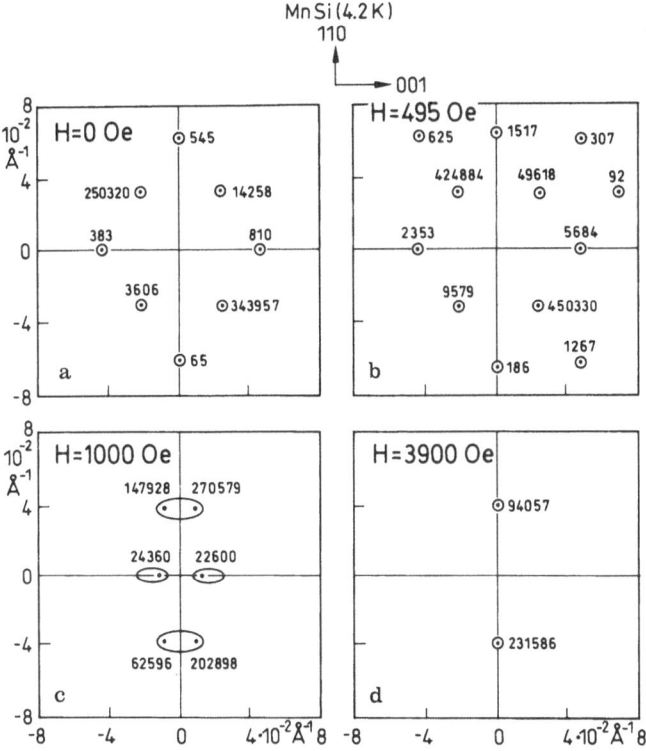

Fig. 3.3. The magnetic field dependence of the scattering patterns of neutron small angle scattering for MnSi. Numerals and circles in the figure indicate the peak intensity and the full width of half maximum respectively [3.11]

Here we review some recent results on the typical metal helimagnets (Appendix A.15) MnSi and Cr. Small angle scattering patterns of MnSi [3.11] around (000) at 4.2 K are shown in Fig. 3.3. Figure 3.3a is the result in zero magnetic field, which shows that MnSi exhibits the helimagnetic spin structure with the long period of $R = 180$ Å ($q = 0.035$ Å^{-1}) along the $\langle 111 \rangle$ direction. By applying a magnetic field above 6.2 kOe the magnetic structure becomes a cone helix and the magnetic moments align parallel with each other (induced ferromagnet (Appendix A.17), Fig. 3.3b–d). The magnetic phase diagram is shown in Fig. 3.4. This induced ferromagnetic state of MnSi shows the various characteristics of a weak itinerant electron ferromagnet (Appendix A.19), which will be discussed in detail later. Recently, the origin of the helimagnet in MnSi is shown by Nakanishi et al. [3.12] and Bak et al. [3.13] to be due to the existence of an antisymmetric exchange interaction $D \cdot [S_i \times S_j]$ (Appendix A.2). This type of interaction originates from the crystal structure of MnSi which has no inversion symmetry. This means that only a right handed or a left handed screw structure

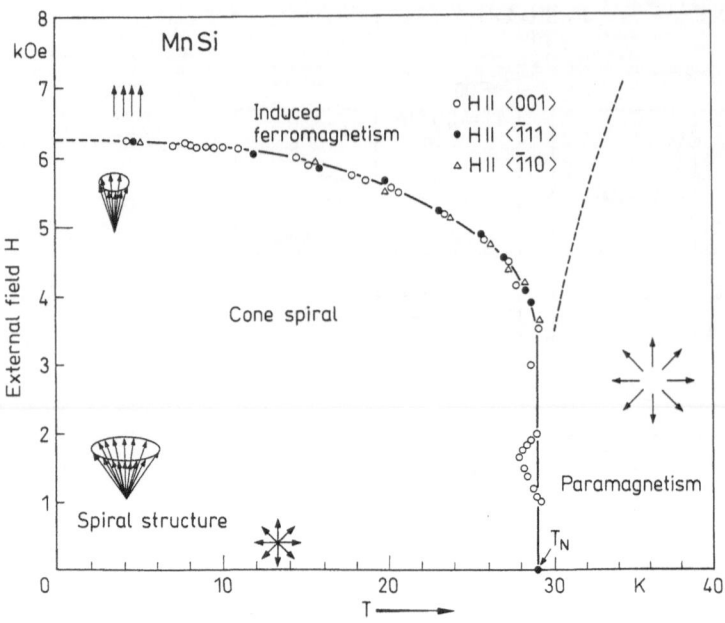

Fig. 3.4. The magnetic phase diagram of MnSi [3.28]

should exist, and actually, polarized neutron scattering experiments have confirmed that the screw structure of MnSi is right handed [3.14]. In general, this
interaction comes from spin-orbit interactions and is as small as $\sim 1/100$
compared with ordinary exchange interactions. Therefore, the long period of the
helimagnetism can be explained by this mechanism.

Cr metal has been known for a long time as a typical itinerant electron
antiferromagnet. It exhibits a spin density wave (SDW) (Appendix A.30) structure which is thought to be stabilized by a peculiar structure of the Fermi surface
[3.4]. The cross section of the Fermi surface of the (100) reciprocal lattice plane
of Cr has an electron and a hole octahedron at the Γ and H point, respectively.
The hole octahedron is slightly larger than that of the electron and these two
octahedra overlap with each other on one side by displacing $Q^\pm = 2\pi/a \pm q$
(Fig. 3.6b). At the Γ and H points, two states mix and produce a new state (SDW
with wave vector q). Thus, the SDW is stabilized by making a gap at the Fermi
surface, (Fig. 3.6b). This is known as the two band theory [3.4].

One of the current studies on Cr is the magnetism of Cr alloys. Phase
diagrams of Cr–V, Mn, Fe, Co have been determined by neutron scattering and
are shown in Fig. 3.7 (upper). These phase diagrams exhibit very different
characters from one another. "I" and "C" in the figure indicate incommensurate
and commensurate spin structures, respectively. In the Cr–Mn alloy system, the
volume of the electron octahedron at the Γ point increases by introducing a Mn

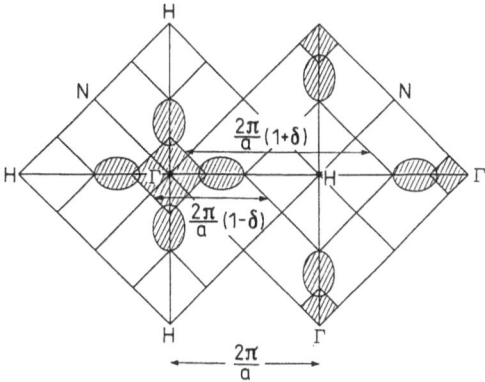

Fig. 3.5. The (100) cross section of the Fermi surface in Cr

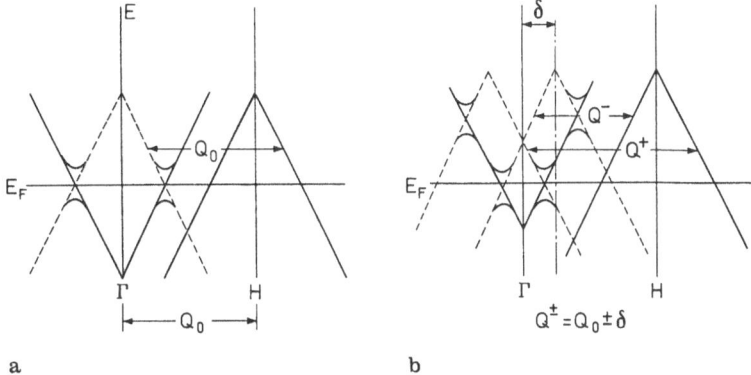

Fig. 3.6a, b. A one-dimensional schematic diagram for the stabilizing mechanism of SDW in Cr. The band gap at E_F appears through the interaction of electrons near the Γ and H points of the Fermi surface. **a** Commensurate structure (CrMn). **b** Incommensurate structure (Cr, CrV)

atom and is equal to the volume of the hole octahedron at the H point. Therefore, the mixture of states of $Q^{\pm} = 2\pi/a$ occurs as is shown in Fig. 3.6a and this mixed state is stabilized by making an energy gap at the Fermi surface. This is the appearance of the commensurate spin structure. On the other hand, in the Cr–V alloy system, the number of d-electrons decreases with the introduction of V atoms and the volume of the electron octahedron becomes smaller. Therefore, only the incommensurate structure appears.

As the stability of this structure depends on the area of the overlapping of the electron and hole octahedra (the energy gap appears at the overlapping region), the decrease and the increase of T_N with the addition of V and Mn, respectively,

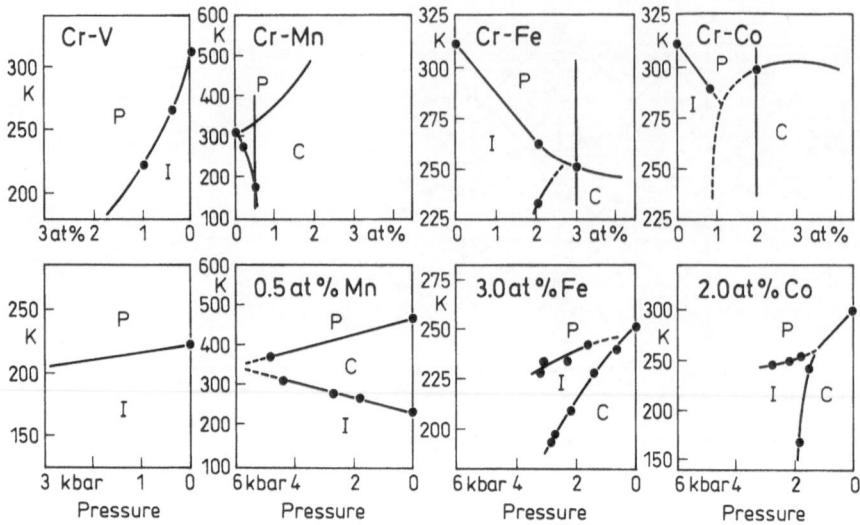

Fig. 3.7. The magnetic phase diagrams of Cr alloys (upper) and their pressure dependence (lower) [3.15]

can be explained by this model. In the CrFe alloy, however, T_N decreases with adding Fe, although the number of d-electrons increases. Moreover, a commensurate structure appears at low temperatures and an incommensurate structure at high temperatures in the 2% Fe alloy, but the situation is reversed in the 0.5% Mn alloy. The change of structure with temperature in the CrFe alloy can not be explained by the simple two band model. In order to explain the complex properties of Cr alloys, neutron scattering experiments under pressure have been performed recently and the magnetic phase diagram has been determined as is shown in the lower column in Fig. 3.7. These pressure phase diagrams are determined for the alloy compositions which are marked by vertical lines in the upper figures. Figures in both rows are very similar to each other if we compare the left hand side of the vertical line in the upper figure with the lower figure. These results indicate that the decrease in the number of d-electrons is equivalent to the decrease in the volume due to pressure. The pressure dependences of T_N, $(1/T_N)\partial T_N/\partial p$, and of the wave vector Q of the SDW at T_N, $(1/Q)\partial Q/\partial p$, have been measured for various alloys which exhibit incommensurate-para transitions. The relationship between $\Delta T_N/T_N$ and $\Delta Q/Q$ thus obtained for various alloys is plotted in Fig. 3.8 (left), and can be expressed well by a universal line except for the CrAl alloy. On the other hand, the same $\Delta T_N/T_N - \Delta Q/Q$ relation obtained from the alloying effect can not be represented by such a line, as is shown in Fig. 3.8 (right). For details of these experiments and the theory one should refer to refs. 3.15 and 3.16.

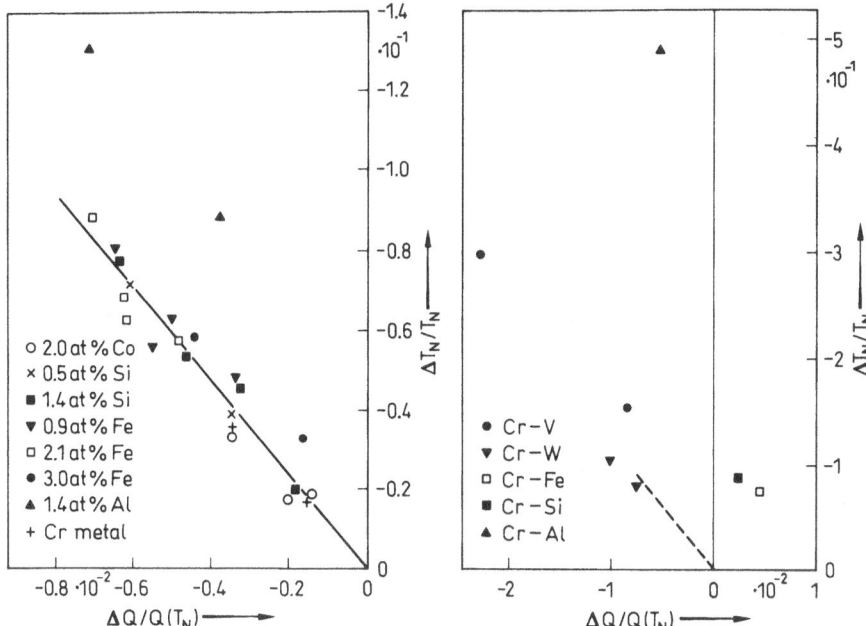

Fig. 3.8. The relation betweeen T_N and $Q(T_N)$ with changing pressure (left) and upon alloying with various elements (right) in Cr alloys [3.15]

3.3 Ferromagnetic Metals Studied by Inelastic Scattering (Dynamical Character)

The investigation of the temperature variation of the physical quantities in metallic ferromagnets is quite important for understanding the origin of magnetism in metals [3.17]. The study of spin fluctuations is thus indispensable and neutron inelastic scattering experiments are quite important for this study. The spontaneous magnetization $\langle M_z \rangle$ and the value of the squared magnetic moment $\langle M_L^2 \rangle$ at an atomic site, where $\langle . . \rangle$ means the thermal average, are important temperature dependent physical quantities. The temperature dependence of $\langle M_L^2 \rangle$ depends on the electron correlations, which has been confirmed recently both theoretically [3.2, 8] and experimentally [3.17]. In Fig. 3.9, the temperature dependences of $\langle M_z \rangle$ and $\sqrt{\langle M_L^2 \rangle}$ are illustrated schematically for various cases, (a) being the strong limit for electron correlation, where $\langle M_z \rangle$ vanishes at T_C while $\langle M_L^2 \rangle$ remains constant even above T_C, as in the case of insulators. The paramagnetic susceptibility χ_p arises from the transverse fluctuation of the atomic magnetic moment. The effective magnetic moment M_{eff} obtained from the Curie constant agrees with $M_z(0)$ and we call this system the

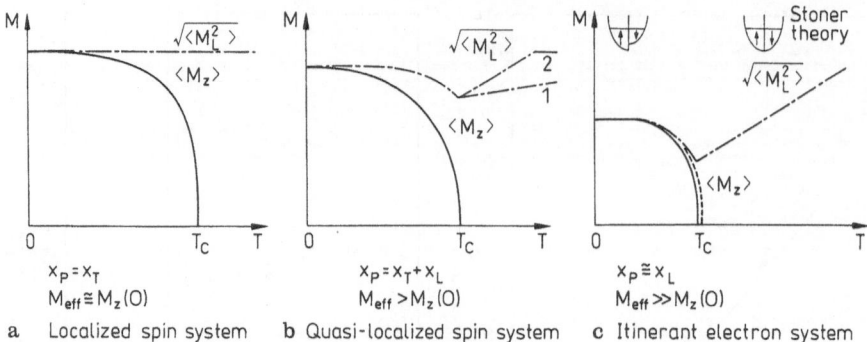

Fig. 3.9a–c. The temperature variations of $\langle M_z \rangle$ and $\langle M_L^2 \rangle$ in three systems classified with different electron correlations [3.17]. **a** Localized spin system. **b** Quasi-localized spin system. **c** Itinerant electron system

localized spin system. The opposite limit to this case, the weak limit of electron correlation, corresponds to the itinerant electron spin system (c). Simple band theory (Stoner theory) (Appendix A.34) predicts that the temperature variation of the spontaneous magnetization arises from the shift of the $+$ and $-$ bands. Therefore, $\langle M_L^2 \rangle$ decreases with increasing temperature and vanishes above T_C, (broken line in (c)). In the spin fluctuation theory (Appendix A.31) proposed by Moriya, however, $\langle M_L^2 \rangle$ decreases with increasing temperature and vanishes above T_C, (dot-chained line in (c)). The Curie–Weiss law in an itinerant electron spin system is due to this increase of $\langle M_L^2 \rangle$ with temperature, i.e., the increase of the amplitude of longitudinal spin fluctuations. Therefore, there is no connection between the M_{eff} obtained from the Curie constant and $M_z(0)$. In the case of intermediate electron correlation (b), the characters of both (a) and (c) mix and the behavior of the magnetic moment is complicated as is illustrated in (b). One of the interesting conclusions of the spin fluctuation theory for this system is that $\langle M_L^2 \rangle$ above T_C tends to saturate at higher temperature as is shown by the chained line (2) in (b), which is really observed in $Co(S_x Se_{1-x})_2$ system [3.18]. This is described in detail in Sect. 4.4.

As discussed above, the direct measurement of $\langle M_L^2 \rangle$ is quite important for studying magnetism in metals. A neutron inelastic scattering experiment gives the q component of the amplitude of the spin fluctuation $\langle M_q M_{-q} \rangle$ and $\langle M_L^2 \rangle$ can be obtained by summing up all the q components over the Brillouin zone;

$$N \langle M_L^2 \rangle = \sum_q \langle M_q M_{-q} \rangle \tag{3.10}$$

and $\langle M_q M_{-q} \rangle$ is related to χ_q through the fluctuation-dissipation theorem,

$$N \langle M_L^2 \rangle \simeq 3kT \sum_q \chi_q = \frac{3kT}{\pi} \sum_q \int \frac{\text{Im}\{\chi(q, \omega)\}}{\omega} \, d\omega \ . \tag{3.11}$$

The unified spin fluctuation theory by Moriya gives $\langle M_L^2 \rangle$ self-consistently from (3.11) using the calculated results of $\chi(q, \omega)$ as a function of $\langle M_L^2 \rangle$. The $\langle M_L^2 \rangle$ illustrated in Fig. 3.9 is obtained in this way.

The observation of $\langle M_q M_{-q} \rangle$ or Im $\chi(q, \omega)$ for all q by means of neutron scattering is, however, almost impossible at present. In theory it should be possible to separate the magnetic scattering in the paramagnetic region from other scatterings using the polarization analysis technique [3.43], but the intensity of the polarized neutron beam is very weak for this purpose. In Fe and Ni, the spectrum extends widely in energy space and epithermal neutrons are needed for these measurements. This is the reason why the high flux spallation neutron source is necessary for the study of magnetism in metals.

The information about $\langle M_L^2 \rangle$ in the ferromagnetic region can be obtained by subtracting the contribution of the spin wave $\langle M_z^{sw} \rangle$ from the total magnetization $\langle M_z \rangle$. The spin wave dispersion relation (Appendix A.33) at the lowest temperature $\hbar\omega_q(0)$ is expressed by the Heisenberg model as,

$$\hbar\omega_q(0) = 2S[J(0) - J(q)] . \tag{3.12}$$

The neutron scattering experiment gives the exchange interaction parameter J_{ij} $[= \sum_q J(q) \exp(iqr)]$ through fitting the experimental results with the dispersion relation (3.12), if the measurement is performed out to the Brillouin zone boundary. Then, the dispersion relation at finite temperature can be calculated from the two-magnon theory. The following two equations are solved self-consistently,

$$\hbar\omega_q(T) = \hbar\omega_q(0) - \frac{2}{N} \sum_k \langle n_k \rangle [J(0) - J(q) - J(k) + J(k-q)] , \tag{3.13}$$

$$\langle n_k \rangle = [\exp(\hbar\omega_k(T)/kT) - 1]^{-1} . \tag{3.14}$$

Thus, the contribution from the spin wave $\langle M_z^{sw} \rangle$ can be obtained as

$$\langle M_z^{sw} \rangle = M_z(0)\left[1 - (NS)^{-1} \sum_k \langle n_k \rangle \right] . \tag{3.15}$$

If the dispersion relation $\hbar\omega(0)$ can not be observed out to the zone boundary so that J_{ij} is not obtained, $\langle M_z^{sw} \rangle$ can be calculated combining (3.14) and the direct observation of $\hbar\omega_q(T)$ at various temperatures in a small q range. $\langle M_L^2 \rangle$ may be considered to be temperature independent if the agreement between $\langle M_z \rangle$ and $\langle M_z^{sw} \rangle$ is good in a temperature range up to $T \sim 0.8\ T_C$.

The temperature variation of $\langle M_L^2 \rangle$ is also obtained from the measurement of the magneto-volume effect. The volume change caused by magnetic ordering $\omega_m (= \Delta V/V)$ is proportional to $\langle M_L^2 \rangle$. Details on this topic are discussed in Chap. 5.

In the study of the dynamical behavior of the metallic ferromagnet by means of neutron scattering, it is very important to choose typical examples which correspond to the three cases shown in Fig. 3.9, and to understand correctly the characteristic properties for each case from careful studies. Results of dynamical behavior measured for typical materials by means of neutron scattering are described in the following sections.

3.4   Spin Dynamics in Localized Spin Systems

A Heusler alloy such as Cu_2MnAl has the chemical formula X_2MnY and is a ferromagnetic alloy in which only the Mn atoms carry the magnetic moment. As the atomic distance between the Mn atoms is more than 4 Å, the magnetic moment localizes on the Mn site and the ferromagnetic coupling is thought to arise via the conduction electrons, i.e. s–d interactions [A28].

Therefore, this alloy system is considered to be a typical localized spin system and it is quite interesting to study the behavior of $\langle M_L^2 \rangle$ and to compare the magnetism at finite temperature with the statistical theory based on the Heisenberg model. For this purpose, neutron spin wave and paramagnetic scattering experiments were carried out for Pd_2MnSn ($T_C = 189$ K), Ni_2MnSn ($T_C = 342$ K) and Cu_2MnAl ($T_C = 610$ K). Only the results for Pd_2MnSn will be described below, because these three alloys exhibit very similar behavior.

The spin wave dispersion relation (Appendix A.33) at the lowest temperature is illustrated in Fig. 3.10. Open circles are for the neutron spin wave scattering

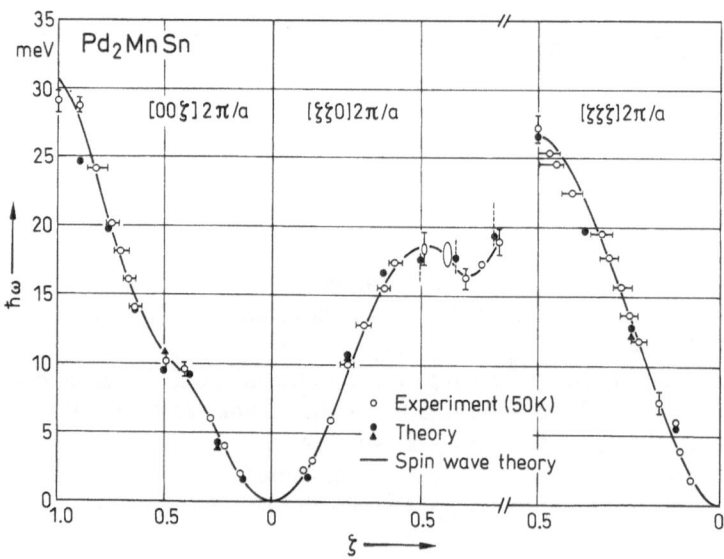

Fig. 3.10. The spin wave dispersion relation of Pd_2MnSn. $\bigcirc$: experimental results [3.19]. $\bullet\,\blacktriangle$: calculations from band theory [3.23]. —: calculations from Heisenberg theory [3.19]

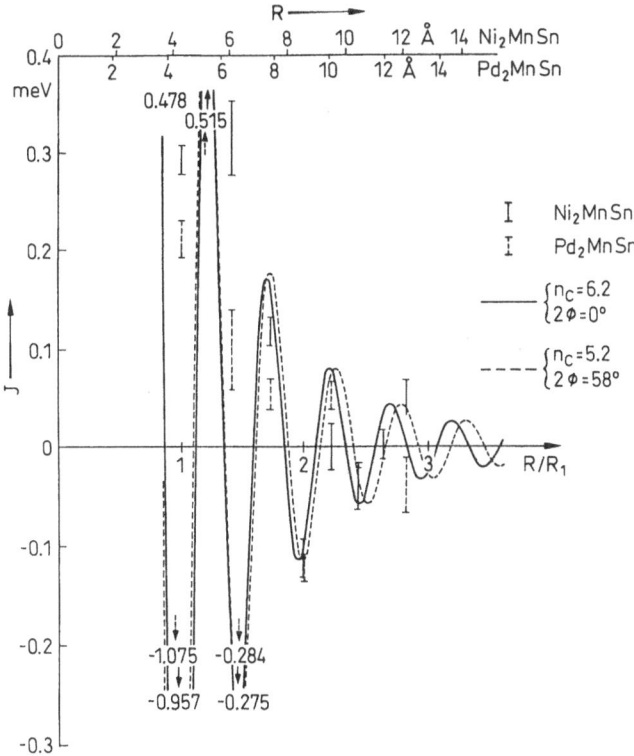

Fig. 3.11. The distance dependence of the exchange parameter $J(r)$ for Pd_2MnSn and Ni_2MnSn obtained from the spin wave dispersion relation in Fig. 3.10. Solid and broken lines are calculated results from te s–d model [3.19]

experiment and the solid lines are the calculated results from (3.12). The exchange interaction parameter J_{ij} thus obtained is plotted against the atomic spacing r_{ij} in Fig. 3.11. Long range exchange interactions, further than eighth neighbor interactions, are necessary to explain the dispersion curve. The result for Ni_2MnSn is also plotted in the figure.

The magnetism of this system at finite temperatures has been studied by calculating $\hbar\omega_q(T)$ and $\langle M_z^{sw} \rangle$ with the two magnon theory (3.13–15) and comparing these with observations. The comparison of calculated and observed $\langle M_z \rangle$ is shown in Fig. 3.12, and good agreement is found in the temperature range $T < 0.8\,T_C$. Above T_C, the energy distribution of the paramagnetic scattering cross section in the localized spin system near the Brillouin zone boundary is expressed by the Gaussian function $(d^2\sigma/d\omega\,d\Omega \sim \exp(-\omega^2/\langle\omega^2\rangle))$ and the width $\langle\omega^2\rangle$ is given as $\langle\omega^2\rangle = (8/3)S(S+1)\sum_n z_n J(R_n)^2 (1 - \cos \mathbf{Q}\cdot\mathbf{R}_n)$, where z_n is the number of n-th neighbor atoms. These relations are confirmed experimentally in Pd_2MnSn. The line

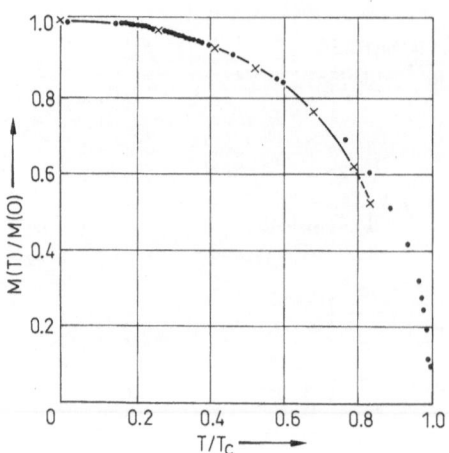

Fig. 3.12. The temperature dependence of the magnetization of Pd_2MnSn. Crosses ($\times$) indicate the calculated results [3.19]

width Γ of the spin wave spectrum in a localized spin system is calculated as [3.20],

$$\Gamma \sim T^2 q^4 \left[\ln\left(\frac{kT}{\hbar\omega(q)} \right) \right]^2 , \tag{3.16}$$

which is in agreement with observations in Pd_2MnSn [3.17].

From the above results, it can be concluded that the $\langle M_L^2 \rangle$ of the Heusler alloy is constant with temperature and its magnetism, including the dynamical behavior at finite temperature, is well explained by the Heisenberg theory. Hence, the role of the conduction electrons, which are the medium of interaction between the local moments, is put into J_{ij}. The dynamical behavior of the conduction electrons is neglected here. Recent measurements of $\langle M_q M_{-q} \rangle$ by means of polarized neutron paramagnetic scattering have confirmed that $\langle M_q M_{-q} \rangle$ at high temperatures ($T = 4T_C$) is independent both of q and temperature, which supports the above conclusion [3.21].

It should be noted that the spin wave dispersion relation for the Heusler alloy has been calculated [3.22, 23] from the band theory based on the random phase approximation (RPA), Appendix A.26. The calculated results are shown by the closed circles and triangles in Fig. 3.10, and are in good agreement with experiment. This result suggests that, even in the localized spin system, band theory can describe correctly the magnetism at the lowest temperature. Therefore, for the description of the magnetism, it may not be unreasonable that we use band theory at the lowest temperature and the Heisenberg theory at finite temperatures.

Another interesting aspect of the Heusler alloy is that information about J_{ij} can be obtained from the band calculation. As shown in Fig. 3.11, J_R exhibits an oscillatory character at large distances which is a characteristic of the s–d interaction (Appendix A.28). Actually, the oscillatory charactor at large distances can be well reproduced by the form of the s–d interaction $J(R) \propto \cos(k_F \cdot R + \phi)/(k_F R)^3$ assuming an appropriate value for the Fermi wave vector k_F and the phase ϕ, as is illustrated by the solid and dashed lines in Fig. 3.11. However, the nearest and 2nd nearest neighbor interactions of this model, which are the most important for the appearance of the ferromagnetism in a Heusler alloy, are negative in sign contrary to observation. Therefore, the s–d interaction model cannot be applied for these near neighbor interactions. These exchange interactions in the Heusler alloy $(X_2 Mn Y)$ are determined by the X atom [3.18]. The interaction is maximal when the X atom is Cu $(T_C \sim 600 \text{ K})$, and minimal in the Pd-Heusler alloy $(T_C \sim 200 \text{ K})$. This may be explained by recent band calculations and readers should refer to ref. 3.23. The study of the exchange interaction based on band theory is a very important problem for localized spin systems in metals.

3.5 Spin Dynamics in Itinerant Electron Systems

The spin dynamics of itinerant electron systems has been studied theoretically for the simple band structure based on the RPA resulting in excitation spectrum shown in Fig. 3.13. The imaginary part of the generalized susceptibility, (Appendix A.14), $\chi(q, \omega)$ in RPA (Appendix A.26) is expressed as,

$$\text{Im } \chi(q, \omega) = \frac{\text{Im } \chi_0(q, \omega)}{(1 - \text{Re}\chi_0(q, \omega)I)^2 + (\text{Im } \chi_0(q, \omega)I)^2} , \quad (3.17)$$

where I is the intra-atomic Coulomb interaction parameter and χ_0 is the non-interacting susceptibility. The region I in Fig. 3.13 is the region of $\text{Im }\chi_0(q, \omega) = 0$ where Stoner excitation (Appendix A.34) from the $+$ spin band to the $-$ spin band does not occur and the spin wave excitation with infinite life time is observable, since $\text{Im}\chi(q, \omega) \propto \delta(1 - \text{Re }\chi_0(q, \omega)I) = \delta(\omega - \omega(q))$. The dispersion relation in the small q region is expressed as $\hbar\omega_q = Dq^2$ which is the same q dependence as in the Heisenberg model. The region II is the region of $\text{Im }\chi_0(q, \omega) \neq 0$ where the transition of electrons from the $+$ spin band to the $-$ spin band occurs. This is well known as the Stoner excitation. The neutron scattering intensity of the spin wave decreases abruptly as the spin wave dispersion curve touches the Stoner boundary and the dispersion curve merges into the Stoner continuum (Fig. 3.13b). The intensity variation obeys the sum rule of the magnetic response,

$$\int_{-\infty}^{\infty} \text{Im } \chi^{-+}(q, \omega)d\omega \propto \langle M_z(q) \rangle \quad (3.18)$$

and the contribution from the Stoner continuum increases when the excitation energy approaches the continuum boundary.

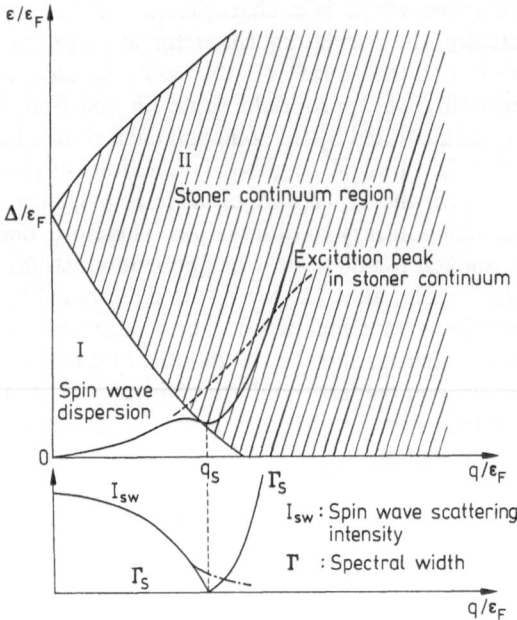

Fig. 3.13. The magnetic excitation spectrum in an itinerant elctron ferromagnet obtained from the RPA

Many attempts have been made to observe the Stoner continuum because this is the clearest demonstration of the itineracy of the electron. However, only in Fe and Ni, has the scattering intensity of the spin wave been observed to decrease abruptly around the transfer energy of 100 meV [3.25, 26], and this is considered as an indication of the existence of the Stoner continuum. However, from the results of band calculations, it was predicted that the decrease of the spin wave scattering intensity along the [100] direction in Ni was due to the crossing of the optical and acoustic modes of the spin wave [3.6] and in fact, a gap around the crossing point was observed [3.27] as shown in Fig. 3.14. Moreover, a recent measurement on a spallation neutron source confirmed that spin waves exist up to 180 meV in Fe [3.45].

At present, magnetic excitations in the Stoner continuum have only been observed in MnSi [3.28]. The intensity of the magnetic excitation in magnetic fields of 5 kOe and 10 kOe are shown with equi-intensity lines in Fig. 3.15. Spin waves with infinite life times which have the dispersion relation $\hbar\omega_q = Dq^2 + C$ have been observed at low temperatures in the low energy transfer region. The width of this excitation spectrum broadens discontinuously above 2.5 meV, which suggests the penetration of the spin wave dispersion into the Stoner continuum. The Stoner boundary moves in the high energy direction with an increase of magnetic field [3.29]. It is interesting to note that the energy of the Stoner boundary (2.5 meV) is nearly equal to the Curie temperature of MnSi.

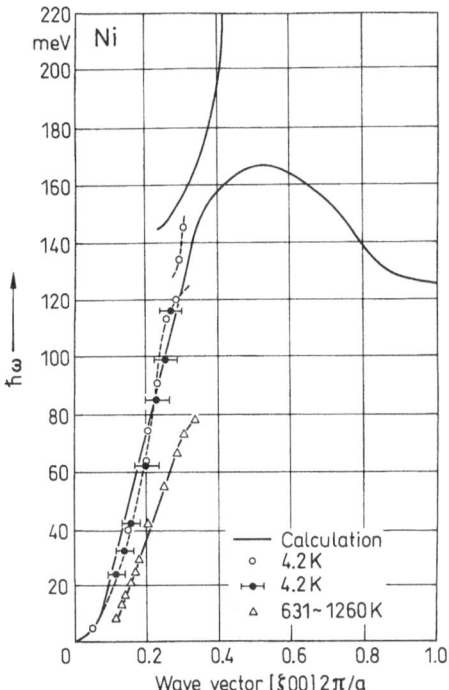

Fig. 3.14. The spin wave dispersion of Ni [3.27]. The experimental result and the calculation based on the band model [3.6]

The spin wave stiffness constant D (Appendix A.8) decreases with increasing temperature and critical scattering is observed above the Curie temperature, while the Stoner continuum is almost independent of temperature and remains observable even at 290 K ($T/T_C \approx 9$), as shown in Fig. 3.15b and c. It is very impressive that this behavior of the magnetic excitations in MnSi, which is considered an itinerant system, agrees at least qualitatively with the result of the band calculation based on the RPA. The temperature dependence of the magnetization in MnSi may be due to the Stoner excitation. Actually, the spin wave contributes only partly to the temperature variation of the magnetization measured at 10 kOe as shown in Fig. 3.16, and it is the change of $\langle M_L^2 \rangle$ with temperature that is expected to explain the major temperature dependence of the magnetization. In fact, a large volume magnetostriction, which corresponds to the temperature dependence $\langle M_L^2 \rangle$, is observed in MnSi [3.30]. From this experimental result, one may conclude that $\langle M_L^2 \rangle$ does not vanish at T_C and that it has a value of $(2.7/5)M(0)^2$ which is very close to the theoretical value of $(3/5)M(0)^2$.

The observation of magnetic excitations in MnSi above T_C is very important in proving the validity of the spin fluctuation theory proposed by Moriya. According to the theory, $\langle M_L^2 \rangle$ should increase above T_C with increasing

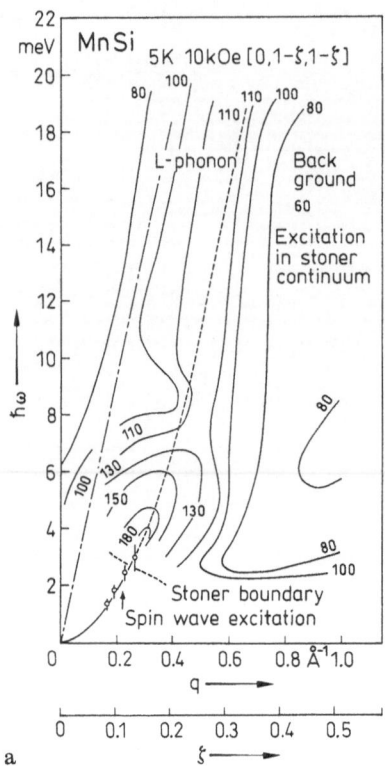

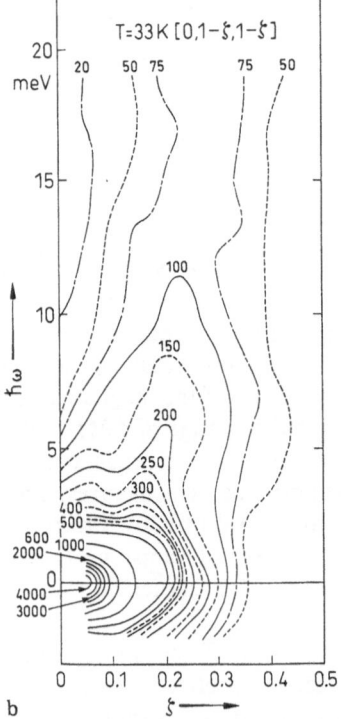

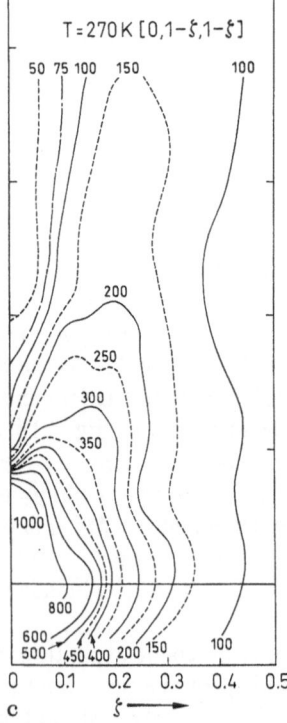

Fig. 3.15a–c. The magnetic excitation spectrum in MnSi measured by neutron scattering. **a** $T = 5$ K, $H = 10$ kOe (ferromagnetic state) [3.28]. **b** $T = 33$ K, $H = 0$ (paramagnetic state) **c** $T = 270$ K, $H = 0$ (paramagnetic state) [3.32]

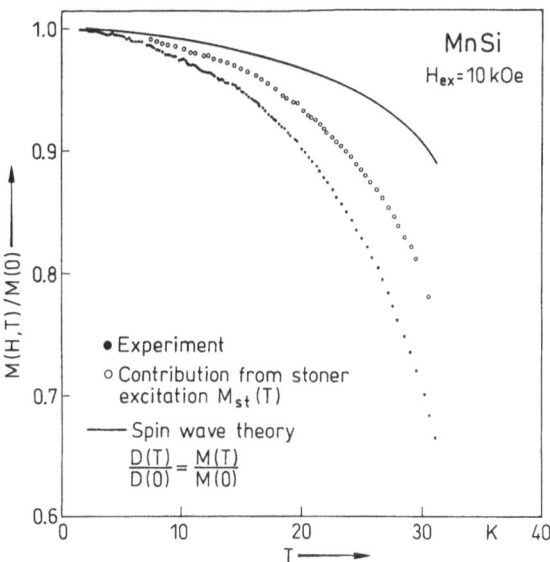

Fig. 3.16. The temperature variation of the magnetization ($H = 10$ kOe) in MnSi (1) and the contribution from the spin wave $\langle M_z^{sw} \rangle$ (3). Open circles (2) correspond to $\sqrt{\langle M_L^2 \rangle}$ [3.30]

temperature, (Fig. 3.9c) and this is the origin of the Curie–Weiss behavior of the susceptibility of this material. This is expected from the experimental result of the magneto-volume effect on MnSi [3.30] and can be studied in more detail by neutron scattering. From the theory of Moriya [3.2, 8], $\chi(q)$ at $T > T_C$ in the small q range is given as,

$$\frac{1}{\chi(q)} = \frac{1}{\chi_0(q)} - 2I + \frac{5}{3}\frac{\gamma}{N}\langle M_L^2(T)\rangle \tag{3.19}$$

where the third term on the right hand side corresponds to the interaction from mode-mode coupling which is introduced by Moriya. The susceptibility based on the RPA, $\chi(q)_{RPA}$, is obtained by neglecting the third term of (3.19). $\langle M_L^2 \rangle$ is given by $\chi(q)$, as shown in (3.11) and so

$$\frac{1}{\chi(q)} = \frac{1}{\chi_{RPA}(q)} + \frac{5\gamma}{N^2}kT\sum_q \chi(q) , \tag{3.20}$$

and $\chi(0)$ diverges at T_C. Therefore, if $\sum_q \chi(q)$ is independent of temperature,

$$\frac{1}{\chi(0, T)} = \frac{5\gamma k}{N^2}\sum_q \chi_q(T - T_C) \tag{3.21}$$

since the $\chi(0)_{RPA}$ changes only little at T_C. This is the Curie–Weiss law for $\chi(0)$.

The Curie constant of (3.21) is proportional to γ which corresponds to the rate of increase in $\langle M_L^2 \rangle$, and has no correlation with $\langle M_z(0) \rangle$.

From the above discussion, the following conditions are necessary for the new type of the Curie–Weiss law which is due to the increase of $\langle M_L^2 \rangle$ with increasing temperature.

(1) $\sum_q \chi(q)$ is independent of temperature.
(2) In order to exhibit the temperature dependence of $\chi(0)$ to be Curie–Weiss type and satisfy the condition (1), $\chi(q)$ must have a large q dependence around $q = 0$.

Experimental results such as those in Fig. 3.15 can be examined by the spin fluctuation theory (Appendix A.31) mentioned above [3.31]. $\chi(q)$ of (3.19) is extended to $\chi(q, \omega)$ and is expressed as

$$\chi(q, \omega) = \frac{\chi_0(q, \omega)}{1 - I\chi_0(q, \omega) + \lambda(q, \omega)} , \tag{3.22}$$

where $\lambda(q, \omega)$ represents the mode-mode coupling of the spin fluctuations. In weak ferromagnetism (Appendix A.19) as in MnSi, only fluctuations with long wavelengths are important. $\chi_0(q, \omega)$ and $\lambda(q, \omega)$ are expanded as

$$\chi_0(q, \omega) = \chi_0(0, 0)\{1 - Aq^2 + \cdots + iB(\omega/q)\} \tag{3.23}$$

$$\lambda(q, \omega) = \lambda(0, 0) = \lambda_0 T . \tag{3.24}$$

Therefore, (3.17) is expressed as

$$\operatorname{Im} \chi(q, \omega) = \frac{c\omega q}{\Gamma_0^2 q^2 [\kappa(T)^2 + q^2]^2 + \omega^2} \tag{3.25}$$

where $\kappa^2(T) = \kappa_0^2(T/T_C - 1)$, $\Gamma_0 = A/B$ and $\kappa_0^2 = \lambda_0 T_C/I_0 \chi_0 A$. This is equivalent to the double Lorentzian expression for paramagnetic spin fluctuations in a localized spin system

$$\operatorname{Im} \chi(q, \omega) = \frac{c_0}{\kappa(T)^2 + q^2} \frac{\Gamma(q)\omega}{\Gamma(q)^2 + \omega^2} \tag{3.26}$$

if $\Gamma(q)$ is expressed as

$$\Gamma(q) = \frac{cq}{\chi(q)} = \Gamma_0 q(\kappa^2 + q^2) . \tag{3.27}$$

The difference with the localized spin system appears in the expression for $\Gamma(q)$. In the localized spin system, Γ is proportional to the square of q, $\Gamma(q) = cq^2/\chi(q)$.

The experimental results on MnSi obtained by neutron scattering in the temperature range between 33 K and 270 K [3.32] can be well reproduced by (3.25) assuming the values of $\kappa_0^2 = 0.0325$ A^{-2} and $\Gamma_0 = 50.0$ meV A^3. It should be noted that the value of κ_0^2 in MnSi is one order of magnitude smaller than that for other localized spin systems (κ_0^2(EuO) = 0.405 A^{-2}, κ_0^2(Fe) = 1.1 A^{-2}). Therefore, the significant temperature dependence of Im $\chi(q, \omega)$ or $\chi(q)$ is limited to the small q region and the q-dependence of $\chi(q)$ is large even at $T = 10T_C$. The large q and T dependent part of Im $\chi(q, \omega)$ is caused by the term $\lambda_0 T$, which is called the MK (Moriya–Kawabata) fluctuation (Appendix A.31). Except for the MK fluctuation, Im $\chi(q, \omega)$ is almost independent of temperature because $\kappa^2 \ll q^2$, can be seen from Fig. 3.15a–c. This temperature independent part corresponds to the excitation in the Stoner continuum. $\chi(q) = (1/\pi) \int (\mathrm{Im}\chi(q, \omega)/\omega)\, d\omega$ is also independent of temperature. Thus, in MnSi, the criteria (i) and (ii) predicted from the spin fluctuation theory are satisfied. The increase of $\langle M_L^2 \rangle$ with increasing temperature above T_C in MnSi is mostly determined by the increase of $\langle M_q^2 \rangle \approx 3kT\chi(q)$ with increasing temperature in the high q region [3.32]. This is because the contribution of $\langle M_q^2 \rangle$ to $\langle M_L^2 \rangle$ ($= \sum_q 4\pi q^2 \langle M_q^2 \rangle$) is significant for large q. It should be noted that these characteristics are due to the large value of A in (3.23).

3.6 Spin Dynamics in Quasi-Localized Spin Systems

The quasi-localized spin system [3.33] is a system which has relatively large magnetic moments and exhibits a spin dynamics different from that of localized spin systems caused by magnetic interaction due to overlap of d-electrons. Ferromagnetic metals such as Fe, Ni and many other transition metals belong to this type of system. In the following, the spin dynamics of MnPt$_3$, FePd$_3$ and FePt$_3$ will be discussed, as their dynamics are very similar to that of a localized spin system. Each of these ordered alloys has the Cu$_3$Au type ordered structure but their magnetic properties are very different. MnPt$_3$ ($M_{Mn} = 3.64\mu_B$, $M_{Pt} = 0.26\mu_B$, $T_C = 390$ K) and FePd$_3$ ($M_{Fe} = 2.86\mu_B$, $M_{Pd} = 0.34\mu_B$, $T_C = 529$ K) are ferromagnets, while FePt$_3$ ($M_{Fe} = 3.3\mu_B$, $M_{Pt} = 0$, $T_N = 170$ K) is an anti-ferromagnet [3.34] with a wave vector of $Q_0 = (1/2, 1/2, 0)2\pi/a$. Spin wave dispersion relations have been observed for MnPt$_3$ [3.35] and FePt$_3$ [3.36] up to the zone boundary as shown in Figs. 3.17 and 3.18. The dispersion curve of FePd$_3$ is similar to that of MnPt$_3$, but an anomalous broadening of the spectrum has been observed in FePd$_3$ near the zone boundary for the [100] direction [3.37]. These dispersion relations have been analysed by the Heisenberg Hamiltonian $H = -2\sum_r J(r)S_0 \cdot S_r$ and $J(r)$ has been determined [3.33] as shown in Fig. 3.19. The assumption that the magnetic moment is located only on Fe or Mn atoms has been made for the analysis (the magnetic moment at a Pd or Pt site is considered to be polarized by the local moment of an Fe or Mn site). Solid lines in Fig. 3.17 and 3.18 are the calculated results using the $J(r)$ thus obtained. The characteristic behavior of $J(r)$ in Fig. 3.19 is that both the magnitude and the distance dependence of $J(r)$ for the three alloys are rather similar to each other. This indicates that the analysis based on the Heisen-

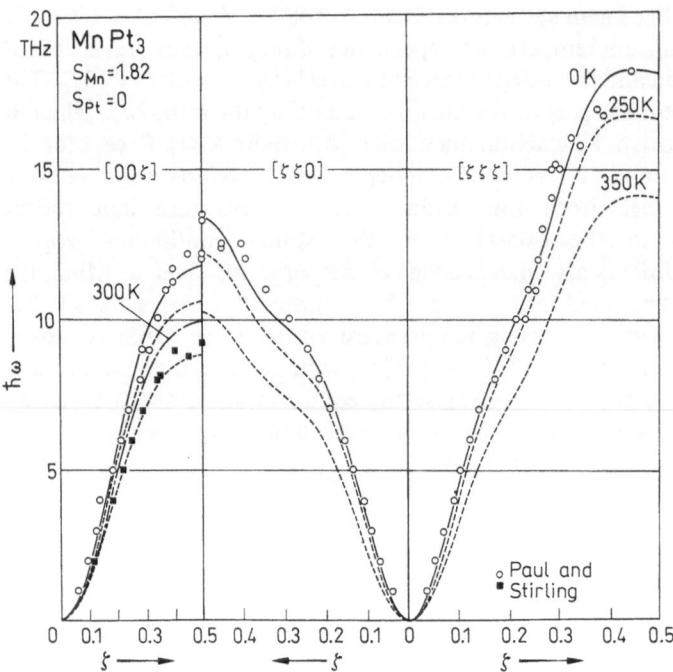

Fig. 3.17. The spin wave dispersion relation in MnPt$_3$ (ferromagnet) [3.35] and the calculation from spin wave theory (solid and broken lines) [3.33]

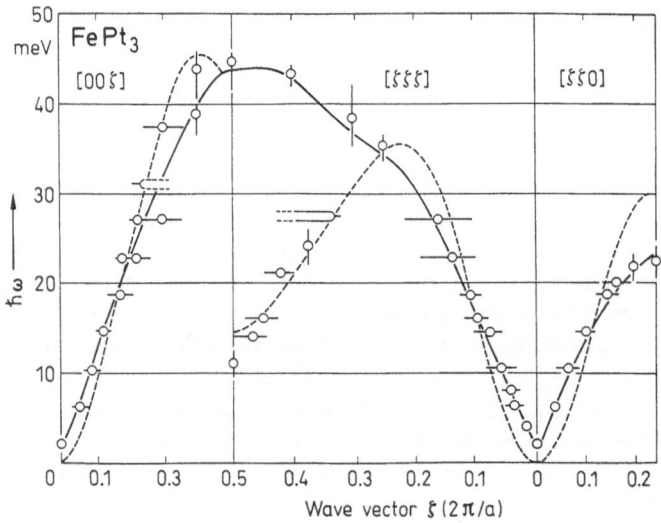

Fig. 3.18. The spin wave dispersion relation in FePt$_3$ (antiferromagnet) and the calculation from spin wave theory (solid and broken lines [3.36]

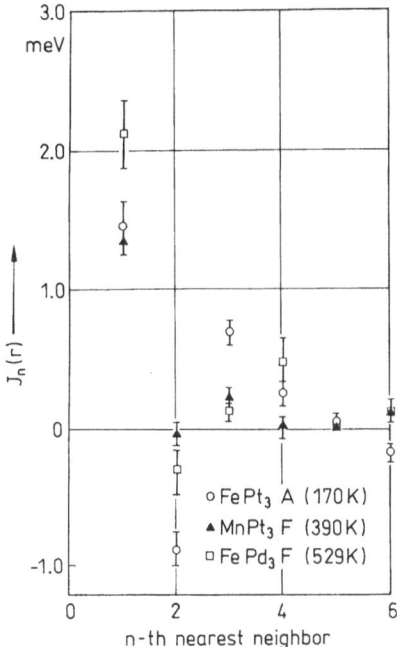

Fig. 3.19. The distance dependence of the exchange parameter $J(r)$ in three Cu_3Au type ordered alloys obtained from the spin wave dispersion relation [3.33]

model may be valid for this type of alloy system, and that the antiferromagnetism of $FePt_3$ may be due to the fact that the second-neighbor interaction parameter has a relatively large value compared with that of the ferromagnetic $MnPt_3$.

The characteristic properties of these materials at finite temperatures can be analysed by the method discussed in Sects. 3.4 and 3.5. In $MnPt_3$ [3.33], the dispersion relation at finite temperatures can be well reproduced by the calculation based on the two-magnon theory as shown in Fig. 3.18. However, a similar calculation for $FePt_3$ did not give good agreement with experimental results. The magnetization $\langle M_q^z \rangle$ of $FePt_3$ observed from neutron scattering is compared with the calculated result based on the two-magnon theory (RSW) or the Green function method (GF) in Fig. 3.20. The agreement between the observed and calculated values is found to be poor [3.36].

On the other hand, the paramagnetic scattering spectrum of $FePt_3$ has a Gaussian shape near the antiferromagnetic wave vector Q_0, which is expected from the Heisenberg theory. The width of the spectrum $\langle \omega^2 \rangle$ is almost in good agreement with the calculation using the value of $J(r)$ shown in Fig. 3.19. The temperature dependence of $\chi(Q_0)$ obtained from the integration of the spectrum obeys the Curie–Weiss law, as is shown in Fig. 3.21a, and the slope of $\chi(Q_0)$ gives

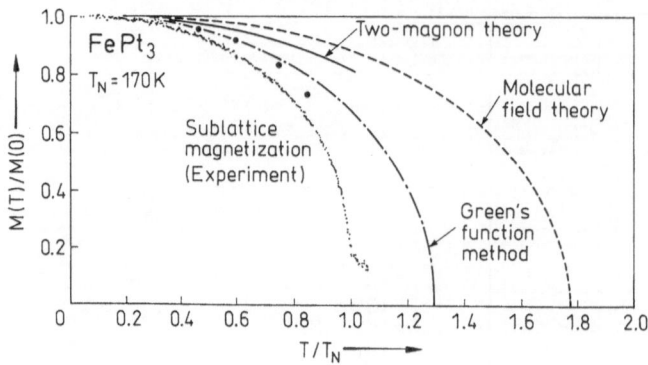

Fig. 3.20. The temperature dependence of the sublattice magnetization in antiferromagnetic $FePt_3$ and calculations based on spin wave theory [3.36]

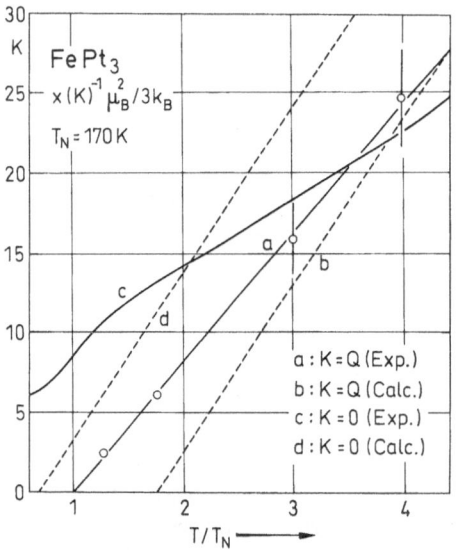

Fig. 3.21. The temperature dependence of $\chi(K)$ for $FePt_3$ in the paramagnetic region [3.36]

the effective magnetic moment M_{eff} of $4.0 \pm 0.3\,\mu_B$ which is close to the value of the saturation moment at $0°$K, $M_{z(Fe)} = 3.3\,\mu_B$. However, M_{eff} obtained from the static susceptibility $\chi(0)$ is $5.7\,\mu_B$ which is very different from $M_{z(Fe)}$. The Q-dependence of $\chi(Q)$ was found to deviate from the expected value of the localized spin model only in the region $Q \cong 0$. These properties can be explained by the assumption that the magnetic moment of Fe behaves as a localized spin whereas the Pt electrons are polarized by the local magnetic moment of Fe [3.36]. An attempt has been made to explain the dynamical behavior of $FePd_3$ using a similar concept with some success [3.38].

3.7 Dynamic Behavior of Invar Alloys

Invar alloys [3.39] such as $Fe_{65}Ni_{35}$ and Fe_3Pt are alloys which exhibit a large positive magneto-volume effect ω_m. The origin of this Invar effect may be essentially due to the volume change being proportional to the temperature variation of the atomic magnetic moment $\langle M_L^2 \rangle$. The temperature dependence of ω_m is indeed very large in MnSi in which the temperature variation of $\langle M_L^2 \rangle$ is very large. The value of ω_m, which can be estimated from $\omega_m = \kappa C \Delta \langle M_L^2 \rangle$, explains quantitatively the temperature and magnetic field dependences of the volume magnetostriction below and above T_C. Therefore, it is important to know what kind of excitation causes the variation of $\langle M_L^2 \rangle$ in Fe_3Pt, $Fe_{65}Ni_{35}$ etc., as well as in MnSi.

The simplest mechanism for the temperature variation of $\langle M_L^2 \rangle$ at low temperatures is the existence of the Stoner boundary located at the low energy excitation region as in MnSi. Neutron scattering experiments of the magnetic

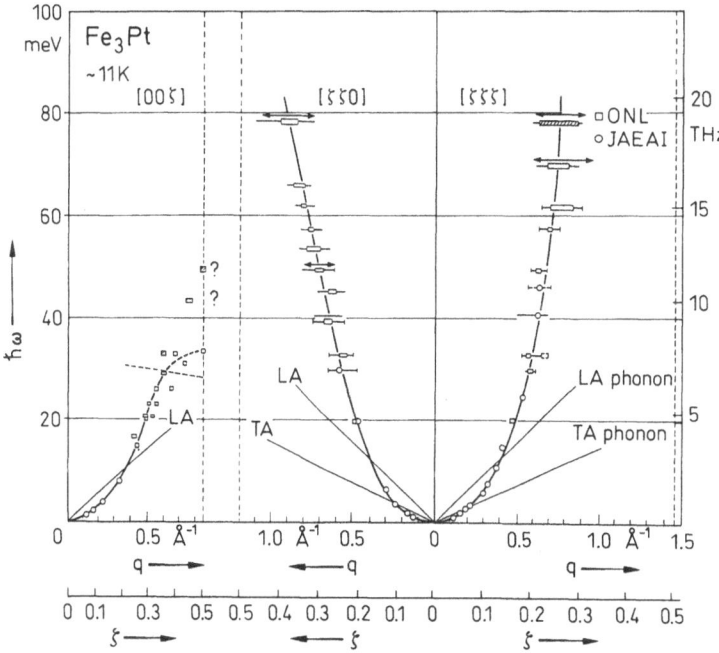

Fig. 3.22. The spin wave dispersion curve of the ferromagnetic Fe_3Pt [3.40]

excitations in $Fe_{65}Ni_{35}$ (disordered alloy) and ordered Fe_3Pt have been performed and well defined spin wave peaks have been observed up to 80 meV in both alloys at low temperatures. As an example, the dispersion curve for Fe_3Pt is illustrated in Fig. 3.22 [3.40]. From the experimental results, it is suggested that the Stoner boundary is located at an energy higher than 80 meV and the simple Stoner excitation may not be the origin of the temperature variation of $\langle M_L^2 \rangle$, in contrast to MnSi. However, no observation has been made near the Brillouin zone boundary and no information exists about the possible low lying Stoner boundary there.

It is also clear that the magnetism at finite temperatures can not be explained by spin wave excitations based on a simple localized spin model. The calculation of $\langle M_z \rangle$ can be made using the observed $\hbar\omega_q(T)$, as has been shown in Sects. 3.5 and 3.6, but the result only explains about a half of the temperature variation of the observed $\langle M_z \rangle$ in $Fe_{65}Ni_{35}$ and Fe_3Pt shown in Fig. 3.23. On the other

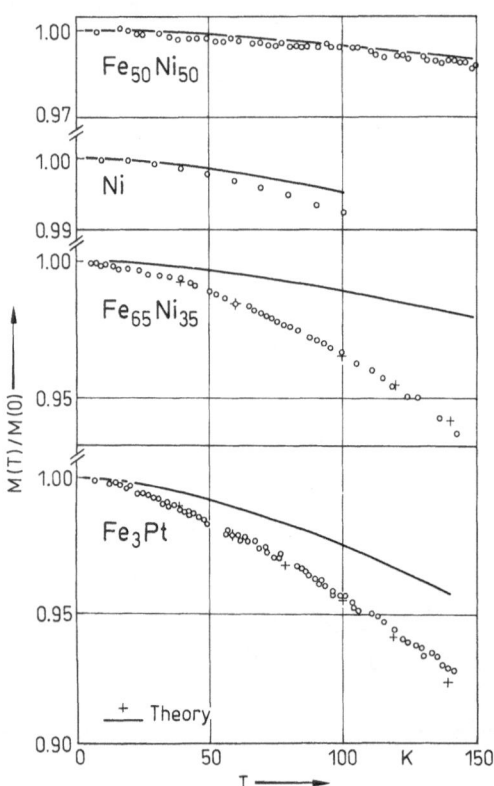

Fig. 3.23. The temperature dependence of the magnetization at low temperatures for various ferromagnetic metals and comparison with the calculated results based on spin wave theory [3.33]

hand, in the non-Invar alloy of $Fe_{50}Ni_{50}$, the temperature variation of the $\langle M_z \rangle$ can be well reproduced by a spin wave excitation. The same situation is also found in amorphous ferromagnetic alloys. In $Fe_{86}B_{14}$ which exhibits the Invar effect, observations of $\langle M_z \rangle$ and calculations by spin wave theory do not agree with each other [3.41], but in the non-Invar alloy of $Fe_{70}Cr_{10}P_{13}$, the agreement is very good [3.42]. These results suggest that the Invar effect is apparently caused by the temperature variation of $\langle M_z \rangle$ due to an excitation different from a spin wave excitation. Moreover, the life time Γ of the spin wave energy spectrum exhibits q and T dependences of $\Gamma \propto q^2(a + bT^\alpha)(\alpha \leq 1)$, which are different from (3.16), which might be due to the interaction of the spin wave with this other excitation.

At present, this excitation cannot be observed, and is thus known as the "hidden excitation". It is speculated that the hidden excitation is a very slow collective motion which is accompanied by the lattice expansion, or it is the Stoner excitation induced by the spin wave excitations. Various experiments for observing the hidden excitation are possible, and if it does exist, it would be a new type of excitation mechanism in magnetic metals.

3.8 Epilogue—The Magnetism of Fe and Ni

In this chapter, we have demonstrated that the problem of the magnetism in transition metal alloys which is complicated and difficult to understand, becomes clear through the classification of metals and alloys into several typical cases with different electron correlations. The study of the dynamical properties of the magnetic moment (spin fluctuation) by the neutron scattering, especially inelastic scattering technique plays an important role for this classification. Then, as mentioned at the beginning of this chapter, how can one understand the magnetism of Fe and Ni, the most typical metallic ferromagnets, at finite temperatures?

In Fe, the temperature variation of $\langle M_z \rangle$ can be confirmed to be due to spin wave excitations at least at low temperatures, as estimated from the calculation of $\langle M_z^{sw} \rangle$ shown in Sects. 3.5 and 3.7. In Ni, however, $\langle M_z \rangle$ and $\langle M_z^{sw} \rangle$ do not agree with each other (Fig. 3.23), and a simple localized model cannot explain the temperature variation of the magnetization. The magnetic property of metals that is essentially different from that of localized spin systems was thought to be the long life time spin wave-like excitation above the Curie temperature (Fig. 3.14). This was regarded as the result of strongly correlated short range order above the Curie temperature, which may be a "special characteristic" of the itineracy of $3d$ electrons, and the mechanism of which has been discussed by many theoreticians [3.1, 2, 44]. However, recently, precise neutron paramagnetic scattering experiments on Fe and Ni have confirmed that the paramagnetic scattering functions of Fe and Ni can be expressed essentially by (3.26), which means that the dynamic behaviour of Fe and Ni is not very different from that of a localized spin system such as Pd_2MnSn in the measured energy ($E < 60$ meV) and momentum ($q < 0.2(2\pi/a)$) transfer range [3.43]. In order to obtain the characteristics of Fe and Ni as metallic magnets, measurements with wider energy and momentum transfer ranges are indispensable. But such measurements are very difficult to perform with present techniques, as mentioned above. However, as studies in this field are now being made extensively, we may have a better understanding of this problem in the near future. In the transition metal compounds which are described in the next chapter, the dynamic behavior of the magnetic moment is also studied by neutron scattering and this may shed more light on the understanding of magnetism in these materials.

Finally, the author (Y.S.) would like to thank Professor S. Chikazumi who gave him the opportunity to begin his studies both on neutron scattering and on the magnetism of metals.

Addendum

KEISUKE TAJIMA

Professor Y. Ishikawa died in February 1986. As he has pointed out, neutron paramagnetic scattering experiments are a very important technique for the study of magnetism in metals, and after this chapter was written, several studies have been made on the neutron paramagnetic scattering of metallic magnets [3.46–49].

In Pd_2MnSn [3.46], measurements were performed up to $T = 4T_C$ and $\chi(q)$ has been obtained from the integration of the constant Q scan [see (3.8)], and this has been compared with the calculation based on the localized spin model,

$$\chi(q) = \frac{(g\mu_B)^2}{(g\mu_B)^2/\chi(0) + 2[J(0) - J(q)]} \ . \tag{3.28}$$

It should be noted that (3.28) contains no adjustable parameter because $J(0) - J(q)$ is obtained from the measurement of the spin wave dispersion at the lowest temperature. The experimental and calculated $\langle M_q^2 \rangle = 3kT\chi(q)$ thus obtained are illustrated in Fig. 3.24. The agreement is very good up to the Brillouin zone boundary. Other properties of the paramagnetic spin fluctuations including the line shape of the energy spectrum can be fairly well reproduced by the localized spin model except for some results at high temperature.

The same method of comparison of $\langle M_q^2 \rangle$ have also been applied to Fe, Ni and their alloys [3.47]. From (3.28), $\langle M_q^2 \rangle$ is expressed as

$$\langle M_q^2 \rangle \simeq 3kT\chi(q) = \frac{\langle M_0^2 \rangle}{1 + \hbar\omega_q \langle M_0^2 \rangle / 3kT(g\mu_B)^2 S} \tag{3.29}$$

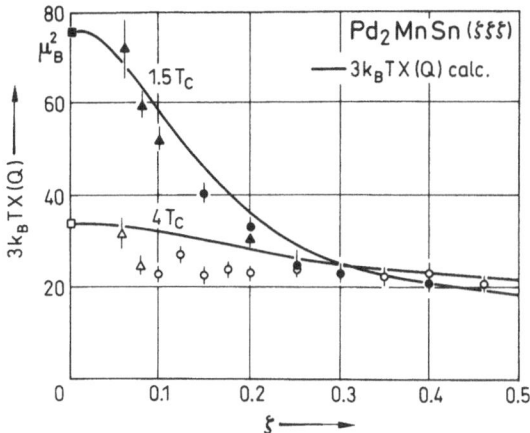

Fig. 3.24. $\langle M_q^2 \rangle$ $(=3kT\chi(q))$ of Pd_2MnSn at $T = 1.5T_C$ and $4T_C$. Solid lines are calculated curves obtained from (3.28)

where $\hbar\omega_q$ is the spin wave energy $\hbar\omega_q = Dq^2(1 - \beta q^2)$ at low temperatures. The parameters D, β, S and $\langle M(0)^2 \rangle$ can be determined from observation. The experimental results for Fe, Ni and $Fe_{65}Ni_{35}$ Invar alloy and the calculated results with (3.29) are illustrated in Fig. 3.25 and are in good agreement with each other. Thus, (3.28) or (3.29) based on the localized spin model describe very well the spin fluctuations not only in the localized spin system but in the itinerant spin system. In the itinerant spin system for Fe, Ni and FeNi alloy, however, the calculation can be applied only for q's less than half way to the Brillouin zone boundary since a well defined spin wave peak is observable only in this q range, whereas in the localized spin system, the calculation can be compared favorably with experiment out to the zone boundary. These results suggest that, in a relatively small q range, the behavior of the spin fluctuation in the itinerant spin system is almost identical with that of the localized spin system and the characteristic behavior in an itinerant spin system may appear only at large q. The conclusion is also consistent with the result mentioned in the Epilogue.

Another interesting study of paramagnetic scattering was made for FeSi [3.48]. It has the same crystal structure as MnSi and exhibits an unusual temperature dependence for the magnetic susceptibility. It shows a broad peak around 500 K, but no magnetic ordering is observed down to low temperature. In order to explain the anomalous paramagnetism in FeSi, a model of the semiconducting band with a narrow band gap was proposed [3.49] and spin fluctuation theory was applied [3.50] to this model. Careful neutron scattering experiments on FeSi confirmed the appearance of a temperature induced paramagnetic moment which has no spacial (instantaneous) correlations. The temperature dependence of the magnetic moment $\langle M^2 \rangle$ thus observed is shown

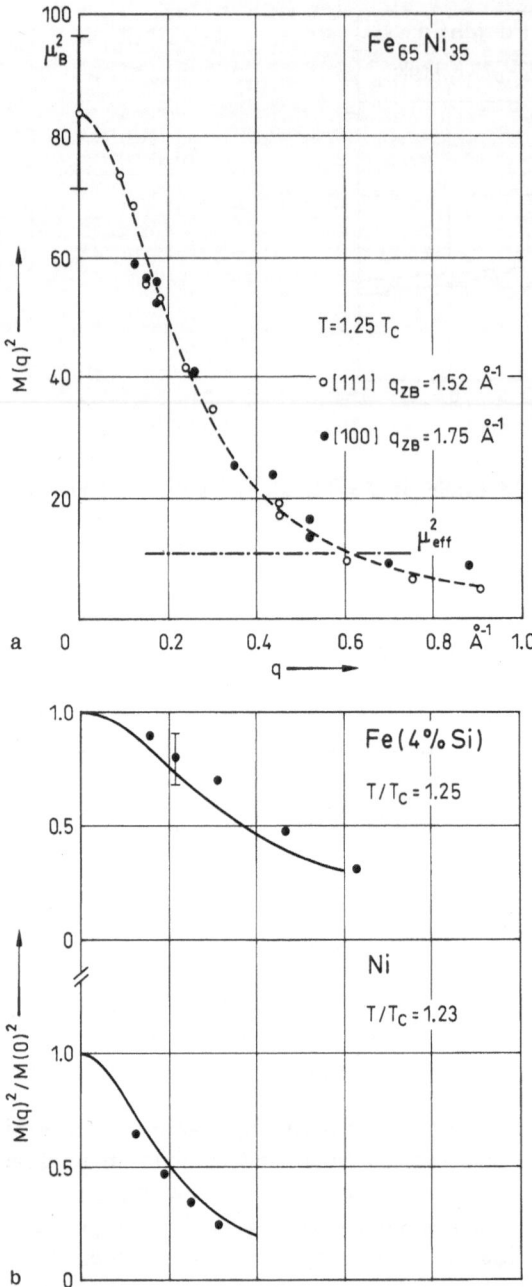

Fig. 3.25. a $\langle M_q^2 \rangle$ of $Fe_{65}Ni_{35}$. The dashed line is the calculated curve from (3.29).
b $\langle M_q^2 \rangle$ of Fe and Ni. Solid lines are calculated curves from (3.29)

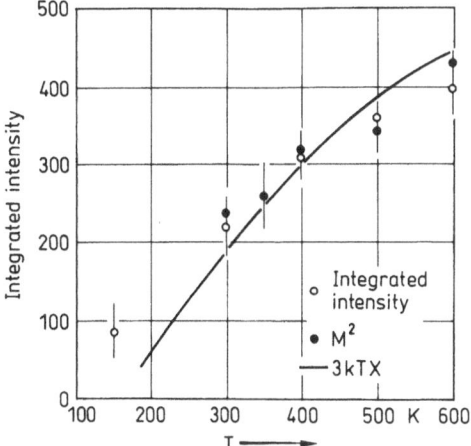

Fig. 3.26. The temperature variation of the integrated intensity of a constant Q scan for FeSi. The solid line is calculated from the static susceptibility

in Fig. 3.26. The solid line is the calculated curve using the measured static susceptibility, and it agrees fairly well with experiment. The behaviour of the life time of the spin fluctuation exhibits a unique feature which may reflect the band structure of FeSi. In an itinerant spin system, the temperature induced magnetic moment is one of the most important characteristics and, actually, is observed also in MnSi at high temperature. Further experiments on FeSi may give important information for understanding the magnetism of metals.

References

3.1. Panel Discussion on 3*d* Magnetism at Finite Temperatures: J. Mag. Mag. Mat. **31–34**, 313 (1983)
3.2. T. Moriya: J. Mag. Mag. Mat. **14** (1979) 1, ibid. **31–34**, 10 (1983)
3.3. A.V. Gold: J. Appl. Phys. **39**, 768 (1968)
3.4. W.M. Lomer: Proc. Phys. Soc. (London) **80**, 489 (1962)
3.5. J. Kanamori: Solid State Physics (in Japanese) **12**, 626 (1977)
3.6. J.F. Cooke, J.M. Lynn and H.L. Davis: Phys. Rev. **B21**, 2095 (1980)
3.7. M. Shimizu: Rept. Prog. Phys. **44**, 329 (1981)
3.8. T. Moriya: *Spin Fluctuations in Itinerant Electron Magnetism* (Springer, 1985)
3.9. G.E. Bacon: *Neutron Diffraction*, 3rd Ed (Oxford Univ. Press, Oxford 1975)
       S. Lovesey: *Theory of Neutron Scattering from Condensed Matter*, Vol. II (Oxford Univ. Press, Oxford 1984)
3.10. Y. Ishikawa: Hyperfine Interactions **17–19**, 17 (1984)
3.11. Y. Ishikawa, K. Tajima, D. Bloch, M. Roth: Solid State Commun. **19**, 525 (1976)

3.12. O. Nakanishi, A. Yanase, A. Hasegawa, M. Kataoka: Solid State Commun. **35**, 995 (1980)
3.13. P. Bak and M.H. Jensen: J. Phys. **C13**, L881 (1980)
3.14. M. Ishida, Y. Endoh, Y. Ishikawa, S. Mitsuda, M. Tanaka: Private communication
3.15. J. Mizuki, Y. Endoh, Y. Ishikawa: J. Phys. Soc. Jpn. **51**, 3497 (1982)
3.16. K. Nakanishi, T. Kasuya: J. Phys. Soc. Jpn. **42**, 833 (1977)
3.17. Y. Ishikawa: Butsuri **36**, 655 (1981) (in Japanese)
3.18. K. Adachi, K. Sato, M. Takeda: J. Phys. Soc. Jpn. **14**, 123 (1979)
3.19. Y. Ishikawa, Y. Noda: AIP Conf. Proc. **24**, 145 (1975)
     Y. Noda, Y. Ishikawa: J. Phys. Soc. Jpn. **40**, 690, 699 (1976)
3.20. V.G. Vaks, A.I. Larkin, S.A. Pikin: Sov. Phys. JETP **26**, 647 (1968)
3.21. P.J. Brown, J. Déportes, D. Givord, K.R.A. Ziebeck: J. Appl. Phys. **53**, 1973 (1982)
3.22. S. Ishida, J. Ishida, S. Asano, J. Yamashita: J. Phys. Soc. Jpn. **45**, 1239 (1978)
3.23. Y. Kubo, S. Ishida, J. Ishida, S. Asano: J. Phys. Soc. Jpn. **48**, 407 (1980)
3.24. Ishikawa: J. Appl. Phys. **49**, 2125 (1978)
3.25. H.A. Mook, J. W. Lynn, R.M. Nicklow: Phys. Rev. Lett. **30**, 556 (1973)
3.26. H.A.J.W. Lynn: Phys. Rev. **B11**, 2624 (1975)
3.27. H.A. Mook, D. Tocchetti: Phys. Rev. Lett. **43**, 2029 (1979)
3.28. Y. Ishikawa, G. Shirane, J.A. Tarvin, M. Kohgi: Phys. Rev. **B16**, 4956 (1977)
3.29. J.A. Tarvin, G. Shirane, Y. Endoh, Y. Ishikawa: Phys. Rev. **B18**, 4815 (1978)
3.30. M. Matsunaga, Y. Ishikawa, T. Nakajima: J. Phys. Soc. Jpn. **51**, 115 (1982)
3.31. Y. Ishikawa, Y. Noda, C. Fincher, G. Shirane: Phys. Rev. **25**, 254 (1982)
3.32. Y. Ishikawa, Y. Noda, Y. Uemura, G. Shirane: Phys. Rev. **B31**, 5884 (1985)
3.33. Y. Ishikawa: J. Mag. Mag. Mat. **14**, 123 (1979)
3.34. G.E. Bacon, J. Crangle: Proc. Roy. Soc. (London) **A272**, 387 (1963)
3.35. D.M.K. Paul, W.G. Stirling: J. Phys. **F9**, 2439 (1979)
3.36. M. Kohgi and Y. Ishikawa: J. Phys. Soc. Jpn. **49**, 985, 994 (1980)
3.37. A.J. Smith, W.C. Stirling, T.M. Holden: J. Phys. **F7**, 2411 (1977)
3.38. H. Yamada, M. Shimizu: J. Phys. **F7**, L203 (1977)
3.39. Y. Ishikawa, S. Onodera, K. Tajima: J. Mag. Mag. Mat. **10**, 183 (1979)
3.40. Y. Ishikawa, K. Tajima, Y. Noda, N. Wakabayashi: J. Phys. Soc. Jpn. **48**, 1097 (1980)
3.41. Y. Ishikawa, K. Yamada, K. Tajima, K. Fukamichi: J. Phys. Soc. Jpn. **50**, 195 (1981)
3.42. Ze Xianyu, Y. Ishikawa, S. Onodera: J. Phys. Soc. Jpn. **51**, 1749
3.43. G. Shirane, O. Steinsvoll, Y.J. Uemura, J. Wicksted: J. Appl. Phys. **55**, 1887 (1984)
3.44. D.M. Edwards: J. Mag. Mag. Mat. **36**, 213 (1983)
3.45. C.K. Loong, J.M. Carpenter, J.W. Lynn, R.A. Robinson, H.A. Mook: J. Appl. Phys. **55**, 1895 (1984)
3.46. M. Kohgi, Y. Endoh, Y. Ishikawa, H. Yoshizawa, G. Shirane: Phys. Rev. **B34**, 1762 (1986)
3.47. K. Tajima, P. Böni, G. Shirane, Y. Ishikawa, M. Kohgi: Phys. Rev. **B35**, 274 (1987)
3.48. G. Shirane, J. Fischer, Y. Endoh, K. Tajima: Phys. Rev. Lett. **59**, 351 (1987)
     K. Tajima, Y. Endoh, J. Fischer, G. Shirane: Phys. Rev. **B38**, 6954 (1988)
3.49. V. Jaccarino, G. Wertheim, J. Wernick, L. Walker: Phys. Rev. **160**, 476 (1967)
3.50. Y. Takahashi, T. Moriya: J. Phys. Soc. Jpn. **46**, 1451 (1979)

4. Magnetic Properties of 3d Compounds with Special Reference to Pyrite Type Compounds

KENGO ADACHI

Since the first investigation on pyrrhotite, Fe_7S_8, by Weiss [4.1] in 1904 in France, an enormous number of magnetic compounds have been studied. Historically, these investigations can be classified into three stages. Those before 1950 were carried out mainly by those interested in crystal chemistry and mineralogy, for example, the typical investigations made by H. Haraldsen and his group in Norway.

After 1950, the intrinsic origin of the ferromagnetism, ferrimagnetism and antiferromagnetism appearing in various oxides and halogenides was made clear by making use of the localized spin model based on the superexchange mechanism (Appendix A.35). The magnetic structures of these compounds were substantiated by neutron diffraction experiments. L. Néel (France) and P.W. Anderson (USA) contributed much to the development of this field.

A similar model was then applied to explain the magnetic properties of other compounds such as chalcogenides (S, Se and Te compounds) and pnictides (P, As, Sb and Bi compounds). These compounds contain metalloid elements with low electronegativities, and do not have typical ionic properties, such as an integer number of spins and insulating behavior. It was therefore difficult to explain their magnetic properties on the basis of a localized spin model.

On the other hand, in order to interpret the magnetic properties of metals and alloys, the Stoner model (Appendix A.34) on itinerant electron magnetism has been utilized extensively and applied to these compounds. However some of the discussions based on this model assumed an artificial density of states curve for the energy band. Therefore, though the localized spin model was successful for ionic compounds, this seems to have been a confused period for many other compounds. The two models, localized spin versus itinerant electrons, were much disputed at this stage.

The latest stage was opened in 1973 by T. Moriya [4.2] and his group (Japan) with the discovery of a new theory called the self-consistent renormalization theory of spin fluctuations, hereafter called the spin fluctuation theory (Appendix A.31). Using this theory, the intrinsic faults of the Stoner theory were greatly improved, such as the linearity of inverse susceptibility for weak ferromagnetism, a reasonable value of the Curie point for band calculations and a temperature variation of local magnetic moments for metallic ferromagnetism. By taking electron correlation effects into account, they also developed a unified theory which connected the two systems, i.e. localized and itinerant electron systems, which appear so widely in insulators and metals. Thus the various magnetic properties for intermediate compounds between these two systems can now be interpreted to test the validity of this theory. Some properties have already been successfully explained from the theory as shown later.

Besides the above mentioned developments, in the second stage, J. Kondo (Japan) discovered in 1964 an effect which explained theoretically the low temperature conduction anomalies found in certain dilute alloys containing $3d$ elements. This is known as the Kondo effect. The theory has been developed in basic physics and applied to electronic processes not only in dilute alloys but also in some intermetallic compounds consisting of rare earth and actinide elements. This phenomenon is called the dense Kondo effect and many interesting magnetic properties have been found and the results interpreted as an effect of valence fluctuations and heavy fermions [4.3]. However, this chapter deals only with $3d$ compounds, so such phenomena are not discussed.

In the next section, a general survey of electronic processes in $3d$ compounds will be mentioned, and some physical properties of pyrite type compounds will be introduced as examples. The results will be interpreted from the standpoint of current theories on magnetism.

4.1   General Survey of $3d$-Magnetic Compounds

Figure 4.1 [4.4] shows a general view of magnetic compounds composed of $3d$ transition metal and some metalloid elements, i.e. nitrides and pnictides (Vb group), oxides and chalcogenides (VIb group) and halogenides (VIIb group), in which the $3d$ atom is surrounded octahedrally by six metalloid atoms. The compounds are arranged according to the number of $3d$ electrons, $3d^n$, the electronegativity of the metalloid ions and their crystal structure.

In this figure, the solid curve means a boundary between the localization and delocalization of $3d$ electrons (Appendix A.20) indicating a metal-insulator boundary. The shaded area signifies a region where the magnetic moments of the $3d$ atom disappear and the compounds show normal metallic behavior, while in the non-shaded region, they show Curie–Weiss type paramagnetism above the transition point. Above the solid line, the $3d$ ions possess in general an integer number of spin moments. In the non-shaded intermediate metallic region, their magnetic moments are more or less non-integer, and some very interesting compounds appear in this region. Compounds with half-filled d levels, d^5, tend to have localized states as in insulators. A similar tendency is also seen in the case of $3d^3$ and $3d^8$ configurations where the d-sublevel is split by an octahedral ligand and is half and fully filled, respectively. In addition, a few ferromagnetic compounds marked in squares appear on both sides of the d^5 configuration, and their behavior will be explained later by the effects of electron correlation in an itinerant electron system.

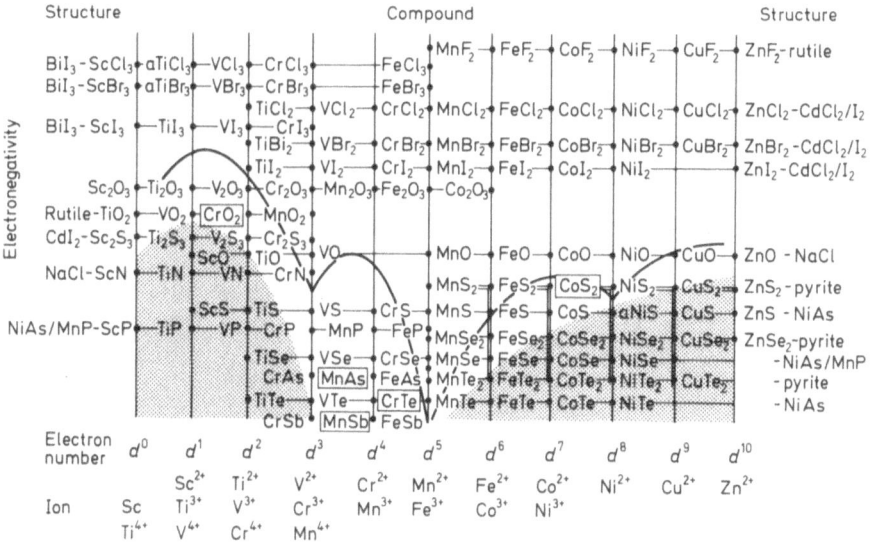

Fig. 4.1. General survey of 3d-compounds with octahedral environments as a function of 3d-electron numbers. d^n, for the ionic state and the electronegativity [4.4]. Solid curve means a boundary between insulating (upper and metallic (lower) regions. In the shaded region, the magnetic moments of the 3d ions disappear. Some intermediate compounds are located between them. Ferromagnetic compounds are marked by □

For many magnetic materials, the value of P_c/P_s can be chosen as a measure of the degree of localization of the 3d electron spins, where P_c is the paramagnetic moment per atom obtained from the effective moment P_{eff} of the inverse susceptibility, and written as $P_c = 2S = \sqrt{1 + P_{eff}^2} - 1$, while P_s is the real atomic moment at 0 K determined from the saturation magnetization in a ferromagnet or from neutron diffraction experiment in antiferro- and helimagnets. The values of P_c/P_s versus the Curie point T_C are shown in Fig. 4.2 [4.5] for some ferromagnetic materials. These values are classified into two branches, the one is $P_c/P_s \simeq 1$ indicating perfect localization of the moments, while the others with $P_c/P_s > 1$ signify some decrease of local moments surviving on the magnetic atom in an itinerant electron system. Weak itinerant ferromagnetism occurs at high values of P_c/P_s and low T_Cs.

It is important to note that in the Moriya theory the local magnetic moment in an itinerant electron system is temperature dependent. On the other hand in the case of ideal localization in insulators, the size of the moment does not change in the temperature range above and below T_C. In the Stoner model, which describes metallic magnetism, the moment changes sharply with temperature and disappears above T_C. According to the new theory, however, neither model is adequate to explain the behavior of real metals because they exhibit an intermediate behavior as shown in Fig. 4.3. The local moment M_{loc} does not

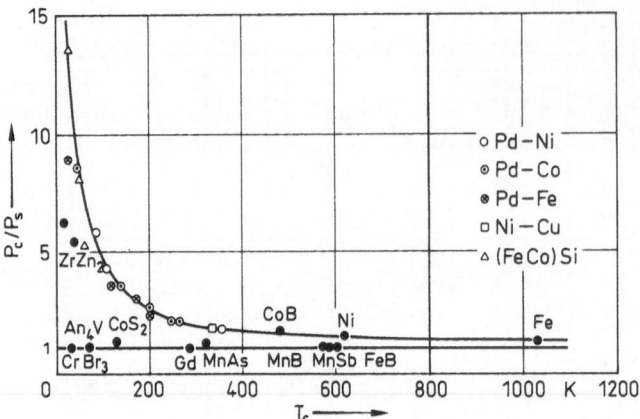

Fig. 4.2. Ratio of magnetic moments P_c/P_s versus T_C for some ferromagnetic compounds, where P_c means that obtained from the inverse susceptibility and P_s the real moment at 0 K

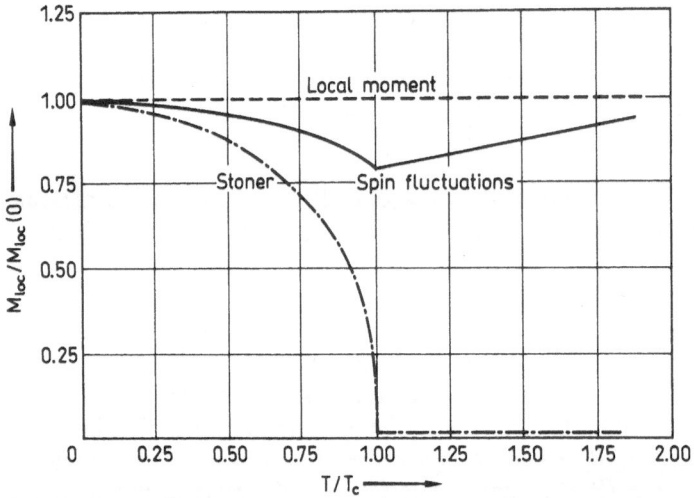

Fig. 4.3. Temperature dependence of the local magnetic moment. Dotted, dashed and solid lines are the localized, Stoner and spin fluctuation models respectively. The last one is the real case

disappear at T_C nor above T_C, and, in particular, the moment remaining above T_C increases slowly with temperature and in some cases, saturates at a certain temperature. Such an increase is called a "temperature induced moment". This effect is seen from the existence of paramagnons above T_C, the change in slope of the inverse susceptibility and the lattice dilatation near T_C relating to the Invar effect, as explained in Chap. 5, in some weak ferromagnets and strong paramagnets.

Many interesting and important phenomena have been reported in chalcogenides, pnictides and their mixed systems which belong to the intermediate region in Fig. 4.1, especially in compounds with FeS_2, NiAs, MnP, Fe_2P, Cu_2Sb type crystal structures. In addition, many ternary compounds with spinel and Heusler type structures have also been investigated. However, we cannot mention all of them within this short chapter. Details on them have been tabulated recently in Landolt–Börnstein [4.6].

In this chapter, we focus our interest on the magnetic and electrical properties of the pyrite (FeS_2) type compounds, which contain phenomena which can be explained by the new theory on itinerant electron magnetism.

4.2 Experimental Results

4.2.1 Physical Properties

The crystal structure of pyrite (FeS_2) belongs to the space group $T_h^6 - Pa3$ as shown in Fig. 4.4. The unit cell is composed of four molecules in which the Fe $= M$ atoms form an fcc lattice. The position of the S $= X$ atom is indicated by a parameter u having the value of 0.38 to 0.40. The center of the S_2 dumbell

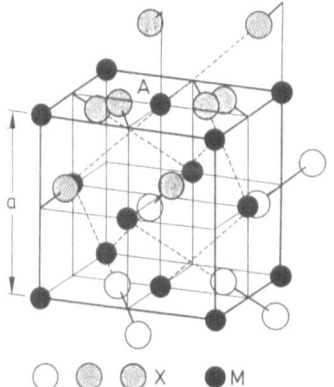

Fig. 4.4. Crystal structure of the pyrite type compound, MX_2. The parameter u indicates the coordinate of the X atoms. The atomic coordinates are: M: (000), (0, 1/2, 1/2), (1/2, 0, 1/2) and (1/2, 1/2, 0), X: $(uuu), (\bar{u}, 1/2 + u, 1/2 - u), (1/2 - u, \bar{u}, 1/2 + u)$ $(1/2 + u, 1/2 - u, \bar{u}), (\bar{u}, \bar{u}, \bar{u}), (u, 1/2 - u, 1/2 + u)$ $(1/2 + u, u, 1/2 - u), (1/2 - u, 1/2 + u, u)$

Table 4.1. Physical properties of pyrite type compounds. Adapted from [4.6]

Compound	Lattice constant a (Å)	Magnetic property	Transition point (K)	Magnetic moment (μ_B)	Electrical conduction	Electron configuration	Remarks
MnS_2	6.101	AF	$T_N = 48$	5	SC	$d\varepsilon^3 d\gamma^2$	spin structure: $2a \times a \times a$
$MnSe_2$	6.417	AF	$T_N = 47$	5	SC		$3a \times a \times a$
$MnTe_2$	6.954	AF	$T_N = 87$	5	SC		$a \times a \times a$ non-col-linear
FeS_2	5.407	WCP	—	—	SC	$d\varepsilon^6$	Temperature-independent susceptibility
$FeSe_2$	5.783^{HP}	WCP	—	—	SC		
$FeTe_2$	6.294^{HP}	WCP	—	—	SC		
CoS_2	5.534	F	$T_C = 125$	0.85	M	$d\varepsilon^6 d\gamma^1$	$1\mu_B$ of the moment from inverse susceptibility
$CoSe_2$	5.861	CWP	—	—	M		
$CoTe_2$	6.318^{HP}	PP or D	—	—	M		
NiS_2	5.680	AF(WF)	$T_N = 40 \sim 150$ ($T_C = 30$)	1.17 (0.03–0)	SC	$d\varepsilon^6 d\gamma^2$	Spin structure: admixture of first and second kinds
$NiSe_2$	5.951	PP	—	—	M		
$NiTe_2$	6.374^{HP}						

CuS_2	5.790^{HP}	PP(S)	$T_s = 1.5$	—	M(S)	$d\varepsilon^6 d\gamma^3$	Lattice anomalies at 50 and 150 K
$CuSe_2$	6.123^{HP}	PP(WF, S)	$T_s = 2.4$ $(T_C = 30)$ $T_s \sim 1.3$	(0.03)	M(S)		Ferromagnetic
$CuTe_2$	6.605^{HP}	(WF, S)	$(T_C = 26)$		M(S)		Superconductor
ZnS_2	5.954^{HP}	D	—	—	SC	$d\varepsilon^6 d\gamma^4$	
$ZnSe_2$	6.290^{HP}	D	—	—	SC		

Note: AF = Antiferromagnet, F = Ferromagnet, WF = Weak ferromagnet, WCP = Weak and constant paramagnet, PP = Pauli type paramagnet, CWP = Curie–Weiss type paramagnet, D = Diamagnet, S = Superconductor, T_C = Curie point, T_N = Néel point, T_s = Transition point of S, SC = Semiconductor, M = Metallic conduction—range depends on the stoichiometry, HP = synthesized under high pressure.

molecule also forms an fcc lattice and its axis is oriented in the (111), $(\bar{1}\bar{1}1)$, $(1\bar{1}\bar{1})$ and $(\bar{1}1\bar{1})$ directions. Thus the M atom is surrounded octahedrally by six nearest neighbor X atoms [4.7].

The main physical data of the MX_2 are summarized in Table 4.1 [4.6] for $M = $ Mn, Fe, Co, Ni, Cu and Zn, and $X = $ S, Se and Te. As seen in this table, from the magnetic and conduction viewpoint the full variety of properties is exhibited by these compounds. With regard to conduction, MnX_2, FeX_2, NiS_2 and ZnX_2 are semiconductive, while metallic conduction appears in CoX_2, $NiSe_2$ and CuX_2. On the other hand, antiferromagnetism appears in MnX_2 and NiS_2, and ferromagnetism only in CoS_2. For the paramagnetic compounds, FeX_2 as an insulator shows a van Vleck type temperature independent susceptibility, $NiSe_2$ a Pauli type and $CoSe_2$ a Curie–Weiss type. $CoTe_2$ and $CuTe_2$ are thought to be Pauli paramagnetic from their metallic conduction. It is possible though that the diamagnetism from the Te ion core may predominate over the intrinsic Pauli paramagnetism.

While all the CuX_2 are superconducting at very low temperatures $CuSe_2$ and $CuTe_2$ possess a weak ferromagnetism above the superconducting transition point, and are regarded as ferromagnetic superconductors. Above the transition point they show Pauli paramagnetism.

It is noticeable that in NiS_2, its weak ferromagnetism appears below a point lower than the Néel point, where a complex magnetic ordering takes place as explained in detail below.

4.2.2 Phase Diagram Constructed from Substitution

With the background for the various properties as above, investigations on substituted systems $M_{1-x}M'_xS_2$ and $M(X_{1-x}X'_x)_2$ were carried out, in order to make clear the effects due to a change of electron numbers for the former, and a change in the intra-atomic distance for the latter. The substitution was done between adjacent elements in the periodic table. The results are summarized as a phase diagram in Fig. 4.5 [4.6], where the one axis, n, means an average electron number per M in the $d\gamma$ band and the other is a change of the $d\gamma$ band width mentioned later.

The main characteristics are as follows:

(1) Near $n = 0$ (FeS_2), ferromagnetism and metallic conduction occur by some percolation mechanism.
(2) Along $n = 1$, the conduction is metallic and the ferromagnetism of CoS_2 disappears rather sharply. The susceptibility in the paramagnetic state is Curie–Weiss type.
(3) Around $n = 2$ ($NiSe_2$), the antiferromagnetic phase extends over both the nonmetallic and metallic phases and weak ferromagnetism appears at lower temperatures. The paramagnetic susceptibility changes from Curie–Weiss type on the NiS_2 side to Pauli type on the $NiSe_2$ side.

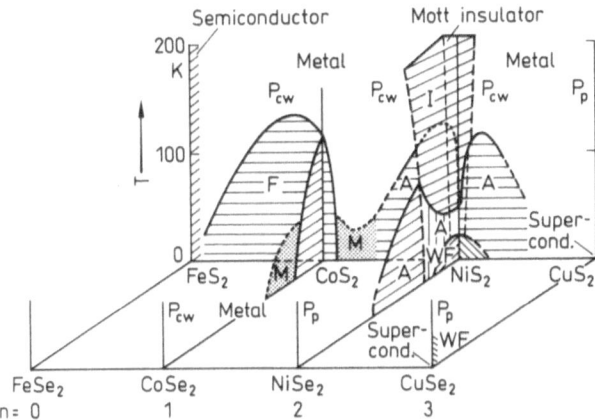

Fig. 4.5. Magnetic and conduction phase diagram for the substituted systems, FeS_2–CoS_2–NiS_2–CuS_2. F: Ferromagnetism, A: Antiferromagnetism, M: Metamagnetism, WF: Weak ferromagnetism, P_{cw}: Curie–Weiss type paramagnetism, P_p: Pauli type paramagnetism. Insulator (Semiconductor) regions are also indicated in the figure

Having given a general view of the phase diagram, in the next subsections we comment on some interesting physical phenomena for each individual compound.

4.2.3 Metal-Insulator Transition of $Ni(S_{1-x}Se_x)_2$

The substituted system, $Ni(S_{1-x}Se_x)_2$, contains five phases (Fig. 4.5) a paramagnetic insulator, an antiferromagnetic insulator, a paramagnetic metal, an antiferromagnetic metal and a weakly ferromagnetic insulator. Here we explain these phases in more detail. Figure 4.6a [4.8] shows a change of electrical resistivity vs. temperature for the various compositions of Se. The resistivity varies from semiconductive (NiS_2) to metallic ($NiSe_2$) viz. a metal-insulator transition takes place. At the same time the magnetic susceptibility changes continuously from Curie–Weiss to Pauli type through the transition.

On the other hand, according to measurements at high pressure on $Ni(S_{0.8}Se_{0.2})_2$, the resistivity occurs with almost similar changes as shown in Fig. 4.6b [4.9]. This suggests that the substitution by the larger Se atom has the same effect as a contraction of the interatomic distance by pressure. Thus, the correspondence rule between pressure p and composition x, 10 kbar $\equiv 0.01$, is found from these data.

The phase diagram of $Ni(S_{1-x}Se_x)_2$ is shown in Fig. 4.7 [4.9, 10], which include the pressure data. The metal insulator boundary appears at $x \simeq 0.25$ at 0 K and the top of this line represents the critical point of this transition. The

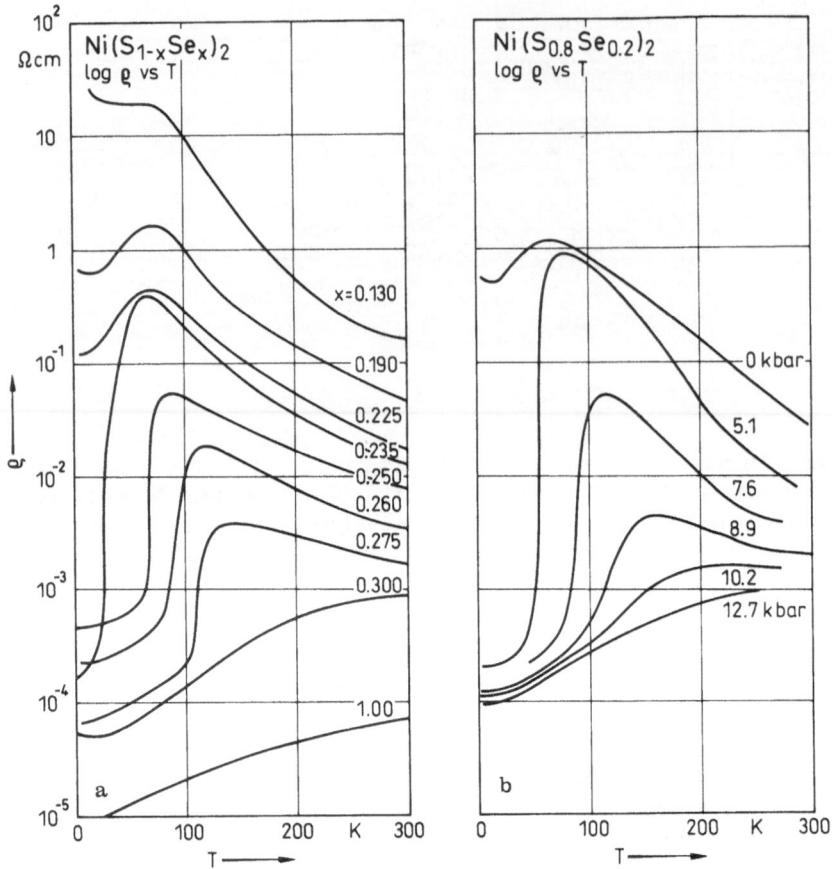

Fig. 4.6. a Electrical resistivity vs. temperature of Ni $(S_{1-x}Se_x)_2$. **b** Electrical resistivity vs. temperature of $Ni(S_{0.8}Se_{0.2})_2$ under high pressure

Néel point increases with x in the insulator phase, reaches a peak at the boundary, then decreases with x and disappears at $x \simeq 0.5$ in the metallic phase. There is also a Curie point from the weak ferromagnetism existing in the antiferromagnetic insulator phase. These phenomena will be interpreted in the next section.

4.2.4 Spin Structure of NiS_2 and its Weak Ferromagnetism

The spin structure of NiS_2 [4.11, 12] has been observed in both powdered and single crystal samples, although the exact spin structure has not yet been determined for this cubic crystal even in single crystals due to the existence of

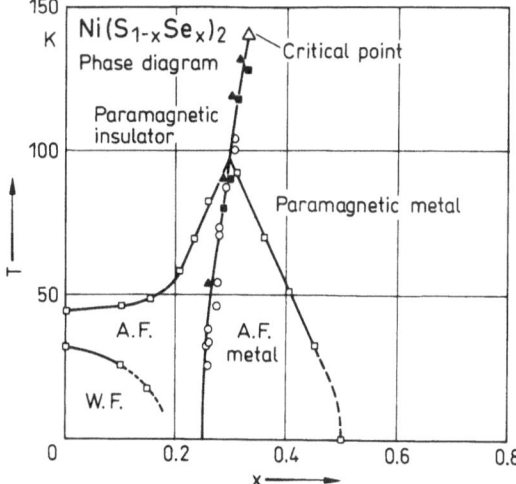

Correspondence rule : 1.0 kbar = 1.0%

Fig. 4.7. Phase diagram of $Ni(S_{1-x}Se_x)_2$ obtained from magnetic and resistivity measurements, including high pressure effects. Data are as follows: ○ Jarrett et al. (1973), ◇ Czjzec et al. (1976) at ambient pressure. ■ $x = 0.15$, ▲ $x = 0.20$ under pressure; ◇ critical point. The correspondance between pressure and composition is: 10 kbar = 1% x

antiferromagnetic domains. Below the Néel point, T_N, down to the Curie point T_C for the weak ferromagnetism, the spin structure is of the first kind in the fcc magnetic lattice (or the noncollinear first kind explained in the next section). A strange phenomenon takes place below T_C, where two structures, the first and second kinds, coexist with each other with different moments of $1.0\mu_B$ and $0.6\mu_B$ respectively [4.11, 12].

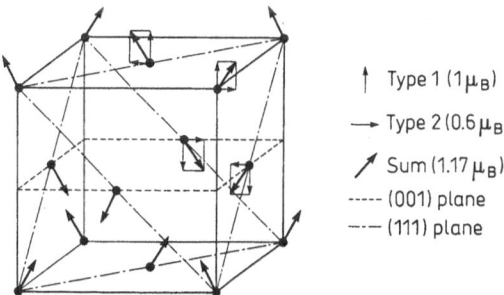

Fig. 4.8. A possible spin structure for NiS_2 below T_C obtained from neutron diffraction. ↑; first kind of ordering with a moment of $1\mu_B$, →; second kind of ordering with a moment of $0.6\mu_B$, ↖; composite ordering with $1.17\mu_B$

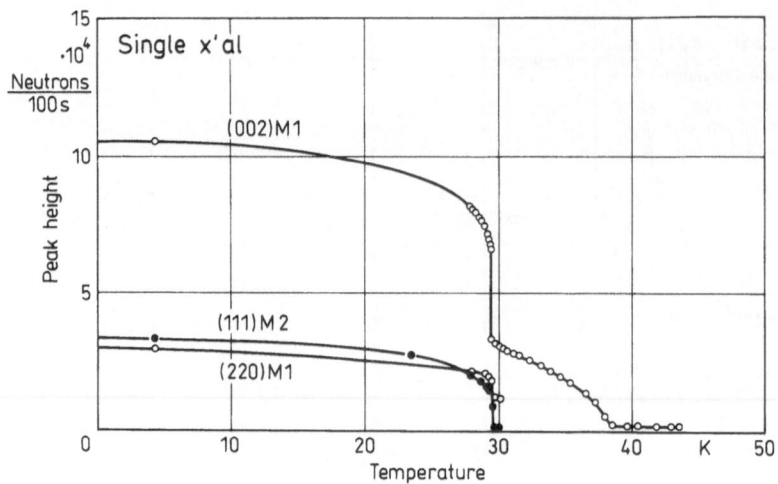

Fig. 4.9. Temperature variation of the magnetic reflections for NiS_2 obtained from neutron diffraction. $M1$ and $M2$ mean the reflection from the first and second kind of ordering respectively

One of the possible structures below T_C is shown in Fig. 4.8 [4.7]. The spin directions are resolved conventionally as two components parallel (μ_B) and perpendicular ($0.6\mu_B$) to the (001) axis for the first and the second kind of orderings respectively. Thus, the composite moment is given as $1.17\mu_B$. This structure is determined from the intensity of the magnetic reflections but is not a unique structure.

Figure 4.9 [4.12] shows the temperature dependence of the magnetic reflection intensity, in which $M1$ and $M2$ are the first and second kind of orderings respectively. The magnetic transition at T_N is of the second kind, while that at T_C is of the first kind [A11], and this has been reconfirmed by specific heat measurements.

4.2.5 Effect of Nonstoichiometry in NiS_2

The antiferromagnetic transition and the magnetic moment for the weak ferromagnetism depend sensitively on the stoichiometry, as shown from the $T_N(x)$ and the $M_0(x)$ in Fig. 4.10 [4.13] for NiS_x. It is surprising that T_N increases from 50 K for $x \simeq 1.9$ to 180 K for $x \simeq 2.1$, while M_0 decreases sharply with x and disappears at $2 < x < 2.1$. In the range of $1.9 < x < 2$, however, T_C stays at 30 ± 1 K. In addition to this, from the lattice parameter and density measurements it can be shown that both Ni and S sites contain random vacant

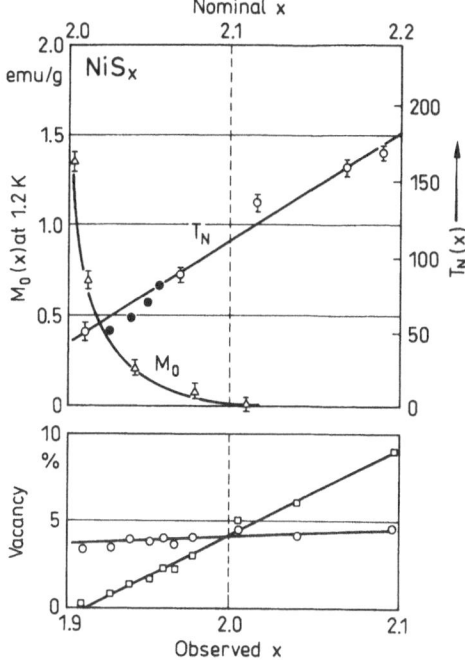

Fig. 4.10. Néel point T_N and magnetization M_0 at 4.2 K for non-stoichiometric NiS_x. The symbols ● and ○ for T_N are obtained from the maximum point for resistivity and susceptibility respectively. The lower figure shows a compositional dependence of vacant sites in Nickel (symbol □) and Sulfur (symbol ○) sites

sites even at the stoichiometry ($x = 2$). The semiconductor estimated from the resistivity decreases with x and the system approaches a metallic state.

These phenomena are interesting but the origin of the vacancy formation and its effect on the magnetic properties have not yet been clarified.

4.2.6   Metamagnetism of $Co(S_{1-x}Se_x)_2$

Figure 4.11 [4.14, 15] shows the change in the Curie point T_C, the critical point T_{cr} of metamagnetism (Appendix A.23) mentioned below and the saturation magnetization M_0 at 4.2 K in the ferromagnetic as well as metamagnetic regions as a function of composition x. The Curie point T_C disappears at $x = 0.12$ obtained from an extrapolation to zero field. The metamagnetism occurs below $x < 0.4$ at 0 K in the paramagnetic state under an applied field. Magnetization takes place suddenly with a hysteresis depending on composition x and temperature T. The magnetization curves of $Co(S_{1-x}Se_x)_2$ at 4.2 K and 77 K are

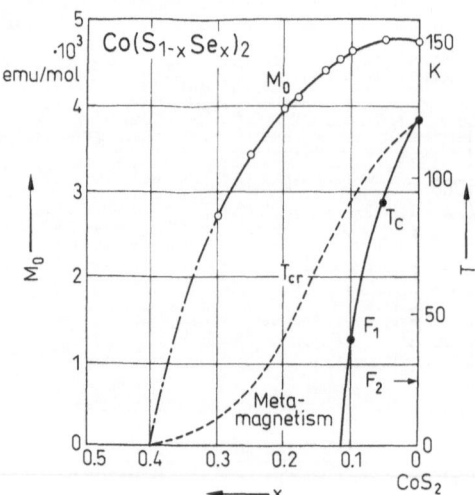

Fig. 4.11. Magnetic phase diagram and saturation moment M_0 at 4.2 K for $Co(Si_{1-x}Se_x)_2$. In this phase diagram, F_1, F_2, T_C and T_{cr} signify the ferromagnetic state with a first order transition, that with second order, the Curie point and the critical point of metamagnetism respectively

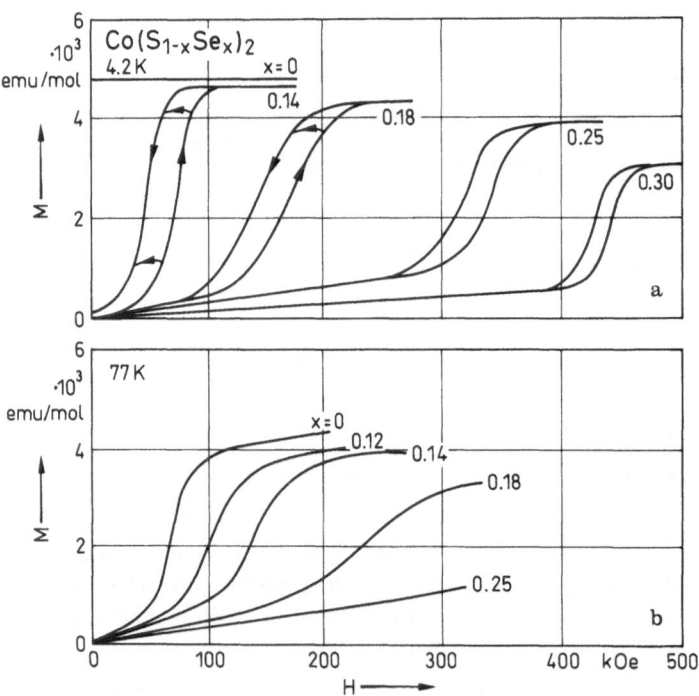

Fig. 4.12a, b. Metamagnetic magnetization of $Co(S_{1-x}Se_x)_2$ under high pressure at (a) 4.2 K and (b) 77 K

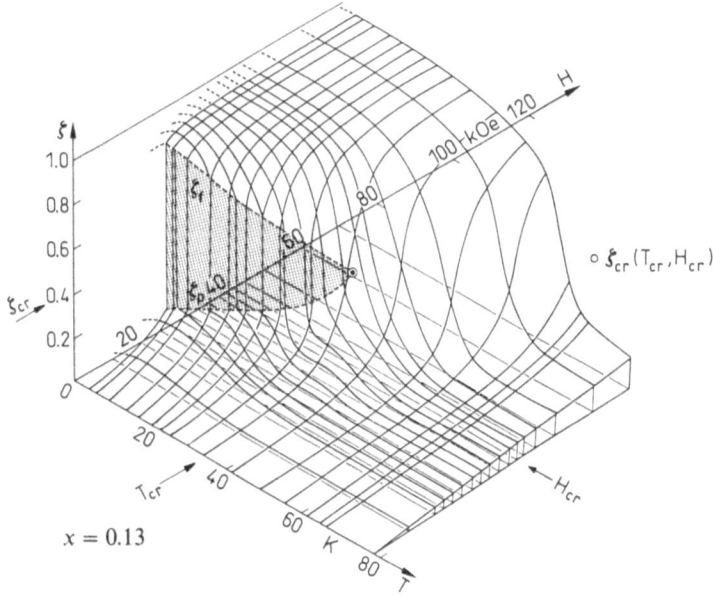

Fig. 4.13. Metamagnetic phase diagram of $Co(S_{1-x}Se_x)_2$ as a function of the relative magnetization ζ, the magnetic field H and the temperature T. The critical point is indicated by $\bigcirc$ at ζ_{cr}, H_{cr} and T_{cr}. ζ_f and ζ_p are the ferromagnetic and paramagnetic relative magnetizations respectively, corresponding to the densities of liquid and gas vs. temperature and pressure (H) in the van der Waals state

shown in Figs. 4.12a and b respectively [4.16] observed in high magnetic fields. The critical composition and field at 0 K are estimated to be $x_{cr} \simeq 0.6$ and $H_{cr} \simeq 600$ kOe by extrapolation. At finite temperatures, for example, the metamagnetic behavior of $Co(S_{0.87}Se_{0.13})$ with a paramagnetic ground state at $T = 0$ and $H = 0$ is shown in Fig. 4.13 [4.17] for a relative magnetization, $\zeta(T, H) = M(T, H)/M_s(0)$, with its saturation magnetization $M_s(0)$ at 0 K. The discontinuous transition in $\zeta(T, H)$ is defined to be a vertical line so as to divide the hysteresis loop into two equal areas.

Thus the critical points for this metamagnetism are determined to be $\zeta_{cr} = 0.43$, $T_{cr} = 34$ K and $H_{cr} = 39$ kOe. On the other hand, $\zeta(T, H)$ near the critical point behaves as a critical phenomenon. Denoting the ferromagnetic and paramagnetic magnetizations by ζ_f and ζ_p respectively, we have the following expressions:

$$\zeta_f - \zeta_p \propto \left\{1 - \frac{T}{T_{cr}}\right\}^\beta \tag{4.1}$$

and

$$\zeta_f - \zeta_p \propto \left\{ 1 - \frac{H}{H_{cr}} \right\}^{\delta} \qquad (4.2)$$

The critical exponents β and δ are estimated to be 0.5 and 0.25 respectively. The values are rather classical as in the case of the van der Waals equation ($\beta = 0.5$, $\delta = 3$).

Several origins of metamagnetism have been proposed based on the localized spin model, however, these are not applicable to this case, since the

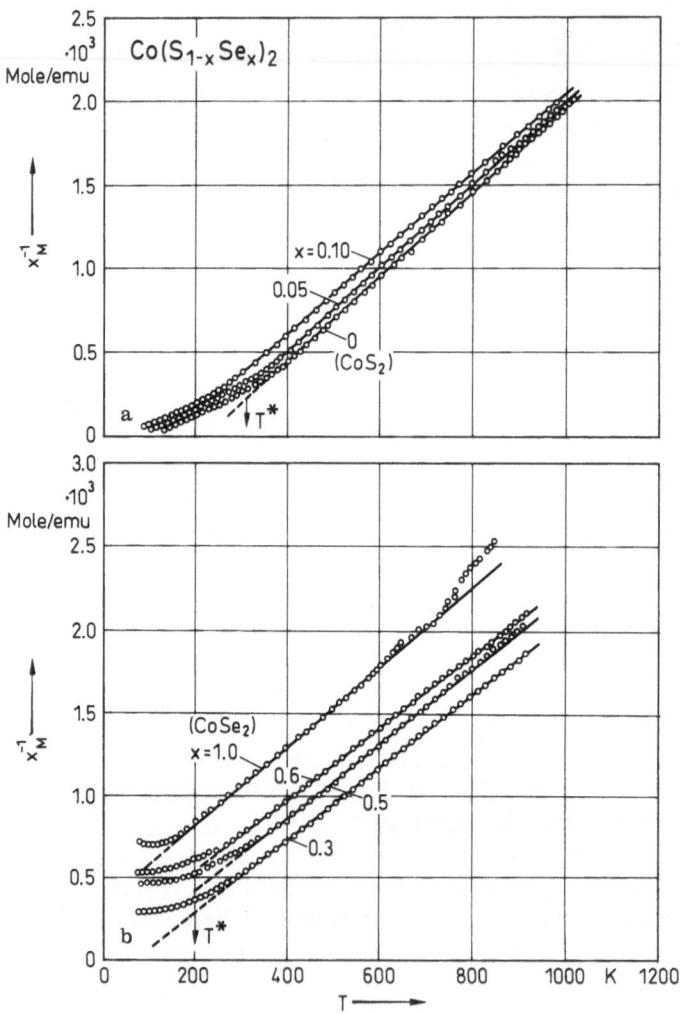

Fig. 4.14a, b. Inverse magnetic susceptibility vs. T of $Co(S_{1-x}Se_x)_2$. The change in slope is denoted by T^*

phenomena appear in the metallic state and must be due to itinerant electron magnetism as shown in the next section.

4.2.7   Paramagnetic Susceptibility and Electrical Resistivity of $Co(S_{1-x}Se_x)_2$

In this section, some characteristic properties of $Co(S_{1-x}Se_x)_2$ are mentioned in connection with a new theory of itinerant electron magnetism.

The inverse magnetic susceptibilities of the system $Co(S_{1-x}Se_x)_2$ are shown in Figs. 4.14a and b [4.18]. The slope of the curves is linear at high temperatures in a Curie–Weiss manner even in the paramagnetic region on the $CoSe_2$ side where there is no local moment. Such behavior cannot be understood by the Stoner theory. It is noted that all the curves show a change in slope above T_C. The temperature of this change is defined here as T^*. Such a change could occur as a result of short range magnetic order above T_C, but it is not expected in the paramagnetic region. The curve for $CoSe_2$ behaves like that of Pd and is regarded as being due to strong exchange-enhanced paramagnetism. The electrical resistivity was measured for the whole system of $Co(S_{1-x}Se_x)_2$ from 4.2 K to 300 K, and the result is shown in Figs. 4.15a and b [4.15]. In the ferromagnetic

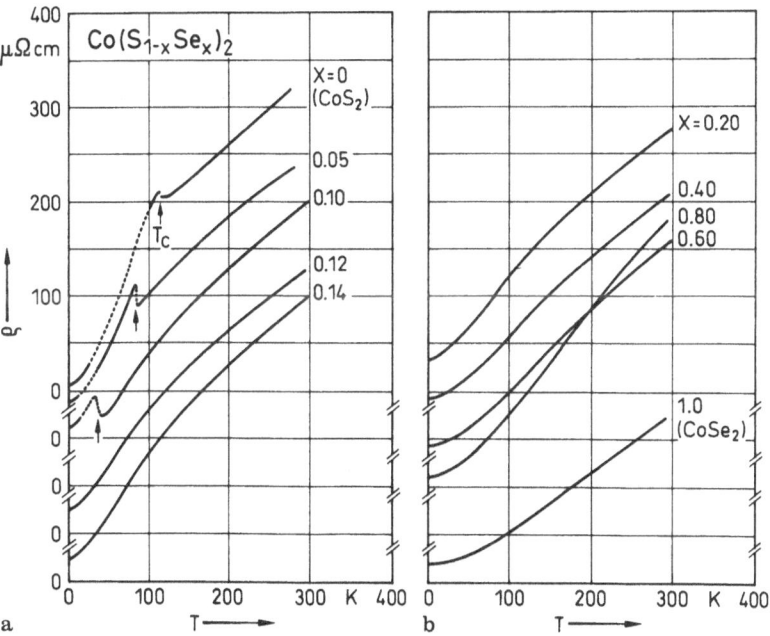

Fig. 4.15a, b. Electrical resistivity vs. T of $Co(S_{1-x}Se_x)_2$

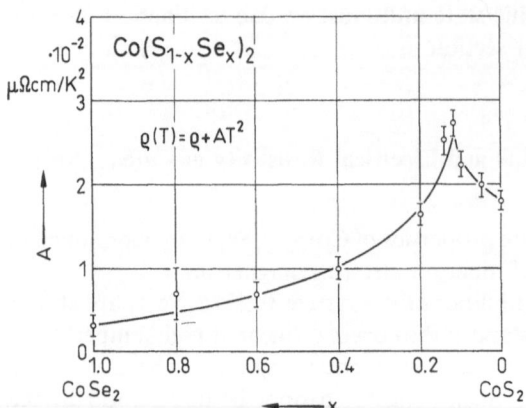

Fig. 4.16. Composition dependence of the temperature coefficient $A(x)$ in $\rho = \rho_0 + A T^2$

range, $0.88 < x \leq 1.0$, a hump appears just below the Curie point. The curves as a whole are not linear at higher temperatures but are concave with respect to the temperature axis. The significance of this is mentioned later. At low temperatures around 0 K, the resistivity is expressed by

$$\rho(T) = \rho_0 + AT^2 \ , \tag{4.3}$$

where ρ_0 is the residual resistivity. The phonon term in the resistivity proportional to T^5 is not important in this low temperature region. The coefficient A depends on x as shown in Fig. 4.16 [4.15]. A noticeable peak appears in $A(x)$ at the boundary between the ferromagnetic and paramagnetic phases at 0 K, $x = x_c = 0.88$.

The Hall effect has been measured in single crystals of CoS_2 and polycrystals of $Co(S_{1-x}Se_x)_2$. The Hall coefficient indicated electron carriers and the Hall number obtained was 0.53 and 1.07 at 4.2 K for CoS_2 and $CoSe_2$ respectively, assuming a 1/4-filled d band.

4.3 Theoretical Interpretation

4.3.1 Electronic Structure

If we adopt the ionic model for the pyrite compounds and consider them formally to be in a $M^{2+}X_2^{-2}$ state, the M^{2+} ion is subjected to a strong cubic crystal field from the octahedral ligands [4.19]. Then its d-level splits into $d\varepsilon(t_{2g})$ and $d\gamma(e_g)$. Thus, as shown in Table 4.1, five electrons in Mn^{2+} for MnX_2 occupy these levels as $d\varepsilon^3 d\gamma^2$ with a moment of $5\mu_B$. In contrast, FeX_2 possesses a smaller lattice parameter than MnX_2, so the six d-electrons are in a "low spin state", due to the stronger crystal field. Thus the $d\varepsilon^6 d\gamma^0$ electron configuration occurs with no moment. These results are consistent with the observations. Accordingly, one would expect CoX_2, NiX_2 and CuX_2 to possess the configurations $d\varepsilon^6 d\gamma^1 (1\mu_B)$, $d\varepsilon^6 d\gamma^2 (2\mu_B)$ and $d\varepsilon^6 d\gamma^3 (1\mu_B)$, respectively, but such a situation is not realized as shown in the experimental results (Table 4.1). Therefore, the ionic scheme is not applicable and their electronic structures should be constructed using energy band formation, especially for their metallic state.

Given this background, band calculations for the disulfides, MS_2, were carried out by some authors [4.20, 21] together with calculations of the density of states (DOS). The calculated band structures are composed of $d\varepsilon$, $d\gamma$ and s–p components, among which a certain amount of mixing occurs. Figure 4.17 [4.21] shows the calculated density of states for FeS_2, CoS_2, NiS_2, CuS_2 and ZnS_2 in the paramagnetic state. The lower occupied states are mainly composed of bonding s–p bands, while the upper empty ones are antibonding. Near the Fermi level, $d\varepsilon$ and $d\gamma$ bands appear except in ZnS_2.

The semiconducting nature of FeS_2 is verified by a gap of 0.7 eV between the filled $d\varepsilon$ and empty $d\gamma$ bands. In CoS_2, the $d\gamma$ band is 1/4-filled corresponding to a metallic state. The Fermi level is located near a peak in the DOS, suggesting the appearance of ferromagnetism. The $d\gamma$ band width is estimated to be about 1 eV. In NiS_2, although its half-filled band shows it to be metallic, this is not so in the real case so its d band must be split due to a strong electron correlation associated with the Ni ion as explained in the next section. In CuS_2, with a 3/4-filled d band, the predicted metallic conduction and Pauli paramagnetism are supported by the observed results except at low temperatures. The diamagnetic and semiconductive behavior of ZnS_2 can be explained by the filled $d\gamma$ and empty s–p antibonding bands separated by a gap of ca. 2 eV.

The most remarkable case here is that of the diselenides. The band widths for $d\varepsilon$ and $d\gamma$ are estimated by some band calculations to be 10 to 20% larger than those of the disulfides [4.20], so that $NiSe_2$ can have a metallic state as seen experimentally. $CoSe_2$ on the other hand has a similar $d\gamma$ band structure and occupation to CoS_2, so that the appearance of paramagnetism is expected, because of a fall in the density of states for the same width. Therefore, $CoSe_2$ can possess a strong (exchange-enhanced) paramagnetic nature as in the case of Pd metal.

Thus, the band calculations are consistent for the pyrite type compounds as far as their magnetic and conduction properties are concerned, except for the case of NiS_2.

4.3.2 Interpretation of the Phase Diagram

In order to explain the phase diagram of the magnetic states in Fig. 4.5, one can apply the Hubbard model (Appendix A.16) for a system with electronic states in

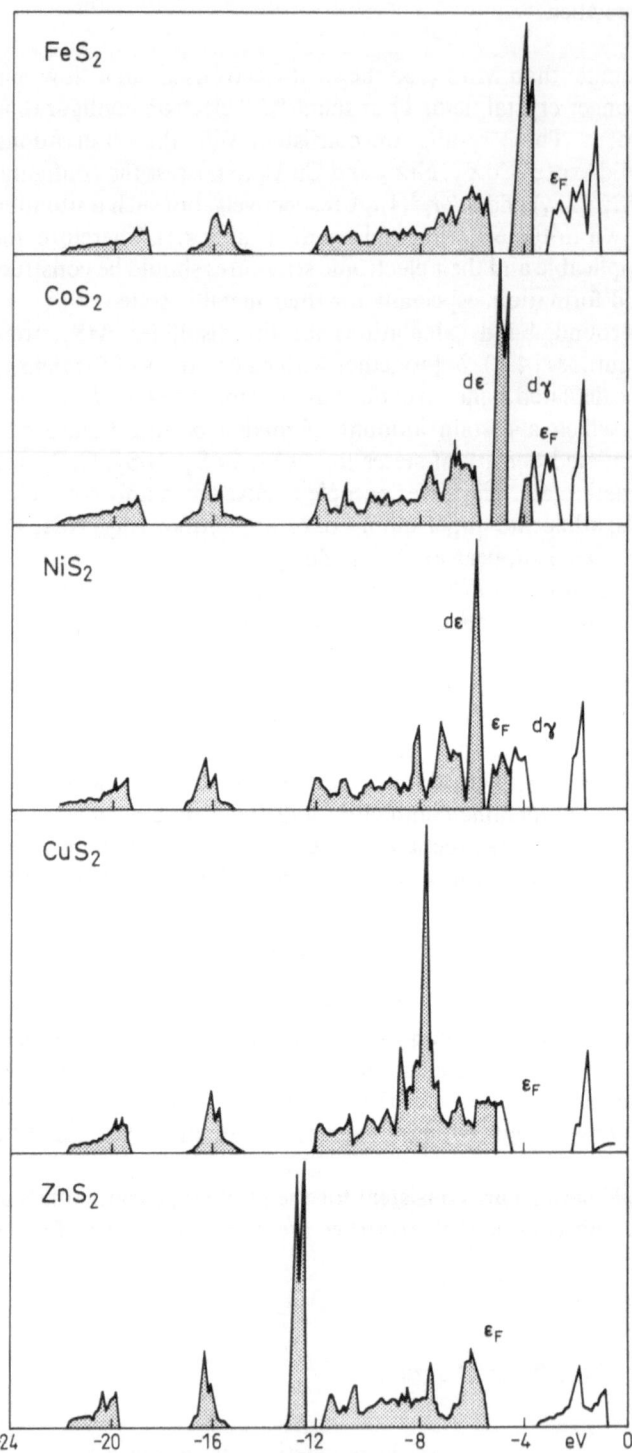

Fig. 4.17. Density of states curve of the pyrite type compounds, FeS_2, CoS_2, NiS_2 and ZnS_2, where ε_F is the Fermi level

a narrow band with strong electron correlation [4.22, 23]. The simplest model consists of a tranfer integral, t_{ij}, for the neighboring z_{ij} atoms, an intra-atomic Coulomb interaction U (repulsive) and an averaged electron number per atom, $n = 0$–2, for a given band. The standard Hamiltonian for this model is described as follows:

$$\mathcal{H} = \sum_{\sigma} \sum_{i,j} t_{ij} a_{i\sigma}^{+} a_{j\sigma} + U \sum n_{i\uparrow} n_{i\downarrow} \qquad (4.4)$$

where $a_{i\sigma}^{+}(a_{j\sigma})$ and $n_{i\sigma} = a_{i\sigma} a_{i\sigma}$ are a creation (destruction) operator on an i (j) site with spin $= \uparrow$ or $\downarrow$ and a number operator on the i-site with spin σ respectively. Then restricting the calculation to only nearest neighbor transfers $t_{ij} = t$ and $z_{ij} = z$, it can be shown that the band width $W = zt$. The important parameters of this model are the ratio W/U and the electron number n.

In real pyrites, however, the $d\gamma$ band has an orbital degeneracy containing $n = 0 \sim 4$ electrons and the transfer is done through the metalloid ion. The appropriate Hamiltonian in this case can be given as follows:

$$\mathcal{H} = t \sum_{\nu, \sigma} \sum_{i,j}^{nn} (a_{j\sigma}^{+} a_{\nu i\sigma} + a_{\nu i\sigma}^{+} a_{j\sigma}) + V \sum_{j,\sigma} n_{j\sigma} + U \left(\sum_{\nu, i} n_{\nu i \uparrow} n_{\nu i \downarrow} + \sum_{j} n_{j\uparrow} n_{j\downarrow} \right)$$
$$+ \sum_{i, \sigma, \sigma'} (K - J\delta_{\sigma, \sigma'}) n_{1i\sigma} n_{2i\sigma'} \qquad (4.5)$$

where ν, i and j mean the degeneracy of $d\gamma$ orbitals ($\nu = 1$ and 2) and the lattice site of M and X, respectively. Here, a nearest neighbor transfer t between M and X is assumed. V, U and $(K - J\delta_{\sigma\sigma'})$ express the Coulomb interaction in X, in M and an orbital interaction in the $d\gamma$ orbitals containing an exchange interaction J in it. Introducing such a model Hamiltonian, we might expect to explain the stability of the observed phases in Fig. 4.5. Unfortunately however, no solution for this Hamiltonian has been found so far.

Leaving aside this complicated Hamiltonian, we return to the simple case (4.4), which has been solved with some approximations for a given crystal. Using these results [4.15, 25], plus one presumed from (4.5) and the experimental results, we give a schematic phase diagram in Fig. 4.18 [4.7]. The asymmetry for the half filled state ($n = 2$) appears because of differences in interatomic distances. Below and above the dashed curve, insulator and metallic states appear respectively. In the metallic state for $W/U \gg 1$, Pauli paramagnetism occurs as seen in $RhSe_2$, $NiAsS$, $NiSe_2$ and CuX_2 in which the itinerancy of the electrons dominates over U. In contrast, in the case of $W/U \ll 1$, Curie–Weiss type paramagnetism appears in the insulator region. An antiferromagnetic state can be stabilized around the half-filled ($n = 2$) state owing to the appreciable electron correlation. This case corresponds to NiS_2.

On the other hand, when W/U takes an intermediate value, some kind of metallic antiferromagnetism can appear in the region near $n = 2$. It is noted here that the high-T_C superconducting state, found recently in some oxides, may occur in this paramagnetic-metallic region near the insulator phase. On both sides of $n = 2$ in the metallic region, a ferromagnetic or paramagnetic state can

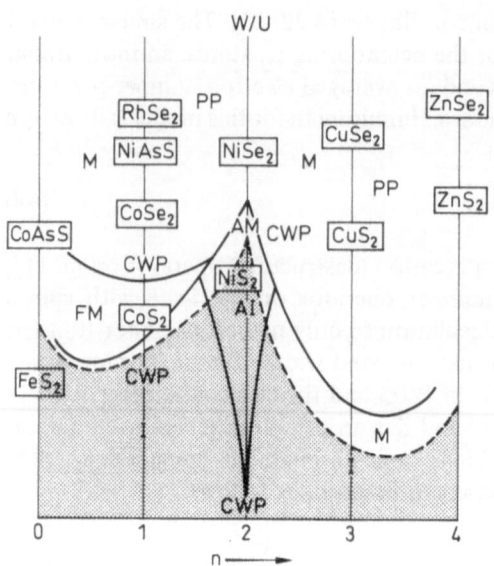

Fig. 4.18. Schematic phase diagram for pyrite type compounds, W/U vs. n. The abbreviations are: M: metallic state, I: insulator state, AI: antiferromagnetic insulator, FM: ferromagnetic metal, PP: Pauli paramagnetism, CWP. Curie–Weiss type paramagnetism

occur under conditions depending on the value of the density of states at the Fermi level and an intra-atomic exchange interaction [J in (4.5)], corresponding to the case of CoS_2 and $CoSe_2$. Weak ferromagnetism or strong paramagnetism is located near this boundary, and the susceptibility in this paramagnetic state shows a Curie–Weiss-like behavior explained by spin fluctuation theory.

The susceptibility in paramagnetic metals, in general, is enhanced by the intra-atomic interaction U, and its value at $T = 0$ is given as,

$$\chi_p(0) = \frac{2\mu_B^2 \rho(\varepsilon_F)}{1 - U_0 \rho(\varepsilon_F)} \, , \tag{4.6}$$

where $\rho(\varepsilon_F)$ is the density of states at ε_F and $\alpha = U_0 \rho(\varepsilon_F)$ is called the exchange enhancement factor (Appendix A.7). The boundary for ferromagnetism is at $\alpha = 1$. The cases for CoS_2, $CoSe_2$ and $RhSe_2$ correspond to $\alpha > 1$, $\alpha \lesssim 1$ and $\alpha \ll 1$ respectively.

4.3.3 Spin Structure and Weak Ferromagnetism Caused by Four-Body Exchange Interactions

In general, collinear spin structures stabilized in an fcc lattice are classified into four kinds [4.28] when the exchange interactions are taken into account up to the second neighbor. They are: the first (tetragonal), the improved first (tetragonal), and the second (rhombohedral) kinds for antiferromagnetism, plus ferromagnetism (cubic). In addition, noncollinear structures derived from the first kind, [4.29] are also possible, in which spins on the four sublattice points couple with each other with an angle $\theta = \cos^{-1}(-1/3)$. This structure is understood to be a three fold degenerate structure of the collinear first kind. The coupling energy is the same as that of the first kind but the symmetry of the magnetic lattice is cubic. This is called, here, the degenerate first kind.

Now let us confine our interest to the weak ferromagnetic spin structure of NiS_2 below the Curie point. As explained above, this structure is composed of the first (or degenerate first) and the second kind. Such a structure cannot be stable theoretically even when neighboring interactions greater than the third are taken into account in the fcc lattice as long as the interaction is based on Heisenberg type exchange coupling. Therefore another type of interaction should be considered.

The Heisenberg type exchange as a superexchange mechanism is produced from a second order perturbation in spin-orbital levels. The other interaction of this type is the Dzyaloshinski–Moriya interaction (Appendix A.2) written as $D|S_i \times S_j|$. When an insulating state approaches a metallic state, effects from higher order perturbations cannot be neglected [4.30], for example, the biquadratic interaction as

$$E_B^{(4)}(i, j) \propto \frac{t_{ij}^2 t_{ji}^2}{U^3}(S_i \cdot S_j)^2 \tag{4.7}$$

is an example of a fourth order one. A more general fourth order perturbation is written as

$$E^{(4)}(i, j, k, l) \propto \frac{t_{ij} t_{jk} t_{kl} t_{li}}{U^3}[(S_i \cdot S_j)(S_k \cdot S_l)$$
$$+ (S_i \cdot S_l)(S_j \cdot S_k) - (S_i \cdot S_k)(S_j \cdot S_l)] \tag{4.8}$$

and is called a "four-body exchange interaction" [4.26, 27]. These interactions are responsible for some of the noncollinear spin structures and for a first order magnetic transition. The famous antiferromagnetic structure found in solid ^{3}He, up-up-down-down, was interpreted in terms of expression (4.8) [4.31].

NiS_2 is located near the boundary of the metallic state, so the four-body interaction is important in explaining its structure. Possible paths for this interaction in the fcc lattice are shown in Fig. 4.19, where the closed path is combined through the first (1) and second (2) neighbors.

Using these ideas for NiS_2, the stability of its spin structure has been explained theoretically, assuming that it is a degenerate structure of the first kind. In addition, it is possible to explain the occurrence of weak ferromagnetism at a T_C below T_N as a first order transition, by taking account of the effect of the magnetic anisotropy energy on the above spin structure [4.25]. Thus the physical origin of the spin structure as well as the weak ferromagnetism derives from a mechanism originating from the many-body problem.

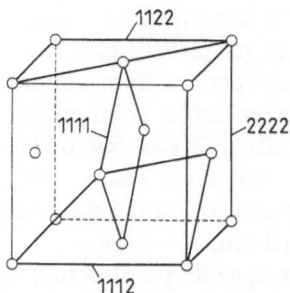

Fig. 4.19. Closed paths for the four-body exchange interaction in fcc lattic (Ni lattice in NiS$_2$). The numbers, 1 and 2, mean the connection through the nearest and next nearest neighbors respectively

4.3.4 Metamagnetism of Co(S$_{1-x}$Se$_x$)$_2$ in the Itinerant Electron Model

As mentioned in Sect. 4.3, the metamagnetism (Appendix A.23), appearing in Co(S$_{1-x}$Se$_x$)$_2$ can be explained on the basis of an itinerant electron state. The occurrence of this state has been explained theoretically by making use of the calculated density of states for CoS$_2$ [4.32, 33].

The free energy per atom in an itinerant electron system can be expressed as follows by using the quantities, $\rho(\varepsilon)$, M (in units of μ_B) and n, as the density of states, the magnetization and electron numbers respectively, and the applied magnetic field H_0. Then

$$F(M, n) = F(H, \mu) + 2\mu n + 2H_0 M + UM^2 \tag{4.9}$$

and

$$F(B, \mu) = -k_B T \sum_\sigma \int d\varepsilon \rho(\varepsilon) \ln\left\{1 + \exp\left(-\frac{\varepsilon - \mu_\sigma}{k_B T}\right)\right\} \tag{4.10}$$

with

$$\left.\begin{array}{l} \mu_\sigma = \mu + \sigma(H_0 + UM) \\[2mm] M = \dfrac{n_\uparrow - n_\downarrow}{2} \\[2mm] n_\sigma = \int d\varepsilon \rho(\varepsilon) f(\varepsilon - \mu_0) \end{array}\right\} \tag{4.11}$$

where $U, \mu, f(\varepsilon)$ and σ denote the intra-atomic interaction, the Fermi level in the paramagnetic state, the Fermi distribution function and the spin, $\uparrow$ or $\downarrow$, respectively. In (4.10), B is contained implicitly, where B^2 means an averaged square amplitude relating to the local spin density expressed according to spin fluctuation theory (Appendix A.31).

The magnetization curve against H and T can be obtained from minimum points of the free energy. The result at 0 K is shown in Fig. 4.20, corresponding

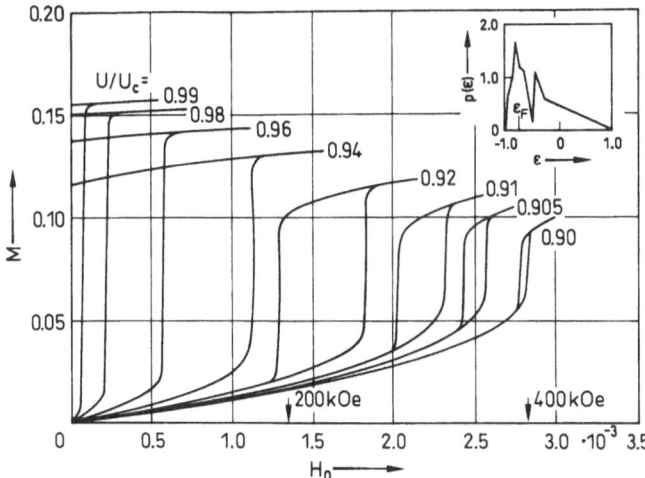

Fig. 4.20. Calculated metamagnetic magnetization curves of $Co(S_{1-x}Se_x)_2$ at 0 K. The insert is a simulated density of states formed from the calculation, ε_F is the Fermi level

to Fig. 4.12 in Sect. 4.3. Here, the insert shows a simulated density of states $\rho(\varepsilon)$ [4.20] and a Fermi level with $n = 1$ in the d band. A metamagnetic behavior with hysteresis is seen for a parameter U/U_c, where U_c is a critical value for the breaking out of ferromagnetism. The value U/U_c corresponds to the composition of $Co(S_{1-x}Se_x)_2$ in the experiment, as $U/U_c = 0.94$, 0.92 and 0.91 to $x = 0.15$, 0.21 and 0.26 respectively.

Thus the calculated results agree with experiment except for a wider hysteresis. At finite temperatures, the spin fluctuation effect becomes important not only for metamagnetic behavior but also for the paramagnetic susceptibility mentioned below.

4.3.5 Magnetic Susceptibility and Electrical Resistivity of $Co(S_{1-x}Se_x)_2$

In this section, we remark first on the change in slope at $T = T^*$ of the inverse susceptibility of this system as shown in Figs. 4.14a and b. In spin fluctuation theory, the magnetic susceptibility in an itinerant electron system is controlled not only by the local moment but also by the temperature induced moment. The former gives the transverse susceptibility χ_t, as the usual susceptibility in the localized model. However, the latter, called the longitudinal susceptibility χ_l, is produced from a change of the length of the local moment with a stiffness different from that in the former. Thus various types of behavior of the susceptibility are possible from both effects. Possible changes of the square of the

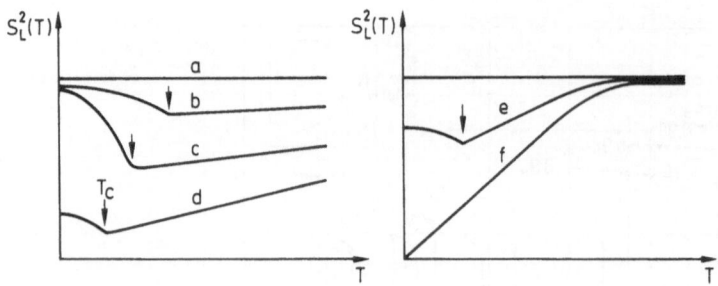

Fig. 4.21. Typical examples for the temperature dependence of S_L^2, the average square amplitude of the local spin fluctuations. The cases are as follows: **a** the local moment limit; **b** the case with a relatively large longitudinal stiffness constant, for example, the case of Fe; **c** the case of the invar alloys; **d** the case of weak ferromagnetism, where $S_L^2(T_e)/S_L^2(0)$ takes a value $3/5$ in the limit; **e** the case of CoS_2 and **f** the case of $CoSe_2$ with a paramagnetic ground state

local moment vs. temperature are classified in Fig. 4.21, where each case is denoted in the figure caption.

In the case of CoS_2 and $CoSe_2$ corresponding to e and f respectively, in Fig. 4.21, the local moment saturates at some temperature, $T = T^*$, under certain conditions mentioned below. Below T^*, the longitudinal susceptibility comes from its soft stiffness together with the ordinary transverse susceptibility. However, above T^*, the longitudinal one disappears and only the transverse one remains. Both inverse susceptibilities are given approximately by a Curie–Weiss type expression in this metallic case, which is expressed as,

$$\chi^{-1}(T) = \frac{T - \theta_p}{C_t + C_1} \tag{4.12}$$

with $C_1 = 0$ for $T > T^*$, and where C_t and C_1 are Curie–Weiss constants.

In this theory, the conditions for the occurrence of a T^* depend on the band structure and its occupation as follows: (1) The Fermi level ε_F must be located near the band edge and a 1/4-filled state is preferred, (2) the local density of states should be rather smooth and there should be a valley situated near ε_F, and (3) in the paramagnetic state, the exchange enhancement must be large. The calculated band structures shown above satisfy these conditions.

In addition, the electrical resistivity due to spin fluctuations in weakly and nearly ferromagnetic metals has been clarified for the whole temperature range [4.34]. The results are summarized as follows:

(1) At low temperatures, $T \ll T_C$, the resistivity begins to increase with an AT^2 dependence from 0 K and its coefficient A depends on the exchange enhancement $(1 - \alpha)^{-1}$, in which A diverges at $\alpha = 1$ corresponding to the boundary of the ferromagnetic and paramagnetic states.

(2) At intermediate temperatures, $T \sim T_C$, its dependence is on $T^{5/3}$.

(3) At very high temperatures, $T \gg T_C$, the resistivity tends to a constant value, that is, it saturates.

As shown in Fig. 4.16, $A(x)$ shows a steep peak at the boundary, $x_c = 0.12$. This is consistent with the theory, though the divergence is not seen due probably to some composition fluctuation in the substituted systems. For the observed resistivity from 4.2 K to 300 K in Fig. 4.15a and b, the $T^{5/3}$ dependence can be fitted in a rather narrow temperature range for all .samples. For some samples, a tendency to the saturation effect can be seen in the measured range of temperature.

Thus the magnetic and electrical properties observed in $Co(S_{1-x}Se_x)_2$ agree with those derived from spin fluctuation theory.

In summary, we have described the magnetic and conduction properties in some pyrite type compounds as being characteristic of many intermediate $3d$ compounds. We have also emphasized that the main characteristics of these materials can only be interpreted from the standpoint of spin fluctuation theory. Further investigations of other compounds are needed to verify and to develop this theory.

References

4.1. P. Weiss: J. Phys. Radium **4**, 459 (1904)
4.2. T. Moriya: Part of his work is summarized, for example. In: J. Magn. Magn. Mat. **14**, 1–14 (1976) *Electron Correlation and Magnetism in Narrow Band Systems* ed. by T. Moriya (Springer, Berlin, Heidelberg 1981) p. 2–28; *Metallic Magnetism* ed. by H. Capellmann (Springer, Berlin, Heidelberg 1985)
4.3. For example, B Cogblin: In *Magnetism of Metals and Alloys* ed. by M. Cyrot (North-Holland, Amsterdam 1982) p. 295–377
4.4. The original figure was presented by J.A. Wilson: Adv. in Phys. **21**, 143 (1972), some other compounds have been added to it
4.5. P.R. Rhodes, E.P. Wholfarth: Proc. Roy. Soc. (London) **273**, 247 (1963)
4.6. K. Adachi, S. Ogawa: In Landolt-Börnstein III/19b (Springer, Berlin, Heidelberg 1988) pp. 1–425
4.7. Some explanation of pyrite type compounds can be found in K. Adachi: JARECT **21**, 19–30 (1986)
4.8. R.J. Bouchard, J.L. Gillin, H.S. Jarrett: Mat. Res. Bull. **8**, 489 (1973)
4.9. M. Kumada, N. Mori, T. Mitsui: J. Phys. C10, L643 (1977)
4.10. G. Czjzek, J. Fink, H. Schmidt, G. Krill, M. F. Lapierre, P. Panissod, F. Gautier, C. Robert: J. Mag. Mag. Mat. **3**, 58 (1976)
4.11. J. M. Hasting, L. M. Corliss: IBM J. Res. Dev. **14**, 227 (1970)
4.12. T. Miyadai, K. Takizawa, H. Nagata, H. Ito, S. Miyahara, K. Hirakawa: J. Phys. Soc. Jpn. **38**, 115 (1975)
4.13. F. Gautier, G. Krill, M.F. Lapierre, C. Robert: J. Phys. C6, L320 (1973)
4.14. K. Adachi, K. Sato, M. Matsumura, M. Ohashi: J. Phys. Soc. Jpn. **29**, 3233 (1970)

4.15. K. Adachi, M. Matsui, M. Kawai: J. Phys. Soc. Jpn. **46**, 1474 (1979)
4.16. K. Adachi, M. Matsui, Y. Omata, M. Mollymoto, M. Motokawa, M. Date: J. Phys. Soc. Jpn. **47**, 675 (1979)
4.17. M. Hattori, K. Adachi, H. Nakano: J. Phys. Soc. Jpn. **35**, 1025 (1973)
4.18. K. Adachi, K. Sato, M. Takeda: J. Phys. Soc. Jpn. **26**, 631 (1969)
4.19. For example, F. Hulliger and E. Mooser: J. Phys. Chem. Solids **26**, 429 (1965)
4.20. S. Asano: Preprint on the annual meeting of Phys. Soc. Jpn. Oct. 1981 and private communication.
4.21. D. W. Bullett: J. Phys. C**15**, 6163 (1982)
4.22. J. Hubbard: Proc. R. Soc. A**281**, 401 (1964)
4.23. S. Doniach: Adv. Phys. **18**, 819 (1964)
4.24. D.R. Penn: Phys. Rev. **142**, 350 (1966)
4.25. T. Moriya, H. Hasegawa: J. Phys. Soc. Jpn. **48**, 1490 (1980)
4.26. A. Yoshimori, S. Inagaki: J. Phys. Soc. Jpn. **50**, 769 (1981)
4.27. Y. Yoshida, S. Inagaki: J. Phys. Soc. Jpn. **50**, 769 (1981)
4.28. J. H. van Vleck: J. Phys. Radium **12**, 262 (1951)
4.29. Y. Yamamoto, T. Nagamiya: J. Phys. Soc. Jpn. **32**, 1248 (1972)
4.30. Y. Yoshimori, S. Inagaki: Magnetism Letters **35**, 69 (1977)
4.31. For a review see M. Roger, J.H. Hetherington, J.M. Delrieu: Rev. Modern. Phys. **55**, 1–64 (1983)
4.32. E. P. Wohlfarth: Philos. Mag. **7**, 1817 (1962)
4.33. Y. Takahashi, M. Tano: J. Phys. Soc. Jpn. **51**, 1792 (1982)
4.34. K. Ueda, T. Moriya: J. Phys. Soc. Jpn. **39**, 605 (1975)

5. Invar Systems

Yoji Nakamura

Towards the end of the 19th Century, in order to encourage the spread of the metric system all over the world, the International Bureau of Standards was making great efforts to discover a new cheap alloy which could be used as a measure instead of the meter standard made from a Pt–Ir alloy which was very expensive. Guillaume who engaged in this project, finally after many trials, found a new alloy with the composition of 65% Fe and 35% Ni which exhibited an almost zero thermal expansion coefficient at room temperature. Since this new alloy had almost constant dimensions independent of temperature, he named it the "invariable alloy", i.e. the "Invar alloy".

Guillaume was clever enough to demonstrate this alloy's remarkable insensitivity to temperature by observing a daily change in the height of the Eiffel Tower due to temperature variation, which had not previously been detected. This material led to a great improvement in land surveying, leading to enormous savings of both time and expense. The most remarkable application of Invar was in the clock and watch industries, where it is used for pendulums and hairsprings. Since then Invar alloys have become important materials for measuring equipment which requires dimensional stability [5.1].

On the other hand, the discovery of Invar had a big impact upon researchers of magnetism in Japan who were just beginning to face the dawn of modernization of their country. In fact many new kinds of Invar-type alloys, such as Superinvar with a thermal expansion coefficient smaller than that of Fe–Ni Invar by more than one order of magnitude, were discovered consequently, mostly by Masumoto and coworkers [5.2].

As shown in Fig. 5.1, the physical origin of the low thermal expansion coefficient of Invar is easily understood phenomenologically as arising from the cancellation of the lattice shrinkage with decreasing temperature by the lattice expansion due to the onset of ferromagnetism below the Curie temperature, which is called the spontaneous volume magnetostriction. However, it took many years after its discovery to make clear the real physical origin of this behavior. Around the 1960s the magnetic moment of Fe–Ni alloys was found to show a sharp deviation from the Slater–Pauling curve with decreasing Ni concentration, as shown in Fig. 5.2 [5.3]. This unusual concentration dependence of the magnetic moment was an attractive problem for physicists, and the magnetism of Invar alloys was studied extensively all over the world, particularly in Japan. In order to explain the unusual magnetic properties of Invar

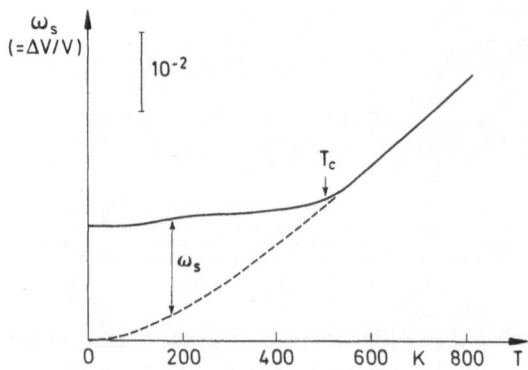

Fig. 5.1. Thermal expansion curve of the Fe–35% Ni alloy. T_C is the Curie temperature. The difference between the observed curve and the fictitious paramagnetic curve shown by a dotted line shows the spontaneous volume magnetostriction ω_s

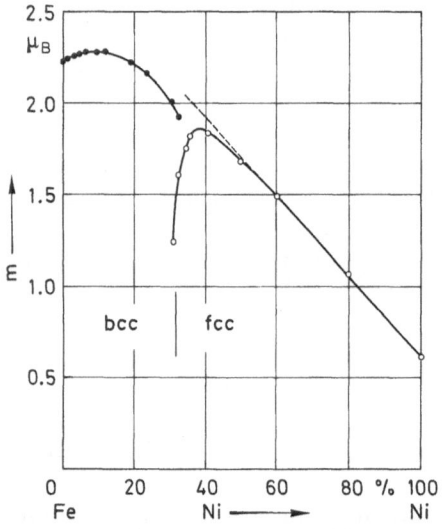

Fig. 5.2. Concentration dependence of the average magnetic moment of Fe–Ni alloys [5.3]. The straight line on the Ni-rich side indicates the Slater–Pauling curve

alloys, many models have been proposed. Following the so-called latent antiferromagnetism [5.4], in which the decrease of the average magnetic moment of the Invar alloys is ascribed to the appearance of antiferromagnetic regions more Fe rich than average, the two-state γ-phase model was proposed by Weiss [5.5] based on thermodynamical data on Fe–Ni alloys accumulated over many years. According to this two-state γ-phase model, the fcc (γ-phase) Fe has two states, one with a large magnetic moment with a large volume and the other with a nearly zero moment with a smaller volume. Since the ratio of these two states depends on both temperature and concentration, the low thermal expansion coefficient as well as the concentration variation of the magnetic moment of the Invar alloys could both be explained. Since this model was so attractive, Chikazumi and his coworkers extended it later in more detail [5.6].

The Fe–Ni Invar alloys show a low thermal expansion coefficient only in the fcc region very near to the fcc-bcc phase boundary. This led to the idea that there is a small bcc region in the Invar alloys. On the other hand, Kachi [5.7] tried to explain the anomaly as being due to inhomogeneities in the concentration, because the magnetic moment and the Curie temperature depend sharply on the concentration in these alloys. Several experiments were carried out to stabilize the fcc structure with much higher Fe concentrations by adding Mn [5.8] or C [5.9] or by suppressing the effect of inhomogeneities by using ordered Fe–Pt alloys [5.10].

In the beginning, the physical origin of the large positive spontaneous volume magnetostriction was discussed on the basis of the localized electron model (Appendix A.20). However, since the magnetism of alloys should be discussed in terms of the itinerant electron model (Appendix A.20), many theories have since been proposed in the framework of itinerant electron ferromagnetism. Originally, the low thermal expansion was limited to fcc Fe alloys. However, very weak itinerant ferromagnets such as $ZrZn_2$ [5.11] have also been found to show a large spontaneous volume magnetostriction, as do the rare earth–Co intermetallic compounds [5.12], antiferromagnets such as Cr–Mn [5.13] and even amorphous alloys such as Fe–B [5.14]. Therefore, the large spontaneous volume magnetostriction has become a more general phenomenon than thought before. Now it is observed in ferro, ferri and antiferromagnets of transition metal alloys including amorphous and weak itinerant electron ferromagnets. As will be described later, the large magnetostriction observed in these various kinds of magnetic alloys can be explained in a very general way as follows: the band polarization causes an increase in the kinetic energy of electrons which can be compensated for by a volume expansion. This idea was first proposed by us [5.15] and was later supported theoretically by detailed band calculations [5.16]. However, the calculated values of the spontaneous volume magnetostriction always give much larger values than those obtained by experiment. This discrepancy between experiment and theory has been solved by the spin fluctuation theory (Appendix A.31) developed recently in Japan [5.17] which indicates that the local magnetic moment still remains above the Curie temperature [5.18].

In addition to the sharp drop of the magnetic moment with decreasing Ni content, the Fe–Ni Invar alloys show several magnetic anomalies such as a rapid decrease in the magnetization with increasing temperature and a very large high field magnetic susceptibility [5.19]. These magnetic anomalies are explainable intuitively as follows: since the Invar alloy is in an unstable state just before its ferromagnetism disappears, its magnetism is very sensitive to concentration, temperature and magnetic field. On the other hand, this instability in magnetism results in a strong mutual dependence between magnetism and volume, leading to the strong dependence of its magnetism on volume, such as the large spontaneous volume magnetostriction, the large forced volume magnetostriction, large pressure effects on both Curie temperature and magnetization etc.

Generally, the phenomena related to the mutual dependence of magnetism and volume are called the magnetovolume effect and the alloys which show strong magnetovolume effects are known as Invar type alloys. The various kinds of anomalous magnetic properties, as a whole, are called Invar effects.

Among the many Invar effects, the magnetovolume effect is the most essential, so that this effect has been studied in detail in many kinds of materials, giving us an almost complete understanding of its physical origin. On the other hand, other Invar effects, such as the anomalous concentration dependence of the magnetic moment, are thought to be secondary Invar effects, so that detailed experiments have not yet been done in many Invar type alloys excluding fcc Fe alloys such as Fe–Ni Invar alloys. Many problems such as the effect of martensitic transformations etc. still remain unsolved. In this review we skip these problems, but readers may refer to [5.1], for example, for further information.

In this review, we first make clear the physical picture of the magnetovolume effect [5.20], then go on to the various kinds of Invar type alloys, including the actinide intermetallic compounds which have recently been found to be Invar type alloys. Finally, we very briefly discuss the elastic anomalies of Invar alloys.

5.1 Magnetovolume Effects

Figure 5.1 shows the thermal expansion curve of Fe–Ni Invar alloy. Above the Curie temperature T_C the thermal expansion coefficient has a value of the order of $10^{-5}/°C$ as in normal metals. However, below T_C the alloy shows an anomaly. The dotted line in the figure shows a fictitious paramagnetic curve. The difference between the observed and dotted curves gives the spontaneous volume magnetostriction ω_s ($= \Delta V/V$) given as

$$\omega_s = AM^2(T) , \tag{5.1}$$

which is proportional to the square of the magnetization. Here, the subscript s means spontaneous and A is the magnetovolume coupling constant indicating the strength of the mutual dependence between magnetization and volume. The Fe–Ni Invar alloy has a large value of A of the order of 10^{-8} cm^6/emu^2. As shown in Fig. 5.3, the value of the spontaneous volume magnetostriction at 0 K, $\omega(0)$ for the 35% Fe Invar alloy, has a large value of about 2×10^{-2} which is ten times as large as that of pure Fe. This large ω_s cancels the thermal shrinkage of the lattice, giving rise to the low thermal expansion in Fe–Ni Invar alloys. If the

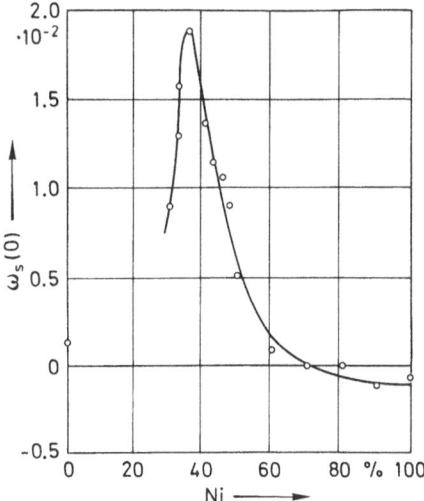

Fig. 5.3. Concentration dependence of the spontaneous volume magnetostriction of the Fe–Ni alloys at 0 K $\omega_s(0)$ [5.21]

spontaneous volume magnetostriction is much larger than the thermal shrinkage, as in Fe–Pt Invar alloys, the alloy shows a negative thermal expansion which means that the volume of the alloy expands with decreasing temperature, contrary to normal metals. Since ω_s is proportional to M^2, a small value of M in a very weak itinerant electron ferromagnet such as $ZrZn_2$ gives a small observed ω_s, even though the magnetovolume coupling constant A is large.

Upon the application of a magnetic field, a ferromagnet is deformed by the magnetostriction, keeping its volume almost constant. If the magnetic field is increased beyond the field necessary to reach technical saturation, the magnetic moment continues to increase slightly, giving rise to a volume change linear in magnetic field. This volume magnetostriction is considered to arise from the same physical origin as the spontaneous volume magnetostriction, because these two kinds of volume magnetostriction are induced by either the external magnetic field or the exchange field. Then we have the forced volume magnetostriction, $\partial\omega/\partial H$, by differentiating (5.1) with respect to the magnetic field

$$\partial\omega/\partial H = 2AM\frac{\partial M}{\partial H} = 2AM\chi_{hf} \ , \tag{5.2}$$

where χ_{hf} is the high field susceptibility. As shown in Fig. 5.4, $\partial\omega/\partial H$ for the Fe–Ni Invar alloys at room temperature reaches values of 3×10^{-8}/Oe which is about one hundred times larger than those for Fe or Ni.

Since the volume of the Fe–Ni Invar alloys increases by an increase in the magnetic moment, one can predict, as a reverse effect to the volume magnetostriction, a decrease of the magnetic moment and the Curie temperature under

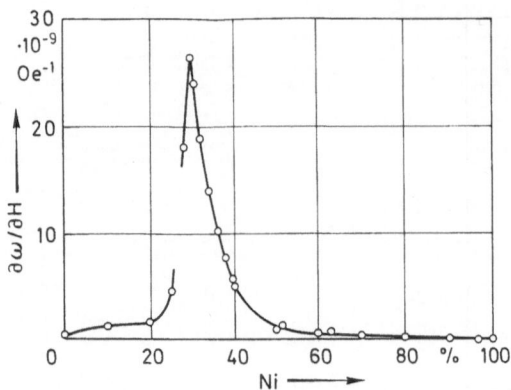

Fig. 5.4. Concentration dependence of the forced volume magnetostriction of Fe–Ni alloys at room temperature $d\omega/dH$ [5.21]

hydrostatic pressure. From the second derivative of the Gibbs free energy of a ferromagnet, we have the thermodynamical relation

$$\left.\frac{\partial\omega}{\partial H}\right)_{T,P} = -\frac{1}{V}\left.\frac{\partial M}{\partial P}\right)_{T,H} , \qquad (5.3)$$

where the subscripts T, P and H mean constant temperature, pressure and magnetic field, respectively. One sees that the forced volume magnetostriction is thermodynamically equivalent to the pressure dependence of magnetization. Therefore, we expect a large decrease of magnetization under hydrostatic pressure for Invar alloys because of their large $\partial\omega/\partial H$. This expectation is fulfilled as shown in Fig. 5.5, where $(1/M)(\partial M/\partial P)$ reaches -4×10^{-2}/kbar for the Fe–Ni Invar composition and its magnitude is more than fifty times as large as that for Fe or Ni.

On the other hand, for the second order magnetic transition from a ferromagnetic to a paramagnetic state, we have the Ehrenfest relation:

$$\frac{dT_C}{dP} = VT_C\frac{\Delta\alpha_m}{\Delta C_m} , \qquad (5.4)$$

where $\Delta\alpha_m$ and ΔC_m are the jumps in the thermal expansion coefficient and the specific heat at T_C, respectively.

As seen in Fig. 5.1, since $\Delta\alpha_m$ is always negative, we expect a decrease in T_C under hydrostatic pressure. This was observed for the Invar composition with a value of dT_C/dP of -3.6 K/kbar. In contrast to this, dT_C/dP is $+0.35$ K/kbar for Ni because of its positive value of ΔC_m. Then we see that (5.4) holds qualitatively. As was discussed previously, the spontaneous volume magnetostriction, the forced volume magnetostriction and the pressure effects on both magnetization and the Curie temperature, are all related to each other through

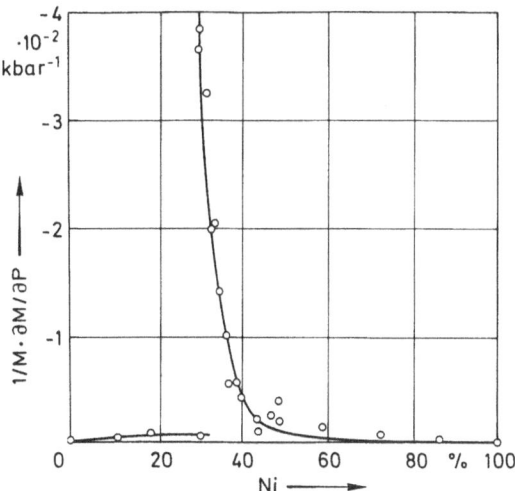

Fig. 5.5. Concentration dependence of the pressure coefficient of magnetization of Fe–Ni alloys at room temperature $(1/M)\partial M/\partial P$ [5.23]

thermodynamical relations. We therefore only discuss here the spontaneous volume magnetostriction which is thought to be the most fundamental of the many magnetovolume effects.

The magnetic properties of transition metals and alloys have been discussed on the basis of two different models, i.e., the localized electron model and the itinerant electron model (Appendix A.20) for many years. The former gives a good account of the high temperature properties such as the Curie–Weiss law, whereas the latter the low temperature properties such as the magnetic moment. However, researchers have tried for many years now to find a unified picture of magnetism. Here, we first discuss the spontaneous volume magnetostriction based upon the localized electron model, because this model gives an easy and intuitive understanding of the origin of the spontaneous volume magnetostriction.

If we use the molecular field approximation based upon the localized electron model and take only the terms related to the magnetization M and the volume change ω, we have the energy of a ferromagnet as

$$E = -\frac{1}{2}WM^2 + \frac{1}{2\kappa}\omega^2 , \qquad (5.5)$$

where W is the molecular field coefficient and κ the compressibility. The first and second terms in (5.5) represent the magnetic energy and the elastic energy, respectively. Therefore, we have the equilibrium volume change ω by differentiating (5.5) with respect to ω, simply because the volume of a ferromagnet changes until the decrease in magnetic energy is balanced by the increase in the elastic energy. Then the spontaneous volume magnetostriction ω_s can be expressed as

$$\omega_s = \frac{1}{2}\kappa\frac{dW}{d\omega}M^2$$

Since the molecular field coefficient is proportional to the exchange integral, we have

$$\omega_s \propto \kappa(dJ/d\omega)M^2 \ . \tag{5.6}$$

Therefore, the magnetovolume coupling constant, A, in (5.1) turns out to be proportional to $dJ/d\omega$. This gives us a physical picture of the origin of the spontaneous volume magnetostriction: the volume of a specimen increases to increase the magnetic energy gain by an increase in the exchange integral. According to the Bethe–Slater curve which relates J to the interatomic distance in ferromagnets, Fe–Ni Invar alloy is expected to have a large ω_s, because Invar is located where the curve has its maximum slope. However, the Bethe–Slater curve has no real physical basis and it is not possible to ascribe the essential origin of the spontaneous volume magnetostriction to $dJ/d\omega$, even though it is convenient to use the Bethe–Slater curve to have an overview of experimental data of the Curie temperature of many alloys. We now discuss the spontaneous volume magnetostriction based upon the itinerant electron model. It is well known that the 3d electrons in transition metals and alloys which form the conduction band also carry magnetic moments. Recent progress in both band calculations using computers and in experimental techniques such as photo-emission have given us a clearer understanding of the band structure of various transition metal alloys and compounds. Here, we discuss the spontaneous volume magnetostriction based upon the Stoner model of a ferromagnet, which is the simplest itinerant electron model for ferromagnets.

As shown in Fig. 5.6, in the paramagnetic state (a) the numbers of electrons with $+$ and $-$ spins are equal, while in the ferromagnetic state (b) the number of electrons with $+$ spin increases to gain the exchange energy. This requires a polarization of the band which in turn induces an increase in the kinetic energy of the electrons, because some of the electrons must move up to the $+$ spin subband from the $-$ spin subband. Therefore, the band is polarized until the magnetic energy gained is balanced by the increase in kinetic energy. This kinetic energy increase can be accommodated by a decrease in the bandwidth, leading to an increase in the interatomic distance as shown in (c). This is the basic physical origin of the magnetovolume effect in the framework of the Stoner model (Appendix A.34). The mechanism can also be understood in the following way: in the ferromagnetic state (b) the number of electrons with high kinetic energy, i.e., the anti-bonding electrons, increases as compared to the para-magnetic state (a), leading to a larger interatomic distance as shown in (c), because the repulsive forces between atoms increase.

If we denote the increase in kinetic energy of electrons ΔE_k and take only the terms dependent on ΔE_s and ω, neglecting the volume dependence of the

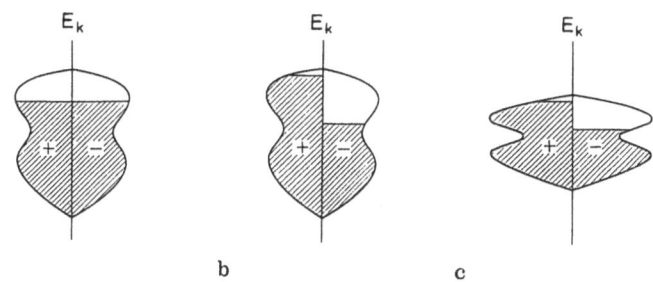

Fig. 5.6a–c. Schematic presentation of ferromagnetism and spontaneous volume magnetostriction based upon the Stoner model. E_k is the kinetic energy of electrons and the abcissa indicates the density of states of electrons. **a** Paramagnetic, **b** Ferromagnetic (large kinetic energy small volume), **c** Ferromagnetic (small kinetic energy large volume)

exchange integral, we have the magnetic energy:

$$E = \Delta E_k + \frac{\kappa \omega^2}{2} \,.$$

giving us ω_s. Here, ΔE_k is, to a first approximation, proportional to M^2, so that ω_s turns out to be proportional to M^2 as in (5.1). The detailed band calculation based on this idea [5.16] has shown large values of $\omega_s(0)$ of a few % for transition metal elements. Therefore, this mechanism seemed to account well for the physical origin of the large spontaneous volume magnetostriction in the Invar alloys. However, the calculation predicted a spontaneous volume magnetostriction for Fe which was much larger than that for the Invar alloys. Thus there is a serious disagreement between theory and experiment, since there is no such large spontaneous volume magnetostriction in Fe. According to the Stoner model the magnetic moment should disappear at the Curie temperature. However, paramagnetic neutron scattering experiments on Fe–Ni Invar alloy have shown that there is a magnetic moment of about $1.4\,\mu_B$ even above the Curie temperature [5.25]. This experiment clearly shows that the Stoner model, which is quite effective for describing the magnetic moment of transition metals at 0 K, fails to describe the high temperature behavior of these magnetic materials and strongly suggests that the $3d$ transition metal alloys have a localized electron character at high temperatures.

As shown in Fig. 5.7a, according to the localized electron model, the magnetization of a sample M becomes zero at the Curie temperature, because the direction of each moment m becomes random. However, each magnetic moment remains constant beyond T_C. On the other hand, the Stoner model gives a different temperature variation of m. As shown in (b) the magnetic moment decreases with increasing temperature, because the band polarization decreases with increasing temperature and vanishes at T_C. Compared to these

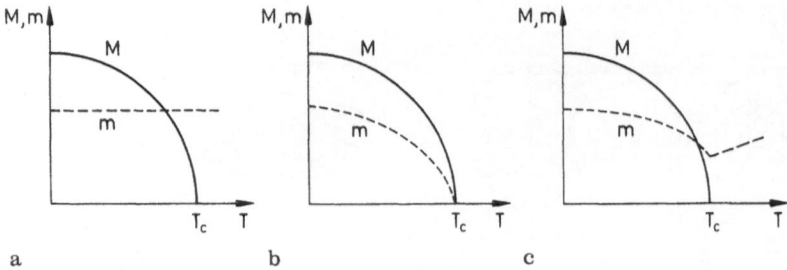

Fig. 5.7a–c. Schematic presentation of the temperature dependence of magnetic moment m and magnetization M. In both cases $M = Nm$, where N is the number of atoms per unit volume. **a** Localized model, **b** Stoner model, **c** Invar model

two cases, Invar alloy is presumably an intermediate case. The magnetic moment m decreases and its direction becomes random with increasing temperature but it still remains beyond T_C where M disappears, as shown in (c) (see also Fig. 4.39).

If the spontaneous volume magnetostriction is given by the square of the magnetic moment m instead of M in (5.1), ω_s observed below T_C is given by

$$\omega_s(T) = AN^2\{m^2(T) - m^2(T_C)\} , \tag{5.7}$$

where N is the number of atoms in the unit volume. This accounts well for $\omega_s(0)$ in Fe–Ni Invar alloy, which was overestimated by the previous calculation. If Fe has nearly localized electrons with almost constant magnetic moments, (5.7) predicts that the spontaneous volume magnetostriction of Fe would be very small, which is what is found experimentally.

As already discussed, the spontaneous volume magnetostriction is directly related to the behavior of the magnetic moment, which is one of the most important problems of alloy magnetism. Many researchers have tried over the years to formulate a unified picture of magnetism in transition metal alloys, which were thought to have an itinerant character at low temperatures whilst behaving as if they had localized electron character at high temperatures. Recently, the spin fluctuation theory (Appendix A.31) of magnetism developed mainly by Moriya and coworkers [5.17] has made a very important contribution to our understanding of the intrinsic nature of alloy magnetism. It has also given a theoretical basis to the magnetic behavior of Invar alloys [5.18] as discussed above.

5.2 Invar Type Alloys

5.2.1 Transition Metal Alloys

The fcc iron alloys such as Fe–Ni, Fe–Pt and Fe–Pd are typical Invar alloys. Fe–Ni–Mn alloys are also Invar type alloys, in which the fcc structure has been stabilized up to a higher Fe concentration by replacing some Ni in Fe–Ni alloys by Mn. These Invar alloys have been studied in detail [5.1], and all show remarkable Invar effects. Among them, the Fe–Pt alloys have the advantage of reducing the effect of inhomogeneities on the Invar effect, because Fe_3Pt is formed in the ordered state. Furthermore, the magnetic moment of most of the fcc Fe Invar alloys decreases sharply with increasing Fe concentration. In contrast, the concentration dependence of the magnetic moment of Fe–Pt alloys follows the Slater–Pauling curve (Appendix A.27) up to a much higher Fe concentration than the Invar composition. Yet the Fe–Pt Invar alloys show a very large spontaneous volume magnetostriction even in the ordered state which overcomes the thermal shrinkage of the lattice, giving rise to a negative thermal expansion coefficient at room temperature [5.26]. These experiments have clearly shown that the idea originally proposed, that the large spontaneous volume magnetostriction is directly related to the sharp decrease of magnetic moment, i.e., ascribing the Invar effect to the transition from a strong ferromagnet to a weak ferromagnet, does not give a consistent explanation for the magnetovolume effect [5.26]. The Fe–Ni–Mn alloys transform from ferromagnetic to antiferromagnetic with increasing Fe concentration, keeping their fcc structure. The spontaneous volume magnetostriction appears again when the alloy becomes antiferromagnetic and it increases with increasing Mn concentration [5.8]. The concentration dependence of the magnetic moment of the Fe–Pd Invar alloys follows the Slater–Pauling curve like that of Fe–Pt alloys [5.27]. These alloys show a transformation of the crystal structure from fcc to face-centered tetragonal at low temperatures [5.28].

Recently, amorphous alloys have been prepared by adding some metalloid elements such as P, B, C or Si to transition metals and alloys, followed by rapid quenching from the melt. Among them, amorphous ferromagnetic alloys such as Fe–P and Fe–B systems also show a large magnetovolume effect [5.29]. As shown in Fig. 5.8 for example, both of these amorphous alloys show a large spontaneous volume magnetostriction comparable to that of Fe–Ni Invar alloys. These alloys also show a sharp decrease of the magnetic moment with

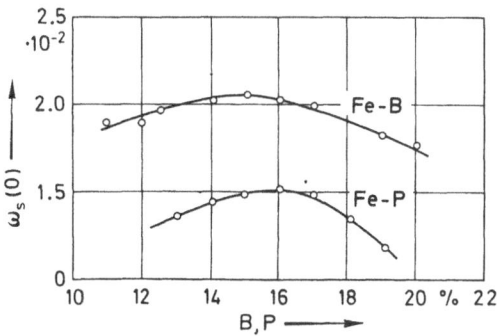

Fig. 5.8. Concentration dependence of the spontaneous volume magnetostriction ω_s of amorphous Fe–B and Fe–P alloys at 0 K [5.29]

increasing Fe concentration [5.29]. These features closely resemble the Fe–Ni Invar alloys. The amorphous structure of these Invar alloys also indicate that the fcc structure is not a necessary condition for the Invar effect.

Among various kinds of ferromagnets, we have very weak itinerant electron ferromagnets such as $ZrZn_2$ [5.11], Ni_3Al [5.30], Ni–Pt [5.30] and MnSi [5.31], which have a small magnetic moment and a low Curie temperature. All show remarkable magnetovolume effects, so that these very weak itinerant electron ferromagnets can be classified as Invar type alloys. Since the magnetic moment of these alloys is very small, the square of the magnetization M^2 should decrease linearly in T^2 with increasing temperature at low temperatures. Therefore, ω_s should decrease linearly in T^2 with increasing temperature, according to (5.1). As shown in Fig. 5.9, this expectation is realized for $ZrZn_2$ at low temperatures where the effect of lattice vibrations is negligible.

Many transition metal alloys with a composition around that at which ferromagnetism disappears show various unusual magnetic behavior such as a maximum in their magnetization–temperature curves and a magnetic field cooling effect. These alloys are called cluster-glasses or mictomagnetic ferromagnets (Appendix A.32). In these alloys, the inhomogeneities existing in the material make the magnetism nonuniform, giving rise to ferromagnetic clusters. The magnetic moments in these clusters behave as superparamagnets at high temperatures, whereas they are frozen in random orientations at low temperatures. If we apply a magnetic field, the moments rotate to the direction of the magnetic field and at the same time the magnetic moment increases due to band polarization induced by the magnetic field, leading to volume magnetostriction. According to a simple calculation [5.33], this volume magnetostriction ω can be

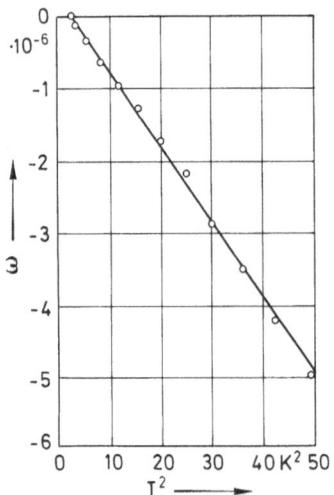

Fig. 5.9. Thermal expansion curve of $ZrZn_2$ at low temperature [5.32]

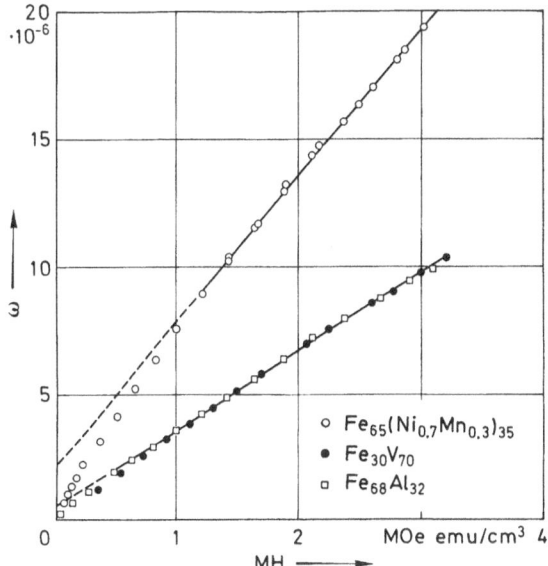

Fig. 5.10. Volume magnetostriction ω versus MH at 4.2 K for Fe–Ni–Mn, Fe–V and Fe–Al cluster glasses [5.34]. The scales of the ω and MH axes in the figure are valid for $Fe_{68}Al_{32}$. For $Fe_{65}(Ni_{0.7}Mn_{0.3})_{35}$ the axis must be multiplied by 2 while the MH axis by 0.5, and for $Fe_{30}V_{70}$ the MH axis by 0.25

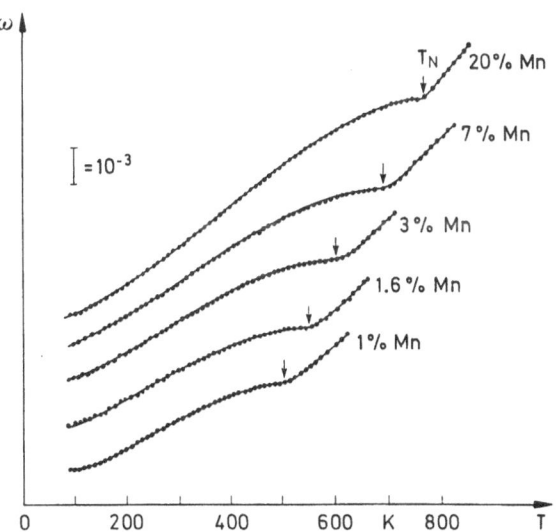

Fig. 5.11. Thermal expansion curves of antiferromagnetic Cr–Mn alloys [5.13]. The arrow shows the Néel temperature T_N

written

$$\omega = A\chi MH \ , \tag{5.8}$$

where χ is the susceptibility of the ferromagnetic clusters. As shown in Fig. 5.10, this relation is almost satisfied for Fe–Ni–Mn, Fe–V and Fe–Al cluster glasses [5.34]. The deviation from (5.8) observed in low magnetic fields is ascribed to the volume magnetostriction due to the exchange interaction between ferromagnetic clusters as given in (5.6).

As already discussed, the large volume expansion in Invar type alloys occurs in order to compensate for the increase in kinetic energy of the $3d$ electrons due to the band polarization. Since this mechanism for the volume magnetostriction may also apply to metallic antiferromagnets, one might expect to have antiferromagnetic Invar alloys. In fact large magnetovolume effects such as a large spontaneous volume magnetostriction have been observed in itinerant electron antiferromagnets such as Fe–Ni–Mn [5.10], Cr–Fe–Mn [5.35] and Cr–Mn [5.13], as shown in Fig. 5.11 for Cr–Mn alloys.

5.2.2   Rare Earth–Transition Metal Intermetallic Compounds

Rare earth metals, R, form various intermetallic compounds with transition metals, T. The $4f$ electrons of the rare earth metals which are responsible for their magnetism behave as localized electrons, because they are well shielded from the outside of each atom by outer closed electron shells of $5s^2 5p^6$. Therefore, $4f$ electrons themselves do not cause any large spontaneous volume magnetostriction, but show a small volume change due to the exchange term as expressed in (5.6). However, rare earth–transition metal intermetallic compounds, such as RCo_2 [5.36, 37], RMn_2 [5.36], R_2Fe_{14} [5.37] and $R_2Fe_{14}B$ [5.38] do show a large magnetovolume effect due to the band polarization of the transition metals.

RCo_2 crystallizes into the cubic Laves phase. The magnetic moments of R and T atoms couple ferromagnetically for light rare earth elements such as Pr, Nd and Sm, leading to ferromagnetism for these compounds, while these moments couple antiferromagnetically for heavy rare earth elements such as Tb, Dy, Ho and Er, giving rise to ferrimagnetism. As shown in Fig. 5.12, these RCo_2 compounds show a large ω_s as well as a large negative dT_C/dP, so that they are classified as Invar type alloys. Here, the magnetic moment of the Co atoms, induced by band polarization due to the exchange magnetic field from the R moments, gives rise to an ω_s nearly proportional to the square of the Co moment as expressed in (5.1). As can be seen in Fig. 5.12, ω_s vanishes discontinuously at the Curie temperature for $R =$ Dy, Ho and Er because of a first order magnetic transition (Appendix A.11) in these compounds. The value of the magnetovolume coupling constant of $GdCo_2$ has the same order of magnitude as that of Fe–Ni Invar alloys. RMn_2 also has the cubic Laves phase structure. In the case

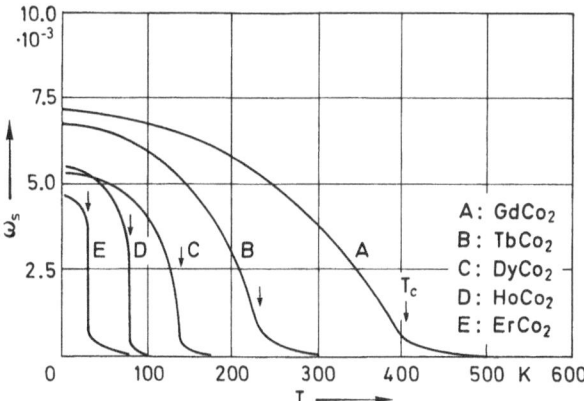

Fig. 5.12. Temperature variation of spontaneous volume magnetostriction ω_s of RCo_2 [5.36]. The arrow indicates the Curie temperature T_C

of a nonmagnetic R such as Y, YMn_2 for example shows a first order magnetic transition from paramagnetic to antiferromagnetic at about 100 K, below which the Mn atoms have a magnetic moment of $2.7\mu_B$ giving rise to an extremely large ω_s of 5% much larger than that of the Fe–Ni Invar alloys and even the largest among the volume changes due to purely magnetic origin, as shown in Fig. 5.13. If R is a light rare earth element such as Pr, Nd or Sm, the RMn_2 are antiferromagnets with a Néel temperature of about 100 K, below which they show a ω_s of about 5×10^{-3}. Among the heavy rare earth compounds $GdMn_2$ is a ferromagnet with a Curie temperature of 40 K, whereas $TbMn_2$ is a ferrimagnet with a Curie temperature of 50 K, both with a large ω_s at the Néel temperature of the Mn sublattice. With a further increase in the atomic number of the R element, the interatomic distance between the Mn atoms decreases, leading to no ordering of the Mn moment for Dy, Ho and Er compounds. Therefore, these heavy rare earth–manganese intermetallic compounds show no volume expansion due to the Mn subband polarization, but do show a small volume shrinkage due to the exchange interaction between localized R moments.

Recently, Sagawa and coworkers [5.38] have discovered a new very strong permanent magnet, $Nd_2Fe_{14}B$. This has a remanent magnetic flux density B of 12 kG and a coercive force H_C between 10 and 20 kOe, leading to an energy product $(BH)_{max}$ of 36 MGOe, over twice as large as that of $SmCo_5$ and its family, which were the strongest permanent magnets until the discovery of $Nd_2Fe_{14}B$. Since then $R_2Fe_{14}B$ and their related compounds have been studied extensively. $R_2Fe_{14}B$ form tetragonal intermetallic compounds which are stabilized by a small amount of B atoms. These intermetallic compounds show a large spontaneous volume magnetostriction of the order of 1% [5.39], as shown in

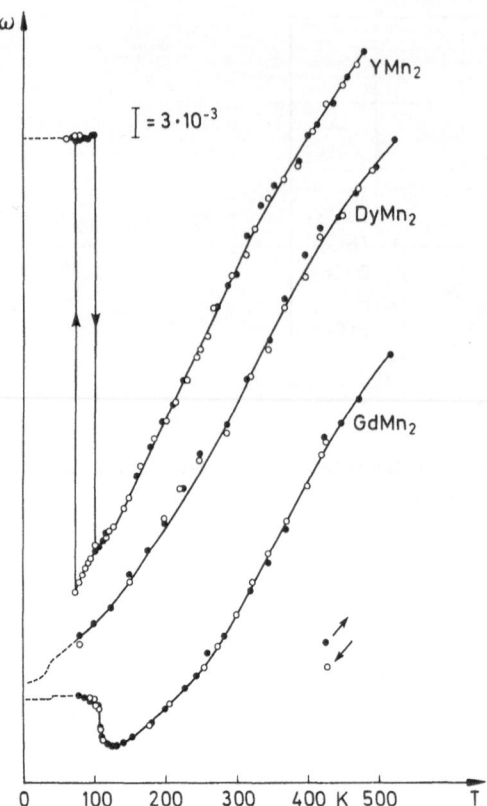

Fig. 5.13. Thermal expansion curves of $R\text{Mn}_2$ as measured by X-ray diffraction [5.36]. The dotted curves at low temperatures were obtained by a dilatometric method

Fig. 5.14 and they reveal a large forced volume magnetostriction as well. These magnetovolume effects can be ascribed to the Fe band polarization.

As already discussed, many kinds of rare earth–transition metal intermetallic compounds show a variety of helimagnetism and some of them are of great importance for practical applications as excellent permanent magnets. Their large spontaneous volume magnetostriction is ascribed to the $3d$ moment of Mn, Fe or Co induced by $3d$ band polarization at the Curie or Néel temperature.

5.2.3 Actinide Intermetallic Compounds

Recently, large pressure effects on both magnetization and Curie temperature have been observed in actinide intermetallic compounds. UPt [5.40] is a weak ferromagnet with a Curie temperature of 25 K and a magnetic moment of $0.5\mu_\text{B}$.

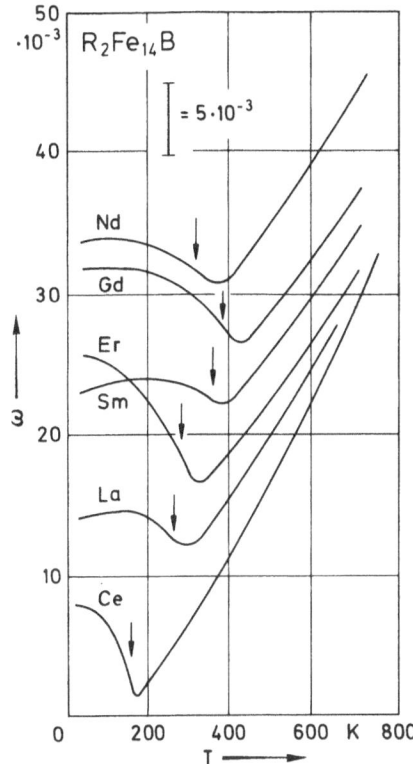

Fig. 5.14. Thermal expansion curves of $R_2Fe_{14}B$ [5.37]. The arrow indicates the Curie temperature T_C

This material shows a large pressure effect on magnetization in such a way that the magnetization becomes one–tenth of its original value by the application of a hydrostatic pressure of 8 kbar [5.40]. Some Np intermetallic compounds also show remarkable magnetovolume effects. $NpAl_2$ and $NpOs_2$ [5.41], both have a Laves phase structure and show very weak ferromagnetism: the Curie temperatures of these two compounds are 56 K and 9 K, and their magnetic moments are $1.4\mu_B$ and $0.5\mu_B$, respectively. Both of them show remarkable pressure effects on both the Curie temperature and the magnetic moment. Figure 5.5 demonstrates the results for $NpAl_2$: the hyperfine field acting on ^{237}Np, as obtained from Mössbauer effect measurements, shows a remarkable pressure effect. This means that the magnetic moment of the Np atoms decreases as a function of hydrostatic pressure. Since U or Np atoms have a $5f$ magnetic moment, the observed large magnetovolume effects suggest that, in contrast to the $4f$ electrons in rare earth metals, the $5f$ electrons of the actinide elements are not localized, but form a $5f$ energy band, which splits below the Curie temperature, giving rise to a magnetovolume effect, similarly to transition metal alloys. As has

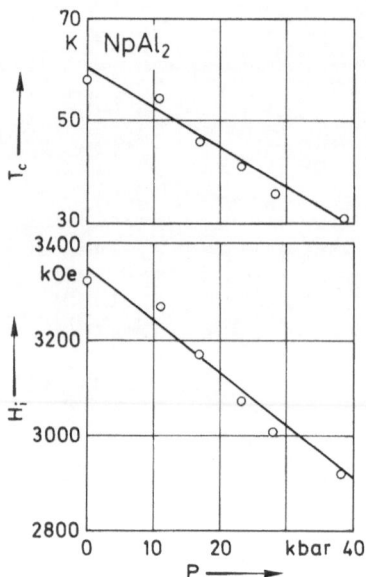

Fig. 5.15. Pressure variation of the Curie temperature T_C of $NpAl_2$ and of the hyperfine field at the Np nuclei [5.41]

been extensively studied theoretically and experimentally, it is of great interest to know whether $5f$ electrons behave as localized or as itinerant electrons. According to our recent understanding, the $5f$ electrons of light rare earth metals are probably itinerant, whereas those of heavy actinide metals have a localized character. For light rare earth actinides, the electric potential for $5f$ electrons is too small to localize $5f$ electrons because of a small positive nuclear charge, whereas it is large enough to show localization of $5f$ electrons for heavy actinides. The experimental results on magnetovolume effects in actinide inter-metallic compounds suggest that the actinide metals have an itinerant electron character up to at least Np.

5.3   Elasticity

So far in this review article, we have only discussed the magnetovolume effect which is directly related to the low thermal expansion coefficient of Invar alloys. Here, we briefly discuss the magnetoelastic effects which are strongly connected to the magnetovolume effect. Generally, in normal metals and ferromagnets the volume decreases with decreasing temperature, leading to a hardening of the

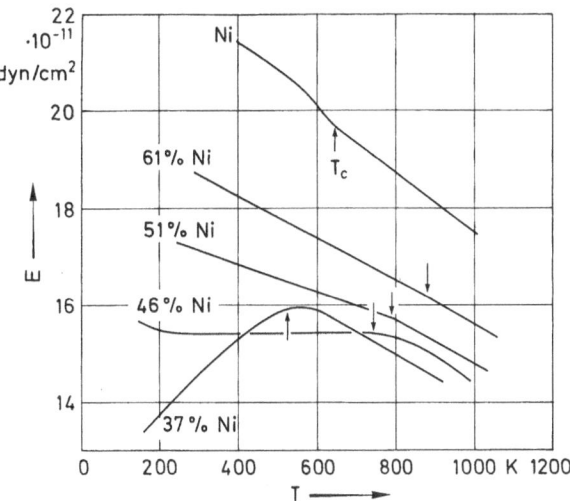

Fig. 5.16. Temperature dependence of Young's modulus of Fe–Ni alloys [5.40]. The arrow shows the Curie temperature T_C

lattice, i.e., an increase in elastic constants. In contrast to this, the Invar alloys expand due to a large spontaneous volume magnetostriction, so that the thermal shrinkage is suppressed. Furthermore, the higher order term in magnetoelastic couplings also reduces the elastic constants. Therefore, the crystal shows a softening with decreasing temperature, leading to a decrease in the elastic constants. The decrease in elastic constants due to magnetic origin, as a whole, is called the ΔE effect (Appendix A.6). For example, the temperature dependence of Young's modulus for Fe–Ni alloys is shown in Fig. 5.16 [5.12]: Young's modulus of the Fe–37% Ni alloy decreases with decreasing temperature below T_C, whereas in the Fe–46% Ni alloy slightly off from the Invar composition the hardening of the lattice and the softening due to magnetoelastic coupling cancel each other, leading to a very small temperature variation of Young's modulus. Guillaume who found this phenomenon named this alloy the "Elinvar alloy" because the elastic properties of this alloy are invariant with temperature. Such phenomena arising from the strong mutual dependence of magnetism and elasticity are called, as a whole, "magnetoelastic effects", and are strongly present in Fe–Ni alloys, Fe–Pt alloys, amorphous Fe–B alloys and antiferromagnetic Cr alloys. The magnetoelastic effect has been discussed theoretically on the basis of both the localized electron model [5.43] and the weak itinerant electron model [5.44] (Appendix A.20). The magnetovolume effects are the phenomena whose physical quantities are expressed by the first volume derivative of more fundamental physical properties, whereas the magnetoelastic effect is related to the second derivative of the energy of a crystal with respect to strain. This makes the understanding of the physical origin of magnetoelastic effects very difficult, so that this problem is still an open question.

5.4 Conclusions

Invar alloys are of considerable practical importance. They are used where dimensional stability is required: standard scales, bimetals, and microwave devices such as wave guides. They are also used on a large scale as transfer tubes for liquefied gas and membranes for containers for liquefied natural gas tankers, whereas the Elinvar alloys are used as spring materials. Both Invar and Elinvar alloys are of major importance in modern technology. For his remarkable discovery of the Invar and Elinvar alloys, Guillaume was awarded the Nobel Prize in physics in 1920. After that time, researchers in solid state physics paid almost no attention to the Invar alloys for many years. Researchers in magnetism now recognize the critical importance of the study of the Invar alloys, because after the 1960s Invar effects, particularly the large magnetovolume effect, were found in a variety of ferromagnets, ferrimagnets and antiferromagnets, and it was recognized that the effects are closely related to the behavior of the $3d$, $4f$ or $5f$ electrons responsible for the magnetism. Today, we are getting very close to having a unified picture of alloy magnetism, due to progress in both experiment and theory, particularly the spin fluctuation theory. However, it is still very difficult to make calculations for each magnetic material, so that it is unnecessary to emphasize the important role of experiments. Theoretically speaking, the magnetic moment above T_C can be obtained directly from paramagnetic neutron scattering. However, excluding Fe–35%Ni [5.25] and Fe_3Pt [5.45], almost no neutron scattering experiments above T_C have been performed because of experimental difficulties. It is extremely necessary that these types of experiments be made.

According to Fig. 5.7c, the magnetic moment of Invar alloys persists above T_C and increases with increasing temperature. Thus we expect to observe a spontaneous volume magnetostriction at high temperatures in the paramagnetic state. In fact, in weak itinerant electron ferromagnets such as $ZrZn_2$ [5.46] and MnSi [5.47] the existence of a spontaneous volume magnetostriction in the paramagnetic state has been suggested from an analysis of thermal expansion experiments. As discussed previously, it is possible to estimate the magnetic moment in the paramagnetic state from the measurements of spontaneous volume magnetostriction at high temperatures. As seen in Fig. 5.13, the thermal expansion curves of RMn_2 show a downward deviation from the straight line at high temperatures. This deviation, which cannot be explained from lattice vibrations at high temperatures, may reflect a variation of magnetic moment in the paramagnetic state. This strongly suggests that precision measurements of thermal expansion at high temperatures up to well beyond the Curie temperature will give us more important information on the physical properties of the Invar alloys as well as the physics of magnetism.

References

5.1.   Honda Memorial Series on Material Science No. 3, *The Physics and Applications of Invar Alloys*, ed. H. Saito (Maruzen Co. Ltd., Tokyo, 1978)
        Y. Nakamura: IEEE Trans Mag. Mag. **12**, 278 (1976)
5.2.   H. Masumoto: Sci. Rep. Tohoku Imperial Univ. **20**, 101 (1931)
5.3.   J. Crangle and G.C. Halam: Proc. Roy. Soc. A**272**, 119 (1963)
5.4.   E.I. Kondorski and V.L. Sedov: J. Appl. Phys. **31**, 331S (1960)
5.5.   R.J. Weiss: Proc. Phys. Soc. **82**, 281 (1963)
5.6.   S. Chikazumi: J. Magn. Magn. Mater. **15–18**, 1130 (1980)
5.7.   S. Kachi, H. Asano: J. Phys. Soc. Jpn. **27**, 536 (1969)
5.8.   M. Shiga: J. Phys. Soc. Jpn. **22**, 539 (1967)
5.9.   G.F. Bolling, A. Arrott, R.H. Richman: Phys. Status Solidi **26**, 743 (1968)
5.10.   K. Sumiyama, M. Shiga, Y. Nakamura: J. Phys. Soc. Jpn. **40**, 996 (1976)
5.11.   S. Ogawa, N. Kasai: J. Phys. Soc. Jpn. **27**, 789 (1969)
5.12.   K.H.J. Buschow, A.R. Miedema, M. Brouha: J. Less–Comm. Met. **38**, 9 (1974)
5.13.   G. Hausch, M. Shiga, Y. Nakamura: J. Phys. Soc. Jpn. **40**, 903 (1976)
5.14.   K. Fukamichi, M. Kikuchi, A. Arakawa, T. Masumoto: Solid State Commun. **23**, 955 (1977)
5.15.   M. Shiga, Y. Nakamura: J. Phys. Soc. Jpn. **26**, 24 (1969)
5.16.   J.F. Janak, A.R. Williams: Phys. Rev. B**14**, 4199 (1976)
5.17.   T. Moriya: J. Magn. Magn. Mater. **45**, 79 (1984)
5.18.   T. Moriya and K. Usami: Solid State Commun. **34**, 95 (1980)
5.19.   O. Yamada, R. Pauthenet, J.C. Picoche: J. de Phys. **32**, C1 119 (1971)
5.20.   M. Shiga: J. Phys. Soc. Jpn. **50**, 2573 (1981)
5.21.   M. Hayase, M. Shiga, Y. Nakamura: J. Phys. Soc. Jpn. **34**, 925 (1973)
5.22.   Ref. [5.1] p. 88
5.23.   J.S. Kouvel, R.H. Wilson: J. Appl. Phys. **32**, 435 (1961)
5.24.   L. Patrick: Phys. Rev. **93**, 384 (1954)
5.25.   M.F. Collins: Proc. Phys. Soc. **86**, 973 (1965)
5.26.   Y. Nakamura, K. Sumiyama, M. Shiga: J. Magn. Magn. Mater. **12**, 127 (1979)
5.27.   M. Matsui, K. Adachi: J. Appl. Phys. **51**, 6319 (1980)
5.28.   M. Matsui, H. Yamada, K. Adachi: J. Phys. Soc. Jpn. **48**, 2161 (1980)
5.29.   K. Fukamichi: In *Amorphous Metallic Alloys* ed. F.F. Luborsky (Butterworths, London 1983) p. 317
5.30.   J.J.M. Franse: Physica **86–88** B + C, 283 (1977)
5.31.   E. Fawcett, J. P. Maita, J.H. Wernick: Intern. J. Mag. **1**, 29 (1970)
5.32.   P.P.M. Meinke, E. Fawcett, G.S. Knapp: Solid State Commun. **7**, 1643 (1969)
5.33.   Y. Muraoka, H. Wada, M. Shiga, Y. Nakamura: Physica **119** B, 174 (1983)
5.34.   Y. Muraoka, M. Shiga, Y. Nakamura: Inst. Phys. Conf. Ser. **55**, 257 (1981)
5.35.   K. Fukamichi, H. Saito: Phys. Status Solidi (a) **10**, K129 (1972)
5.36.   Y. Nakamura: J. Magn. Magn. Mater. **31–34**, 829 (1983)

5.37. D. Gignoux, D. Givord, F. Givord, R. Lemaire: J. Magn. Magn. Mater. **10**, 288 (1979)
5.38. M. Sagawa, S. Fujimura, N. Tagawa, H. Yamamoto, Y. Matsuura: J. Appl. Phys. **55**, 2083 (1984)
5.39. K.H.J. Buschow: J. Less–Comm. Met. **118**, 349 (1986)
5.40. J.J.M. Franse: J. Magn. Magn. Mater. **31–34**, 819 (1983)
5.41. J. Moser, W. Potzel, B.D. Dunlap, G.M. Kalvius, J. Gal, G. Wortmann, D.J. Lam, J.C. Spirlet, I. Nowick: In *Physics of Solids under Pressure*, ed. by J.S. Schilling, R.N. Shelton (North-Holland, Amsterdam 1981) p. 271
5.42. G. Hausch, H. Warlimont: Z. Metallkde **64**, 152 (1973)
5.43. G. Hausch: J. Magn. Magn. Mater. **10**, 163 (1979)
5.44. S.G. Steinemann: J. Magn. Magn. Mater. **7**, 84 (1978)
5.45. K.R.A. Ziebeck, P.J. Webster, P.J. Brown, H. Capellmann: J. Magn. Magn. Mater. **36**, 151 (1983)
5.46. S. Ogawa: Physica **119B**, 68 (1983)
5.47. M. Matsunaga, Y. Ishikawa, T. Nakajima: J. Phys. Soc. Jpn. **51**, 1153 (1982)

6. Magnetic Anisotropy and Magnetostriction

Tokuo Wakiyama

Professor Sōshin Chikazumi chaired an International Conference on Magnetism in 1982 which was held in Kyoto. At this conference, which was very successful, many papers on the physics of magnetism were presented and discussed. First, let us look at the logo of ICM 82 (Fig. 6.1). In this logo, we can see the results of the world-famous experiment carried out in 1926 by Honda and Kaya [6.1], who were leaders in the physics of magnetism in Japan; the experiment which marks the discovery of the magnetocrystalline anisotropy of a ferromagnetic material. On the other hand, the phenomenon of magnetostriction was investigated by Nagaoka at an even earlier time. In 1900, at the first International Conference on Physics held in Paris, Nagaoka gave an invited lecture on magnetostriction at the request of Guillaume, the discoverer of Invar (Chap. 5). Thus, the pioneering work on magnetic anisotropy and magnetostriction was carried out in Japan at a very early stage in the history of modern physics.

Many investigations have been carried out on magnetic anisotropy and magnetostriction from theoretical, experimental and practical points of view. Various magnetic materials have been investigated, including ferromagnetic, ferrimagnetic and antiferromagnetic materials. It is impossible to describe all of these investigations in these limited pages. In this chapter, magnetocrystalline anisotropy and magnetostriction are described briefly. First, the phenomenological theories and origins of the effects are discussed. Next, various experimental investigations on typical materials are mentioned. In addition, an example is presented of the correlation between magnetocrystalline anisotropy, magnetostriction and related characteristics of practical magnetic materials.

Fig. 6.1. Logo of the International Conference on Magnetism 1982. The design depicts the magnetization curves of Fe single crystals in the $\langle 100 \rangle$, $\langle 110 \rangle$ and $\langle 111 \rangle$ directions

Finally, induced magnetic anisotropy is described; this is of both physical and practical interest.

6.1 Magnetocrystalline Anisotropy

There exists a free energy in a ferromagnet which depends on the direction of spontaneous magnetization. The magnetization vector lies in the stable direction in which this energy is minimal. As the magnetization vector rotates towards any other direction, this energy increases. Such a phenomenon is called magnetic anisotropy and this free energy is called magnetic anisotropy energy. In a ferromagnetic crystal there is an intrinsic magnetic anisotropy which reflects the symmetry of the crystal lattice, and it is called crystal magnetic anisotropy or magnetocrystalline anisotropy.

In a cubic crystal, the magnetocrystalline anisotropy energy E_a is expressed as a function of the direction cosines $(\alpha_1, \alpha_2, \alpha_3)$ of spontaneous magnetization with respect to the cubic crystallographic axes. By considering cubic symmetry and by using mathematical relations among the direction cosines, we can express E_a as

$$E_a = K_0 + K_1(\alpha_1^2\alpha_2^2 + \alpha_2^2\alpha_3^2 + \alpha_3^2\alpha_1^2) + K_2\alpha_1^2\alpha_2^2\alpha_3^2$$
$$+ K_3(\alpha_1^2\alpha_2^2 + \alpha_2^2\alpha_3^2 + \alpha_3^2\alpha_1^2)^2 + \ldots , \tag{6.1}$$

where K_1, K_2 and K_3 are the magnetocrystalline anisotropy constants. In a hexagonal crystal, we can express E_a, as a function of the angle θ between the c axis and the spontaneous magnetization and of the angle ψ between the component of the spontaneous magnetization in the c-plane (plane normal to the c-axis) and one of the a-axes, by

$$E_a = K_0 + Ku_1 \sin^2\theta + Ku_2 \sin^4\theta + Ku_3 \sin^6\theta$$
$$+ Ku_4 \sin^6\theta \cos 6\psi + \ldots . \tag{6.2}$$

When we define the anisotropy energy as mentioned above, we consider that the intensity of spontaneous magnetization does not depend on the crystallographic orientation. In most materials, the anisotropy of spontaneous magnetization can be neglected in practice. However there are some materials in which the anisotropy of spontaneous magnetization is important. We can express, by symmetry considerations, the anisotropy of spontaneous magnetization in the same form as the expression (6.1) for the magnetocrystalline anisotropy energy. In such materials, it is important to consider the contribution of the anisotropic magnetization to the anisotropy energy.

The origins of magnetocrystalline anisotropy exist in the mechanisms via which the spin magnetic moments sense the symmetry of the crystal lattice. One mechanism is the anisotropic interaction between two spin magnetic moments and the other is the single-ion anisotropy of a magnetic ion in a crystalline field. We call the former the "pair model" and the latter the "single-ion model".

We can consider two kinds of anisotropic interactions. One is the classical magnetic dipolar interaction between spin magnetic moments. This interaction energy is expressed as

$$-\frac{3\mu_s^2}{4\pi\mu_0 r^3}(\cos^2\phi - 1/3) , \tag{6.3}$$

where ϕ is the angle between the direction of the spin magnetic moment μ_s and the bond direction, and r is the distance between two spins, as shown in Fig. 6.2. In a cubic crystal, summation of the energy (6.3) over all neighboring spin pairs gives a value of zero. Thus, the magnetic dipolar interaction only becomes the origin of the anisotropy energy for crystals having a lattice of lower than cubic symmetry.

The other anisotropic interaction is the anisotropic exchange interaction which originates from the combined effect of the spin-orbit interaction and the isotropic exchange interaction. The physical interpretation of this mechanism is illustrated by Fig. 6.3. Electron charge distributions on magnetic atoms depend on the spin directions through the spin-orbit interaction. Accordingly a rotation of the spin direction leads to a change in the overlap of charge distributions on neighboring atoms, and thus the strength of the exchange interaction changes depending on the direction of spontaneous magnetization. Van Vleck treated the anisotropy energy by expressing such anisotropic exchange interactions in the form of dipole-dipole and quadrupole-quadrupole interactions between spins. These interactions are essentially electrostatic interactions which are stronger than simple magnetic ones and are called the pseudo-dipolar and pseudo-quadrupolar interactions (Appendix A.25). Néel [6.2] expressed the interaction energy of atomic pairs as shown in Fig. 6.2 by expanding it in

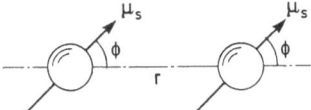

Fig. 6.2. Spin pair

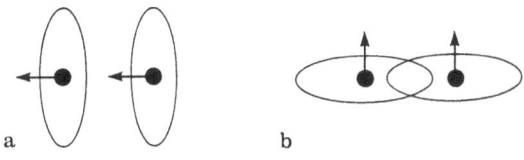

Fig. 6.3a, b. Model of the anisotropic exchange interaction

Legendre polynomials of $\cos\phi$ as follows:

$$w = (l + m\delta r)(\cos^2\phi - 1/3) + (q + s\delta r)(\cos^4\phi - \tfrac{6}{7}\cos^2\phi + \tfrac{3}{35}) + \ldots ,$$
$$(6.4)$$

where l and q are the coefficients of pseudo-dipolar and pseudo-quadrupolar interactions, and m and s are the differential coefficients of l and q relative to r, respectively. δr is the variable part of the interatomic distance r, namely $r = r_0 + \delta r$. We can calculate the magnetocrystalline anisotropy constants on the basis of (6.4).

The orbital state of a single magnetic ion in the crystal "sees" the lattice symmetry by feeling the crystalline field made by the surrounding ions. Thus, the spin of this ion comes to "see" the lattice symmetry through the agency of the spin-orbit interaction. Since the energy of the ion affected by the crystalline field becomes minin um when its electron cloud is directed along preferred crystallographic orientations, the spin also takes stable directions with respect to the crystallographic axes by means of the spin-orbit coupling. When the spin is rotated from i.s stable direction by an applied magnetic field, the electrostatic energy of the ion increases due to distortion of the electron cloud. This gives rise to the anisotropy energy.

On the other hand, in $3d$-transition metals and alloys, $3d$ electrons carrying the magnetic moment have itinerant character in the crystal. Therefore, the mechanism of anisotropy using a band picture of $3d$ electrons is more appropriate than using anisotropic interaction between localized magnetic moments or the single ion anisotropy mentioned above. The origin of anisotropy based on the itinerant electron model was first discussed by Brooks (Appendix A.20). Topics concerning this problem will be described, in the next section, using Ni as an example. Although there are various origins of magnetocrystalline anisotropy, each origin plays a meaningful role depending on individual magnetic materials. It is important in the study of magnetism to clarify which type of mechanism is the most effective in the particular magnetic material under investigation, considering the lattice symmetry of the crystal and the electronic state of the magnetic ion.

Measurements of magnetic anisotropy energy are usually carried out using a torque magnetometer. When a strong magnetic field is applied to the specimen and the spontaneous magnetization is rotated by an angle θ from the reference axis in the measuring plane, the torque L exerted by a unit volume is given by $L = -\partial E_a/\partial\theta$. Methods of torque measurement and analysis are given in [6.3]. For example, a computer-controlled automatic torque magnetometer has been designed and fabricated by Abe and Chikazumi [6.4]. This apparatus contains an electro-mechanical system in which the specimen can be rotated stereographically and its principal crystallographic axis can be set vertically using the symmetry of the magnetic anisotropy. So we can choose the desired plane of the specimen as a measuring plane and keep the specimen there at a given temperature (for example, at low temperatures). Taking advantage of these points, Chikazumi et al. cooled a magnetite specimen down to low temperatures in a magnetic field, rotated the specimen to find the required planes at this low temperature, and then carried out their torque measurements. In this way, they obtained considerable information about the magnetite.

6.2 Magnetostriction

The study of magnetostriction has a long history. When a ferromagnet is magnetized, its shape changes slightly. This phenomenon, known as magneto-striction, was discovered by Joule and reported by him in 1842, one year before the publication of his famous paper on the "mechanical equivalent of heat". In Japan, Nagaoka, who is famous for his proposition of a Saturn-like atomic model, began experiments on magnetostriction and published the first paper on it in 1888. We have mentioned already that Nagaoka made his debut on the international academic scene by his study of magnetostriction, and due to his successes the study of this subject was said to look like a traditional Japanese speciality in physics.

The deformation due to magnetostriction is expressed as $\delta l/l$, where δl is a change of linear dimension of a ferromagnetic specimen with a length l. The magnitude of $\delta l/l$ changes with a change in magnetization and reaches its saturation value when the magnetization reaches saturation. This saturation value is of the order of 10^{-6}, depending on the magnetic material. Thus, the crystal lattice of a ferromagnet spontaneously deforms when spontaneous magnetization occurs. When a field is applied the magnetization vector in each domain rotates together with the axis of spontaneous deformation, causing a deformation of the magnet as a whole.

The magnetostriction of a cubic crystal can be expressed as

$$\frac{\delta l}{l} = \frac{3}{2}\lambda_{100}(\alpha_1^2\beta_1^2 + \alpha_2^2\beta_2^2 + \alpha_3^2\beta_3^2 - 1/3)$$

$$+ 3\lambda_{111}(\alpha_1\alpha_2\beta_1\beta_2 + \alpha_2\alpha_3\beta_2\beta_3 + \alpha_3\alpha_1\beta_3\beta_1) , \tag{6.5}$$

where $(\alpha_1, \alpha_2, \alpha_3)$ are the direction cosines of the spontaneous magnetization and $(\beta_1, \beta_2, \beta_3)$ those of the direction which is observed. λ_{100} and λ_{111} are the material constants and are called the magnetostriction constants. The expression for the magnetostriction of a cubic crystal using higher order terms and that of a hexagonal crystal are given in [6.5].

The origin of magnetostriction, like that of magnetocrystalline anisotropy, is associated with the anisotropic interaction between spins and the single-ion anisotropy of a magnetic ion in a crystalline field. Magnetostriction results from the fact that, when a crystal lattice deforms spontaneously, the decrease of magnetic energy is counterbalanced by an increase in the elastic energy. Néel [6.2] calculated magnetostriction on the basis of the anisotropic interaction between spins of (6.4). It is interesting to note that the coefficients of pseudo-dipolar interaction, l, and its derivative with respect to r, m, can be estimated from measured values of the magnetostriction constants, λ_{100} and λ_{111}. Also, the single-ion anisotropy energy associated with the crystalline field will depend on the strain, because the crystalline field changes when there is a deformation of the crystal lattice. Thus, we can discuss magnetostriction on the basis of the single-ion model.

Magnetostriction is usually measured using a strain gauge or by a three terminal capacitance method. Methods of measurement and analysis are given in [6.5].

6.3   Representative Materials and Topics

The emphasis in this section will be on metals and alloys.

6.3.1   Iron-Group Transition Metals and Alloys

a) Temperature Dependence of Magnetic Anisotropy in Fe

For ferromagnet Fe, the temperature dependences of K_1 and K_2 were obtained by Bozorth [6.6] from analysis of the temperature variations of the magnetization curves measured by Honda et al. [6.7]. Later, torque measurements were also carried out, but there were differences in the K_2 values depending on the type of measurement. Gengnagel and Hofmann [6.8] obtained the temperature dependences of K_1, K_2 and K_3 from torque measurements using precise spherical specimens of Fe single crystals prepared by a recrystallization method. These results are shown in Fig. 6.4, where K_1^∞, K_2^∞ and K_3^∞ are the constants determined by the extrapolation $H \to \infty$ in the plot of the Fourier coefficients of the torque curve as a function of the reciprocal of the applied field, $1/H$.

According to Zener's theory, the following relationship exists between the temperature dependences of the anisotropy constant K_n and the saturation magnetization M_s:

$$\frac{1}{K_n}\frac{\partial K_n}{\partial T} = m_T \frac{1}{M_s}\frac{\partial M_s}{\partial T} . \tag{6.6}$$

In a cubic crystal, $m_T = 10$ for $K_n = K_1$. Comparison between this theory and the experimental results by Klein and Kneller [6.9] is shown in Fig. 6.5, and there is no agreement between them. It is difficult to express K_1 by the tenth power law of M_s or a single value of m_T over the whole temperature range of the investigation. On the basis of band theory, Asdente and Delitala [6.10], and

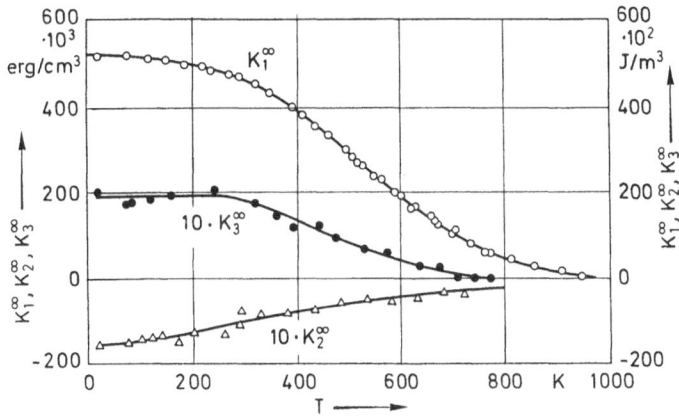

Fig. 6.4. Temperature dependence of the magnetocrystalline anisotropy constants of Fe [6.8]

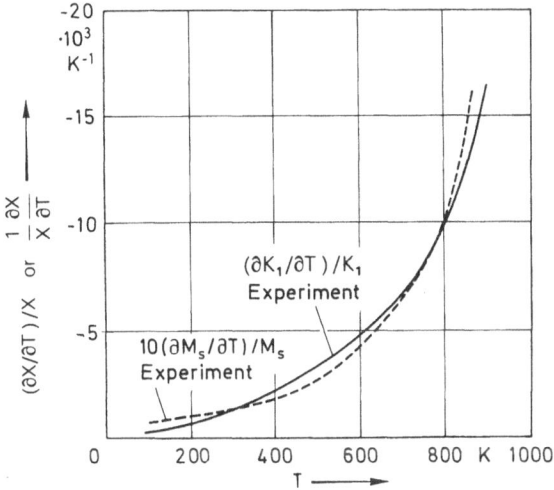

Fig. 6.5. Comparison between K_1 and the tenth power law of M_s in Fe [6.9]

Mori [6.11] calculated K_1, but there were differences between them and the problem remained controversial.

b) Anomalous Torque Curves at Low Temperatures in Ni

Franse and De Vries [6.12] found that higher-order anisotropy constants had to be considered to analyze the data from torque measurements in Ni at low temperatures. This fact is shown clearly as follows. When the spontaneous magnetization rotates by an angle θ from the [001] direction in the (100) plane, the torque L is expressed as

$$L = -\sin 2\theta \cos 2\theta (K_1 + \tfrac{1}{2}K_3 \sin^2 \theta + \dots) . \tag{6.7}$$

Therefore, if there is no contribution from the higher-order anisotropy constants such as K_3 and so on, the plot of the value of $L/(\sin 2\theta \cos 2\theta)$ as a function of θ will give a constant value of K_1. If such contributions do exist however, this plot will show an angular dependence. Figure 6.6 shows the plots of data measured for Ni and Fe [6.13]. In Ni, the presence of contributions from higher-order anisotropy constants can be seen clearly at low temperatures. On the other hand, in Fe, the torque curve for the (100) plane can be described sufficiently by K_1 alone. It is interesting to note that the higher order contributions observed at

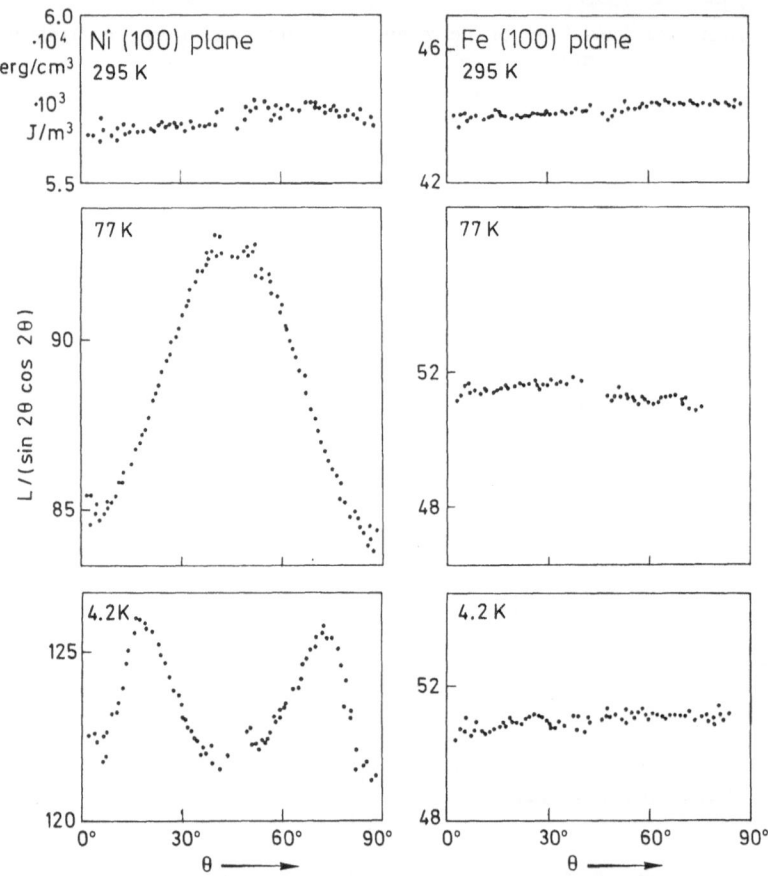

Fig. 6.6. Torque curves of Ni and Fe in the (100) plane. In order to investigate higher order contributions to the torque curves, $L/(\sin 2\theta \cos 2\theta)$ is plotted as a function of θ [6.13]

low temperatures for Ni depend on the purity of the specimen. This is shown in Fig. 6.7, where the high purity specimen contained 20 ppm Fe and its resistance ratio was $R_{273}/R_{4.2} = 610$, whereas the less pure specimen possessed 150 ppm Fe with $R_{273}/R_{4.2} = 230$. Later, Aubert et al. made more detailed investigations and discussed further the contributions of higher-order constants to the aniso-tropy energy of Ni [6.14]. Figure 6.8 shows the temperature dependence of the anisotropy constants K_1, K_2 and K_3 measured by Franse [6.13]. The K_1 of Ni changes more steeply than a tenth power law as mentioned previously.

Furey [6.15] discussed the magnetocrystalline anisotropy of Ni on the basis of band theory as follows: A rotation of the spontaneous magnetization results in an energy shift of the bands due to the spin-orbit interaction. These energy shifts are largest near the symmetry point X of the Brillouin zone, where the bands are degenerate in the absence of a spin-orbit interaction, and go toward the positive side or the negative side depending on the bands. Thus the summation of the energy shifts over the states gives rise to the anisotropy energy, and the calculation results in a reasonable value for K_1. Franse [6.13] pointed out that the observed fine structure in the anisotropy energy may originate from effects due to a band which crosses the Fermi level at a zone boundary (near the point L) with a rotation of the spontaneous magnetization.

Aubert et al. [6.14] carried out high-precision magnetic torque measure-ments on Ni single crystals as a function of the crystallographic direction of the magnetization. Gersdorf [6.16] found that, from analyses of the data obtained by Aubert et al., the magnetic torque curves of Ni at 4.2 K for the (100) and (110) planes can be decomposed into a normal torque curve, which is described by the four constants K_1 to K_4, and an anomalous torque curve with an amplitude of only 2% of that of the normal curve. These results are shown in Fig. 6.9, where a small anomaly in the torque can be seen when the magnetization direction is at

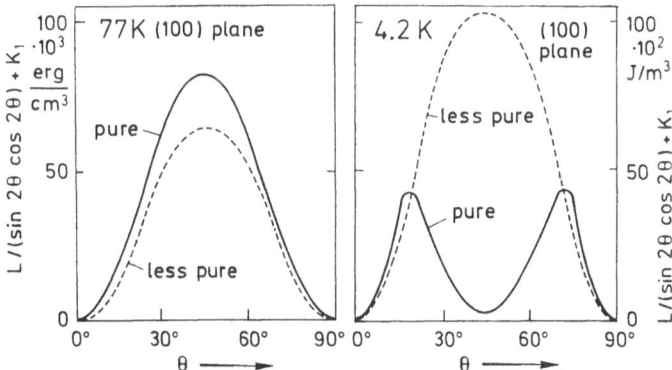

Fig. 6.7. Torque curves of Ni in the (100) plane. Higher order contributions to the torque curves depend on purity of samples. $L/(\sin 2\theta \cos 2\theta) + K_1$ is plotted as a function of θ [6.12]

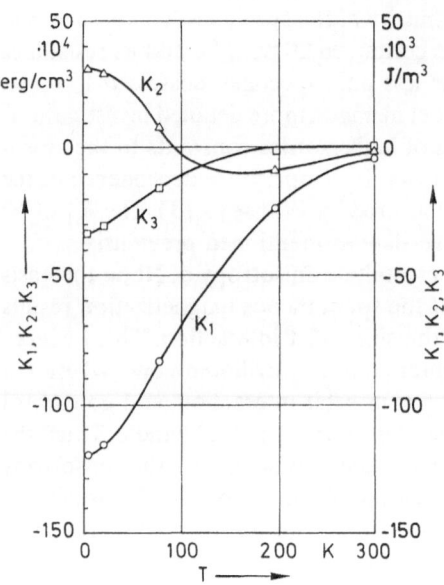

Fig. 6.8. Temperature dependence of the magnetocrystalline anisotropy constants of Ni [6.13]

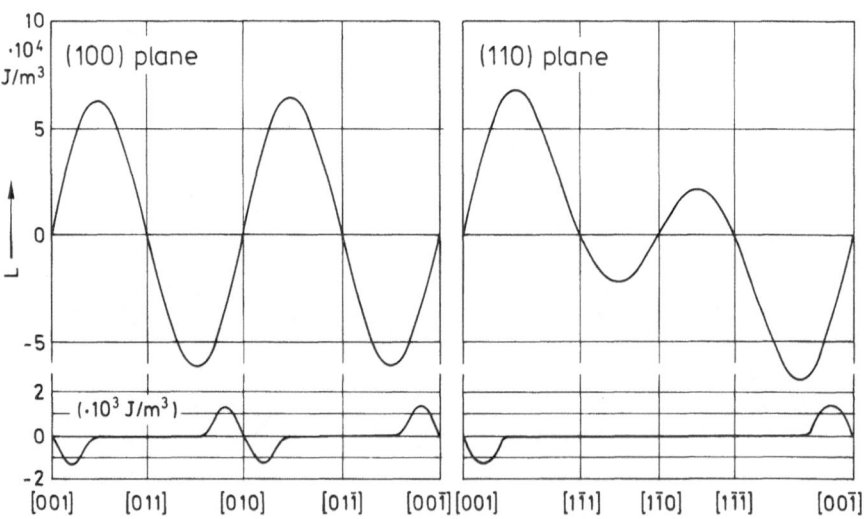

Fig. 6.9. Torque curves of Ni in the (100) and (110) planes. Anomalous torque curves are superposed on regular curves [6.16]

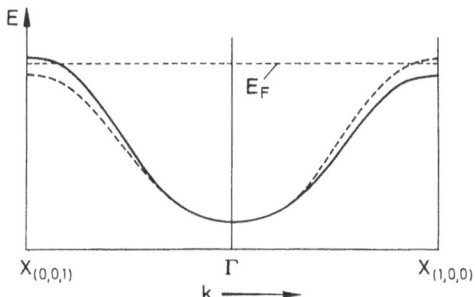

Fig. 6.10. Schematic representation of the $X_{2\downarrow}$ sub-band in Ni. Solid curve: M_s along [001]. Dashed curve: M_s along [100] [6.16]

an angle of about 18° from the nearest cubic axis. Gersdorf [6.16] explained these phenomena by assuming the following: As shown in Fig. 6.10, because of spin-orbit coupling, the top of the $X_{2\downarrow}$ band at $X(0, 0, 1)$ sits slightly above the Fermi energy E_F when the magnetization direction is parallel to the [001] axis. However, it shifts down with a rotation of the magnetization and comes down to the Fermi energy at a magnetization direction of 18° from the [001] axis. Thus, redistribution of electrons occurs from the top of the $X_{2\downarrow}$ band to other parts of the Fermi surface while the magnetization exists within 18° from a cubic axis. This situation results in a negative contribution to the total energy of the crystal. The results obtained by analyzing the torque data using the above assumption are in agreement with the results of band calculations by Wang and Callaway. The studies mentioned here show that high-precision magnetic torque measurements are useful for understanding the details of band structures of ferromagnetic metals and alloys.

c) Direction of Easy Magnetization and Magnetic Domain Structure in Co

Figure 6.11 shows the temperature dependence of the magnetocrystalline anisotropy constants, K_{u1} and K_{u2}, of hcp Co measured by Pauthenet et al. [6.17]. The stable direction of the spontaneous magnetization can be deduced from the expression (6.2) for the uniaxial anisotropy as follows: (a) for $K_{u1} > 0$ and $K_{u1} + K_{u2} > 0$, the c-axis is the easy axis; (b) for $K_{u1} > 0$ and $K_{u1} + K_{u2} < 0$, or for $K_{u1} < 0$ and $K_{u1} + 2K_{u2} < 0$, the c-plane is the easy plane; (c) for $K_{u1} < 0$ and $K_{u1} + 2K_{u2} > 0$, the easy direction lies on a conical plane at an angle θ_0 from the c-axis, and here $\theta_0 = \arcsin\left(-K_{u1}/2K_{u2}\right)^{1/2}$. The temperature dependence of the direction of easy magnetization in Co has been investigated by means of neutron diffraction and NMR. It has also been studied by observation of magnetic domains, but the results are controversial.

Watanabe et al. [6.18] have made dynamical studies of the changes with temperature in domain structures of hcp Co single crystals in the range

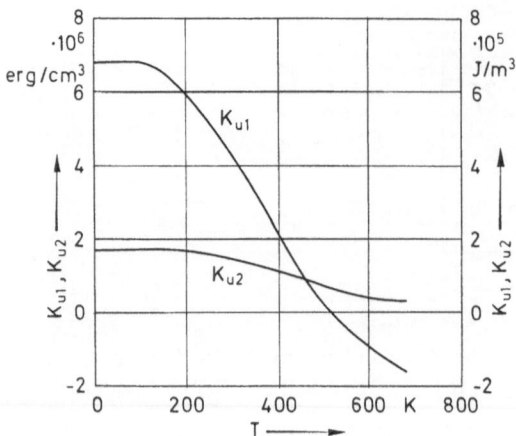

Fig. 6.11. Temperature dependence of the magnetocrystalline anisotropy constants of Co [6.17]

20–380°C, using the Tohoku University 1000 kV electron microscope and a VTR recorder. Figure 6.12 shows a sequence of the images of domain structure in $(11\bar{2}0)$ Co recorded on the VTR in the heating experiment. We see that the direction of easy magnetization in the domain rotates from the axis to a direction in the c-plane with increasing temperature in the range $200 \sim 330\,°C$. The rotation of the easy direction occurs reversibly with increasing and decreasing temperature. The angle of rotation θ_0 was measured on the images and plotted as a function of the specimen temperature in Fig. 6.13, where the solid curve indicates the temperature dependence of θ_0 calculated from the data of K_{u1} and K_{u2}. The experimental results from the domain observations are in good agreement with the calculated curve.

d) Discovery of Double Hexagonal Co–Fe

For Co with the hcp structure at room temperature, the direction of easy magnetization is parallel to the c-axis and the direction of hard magnetization is a direction in the c-plane. This fact was well known from Kaya's famous experiment. However, in contrast with Co, for Co containing only about 1 at. % Fe the direction of easy magnetization becomes a direction in the c-plane and the direction of hard magnetization is parallel to the c-axis. This interesting phenomenon was discovered by Chikazumi and Wakiyama during a study of magnetocrystalline anisotropy in Co-based dilute alloys containing 3d transition metal impurities [6.19]. The crystal structure of this "anomalous" Co–Fe alloy was investigated by X-ray analysis and was found to be a double hexagonal close-packed (dhcp) structure. The stacking sequence of atomic planes is ABAB type for the hcp structure, while it is ABAC type for the dhcp

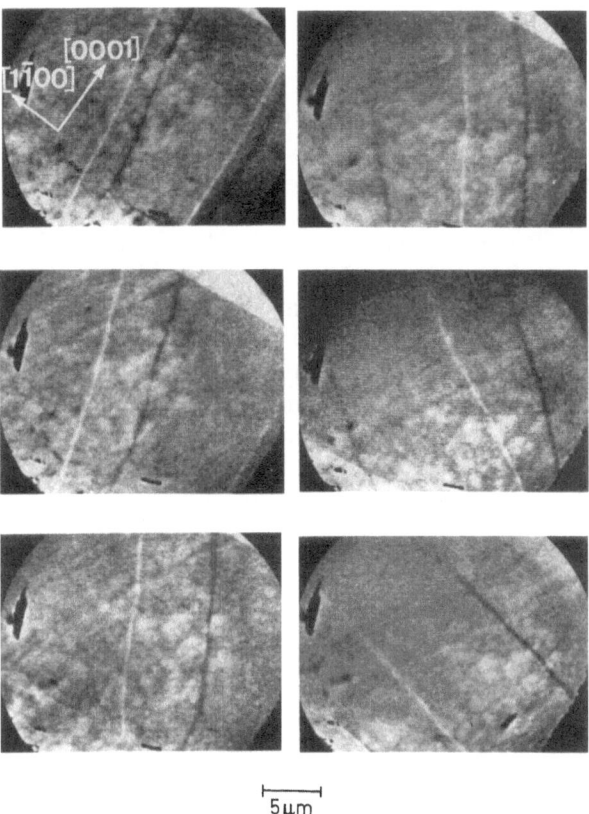

Fig. 6.12. A sequence of the images of domain structures of Co in the (11$\bar{2}$0) plane at various temperatures [6.18]

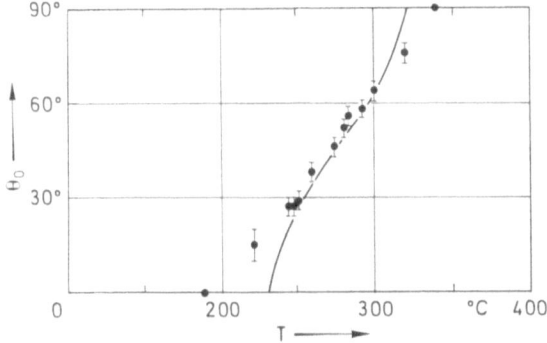

Fig. 6.13. Temperature dependence of the angle θ_0 between the easy direction of magnetization and the c-axis in Co [6.18]

structure. The composition and temperature dependences of magnetocrystalline anisotropy constants for the hcp and dhcp Co–Fe alloys have been studied in detail [6.20, 21].

Tanaka et al. [6.22] found that the direction of easy magnetization for the Co–1.2 at.% Fe alloy was changed from a direction in the c-plane to the c-axis by an applied magnetic field at room temperature. The magnetization curves along the a-axis (dotted lines) and the c-axis (solid lines) for this alloy are shown in Fig. 6.14, where the number on the curves denotes the sequence of the measurements. For the specimen heated up to 300°C and cooled down to room temperature, first the a-axis was easy to magnetize and the c-axis was the hard axis. However, after the specimen was kept at room temperature in a field of 18 kOe applied parallel to the c-axis for a while, the c-axis became easier to magnetize and the a-axis became harder. After the same treatment for more than 3 hours, the easy direction was found to change from the c-plane to the c-axis. We call such a phenomenon "the field-induced transition of the easy axis". Minimum fields required to cause the change of easy direction at room temperature were determined as a function of the angle θ between the c-axis and the field direction. The relationship between the field strength and the angle is shown in Fig. 6.15.

Before and after "the field-induced transition of the easy axis" for the Co–1.2 at.% Fe alloy, the crystal structures were examined by means of electron diffraction. It was found that the structure was the dhcp type before the transition and was the hcp type after the transition. Thus, we understand that "the field-induced transition of the easy axis" is associated with "the field-induced dhcp → hcp transformation". A thermodynamic process that induces

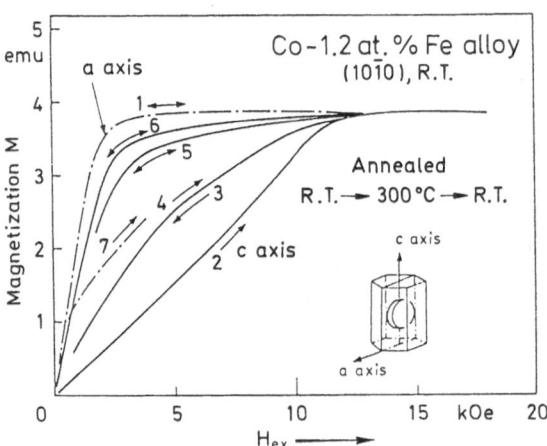

Fig. 6.14. Magnetization curves for the a-axis and c-axis of the Co–1.2 at.% Fe alloy, showing the change of easy direction induced by a field of 18 kOe applied parallel to the c-axis [6.22]

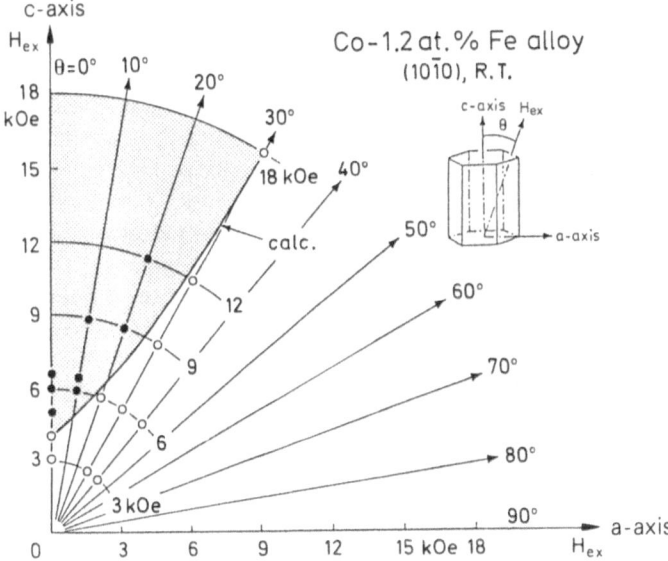

Fig. 6.15. The relationship between the strength and direction of the field to induce (●) and not to induce (○) the change of easy direction. The shadowed area shows the calculated region where the dhcp → hcp transformation occurs [6.22]

such a transformation was considered using the free energy given by the sum of the free energy in the absence of a field and the magnetic energy, consisting of the field term and the magnetocrystalline anisotropy term. As a result, the region where "the field induced dhcp-hcp transformation" occurs was obtained as the shadowed area in the diagram of the strength and direction of the field (Fig. 6.5), showing good agreement with the experimental results.

The dhcp Co–Fe alloy is a unique ferromagnet. In addition to the magneto-crystalline anisotropy, its magnetostriction is also quite different from that of hcp Co. This interesting alloy presents many problems concerning its phase diagram, saturation magnetization, magnetic domains, NMR and magnetic annealing effects [6.23].

6.3.2 Rare-Earth Metals and Alloys

a) Magnetocrystalline Anisotropy of Gd–R

Rare-earth metals and alloys have a very large magnetocrystalline anisotropy energy compared with the iron-group transition metals and alloys. For example, Fe, Ni and Co have an anisotropy energy of 10^3–10^5 J/m^3, while Tb, Dy and Ho have 10^7–10^8 J/m^3. Where does this large difference come from?

In the iron-group transition metals and alloys, the $3d$ electrons carrying the magnetic moments are itinerant in a crystal. In this situation, the orbital angular momenta of the $3d$ electrons are quenched. However, these angular momenta are partly recovered by the spin-orbit interaction and are coupled with the crystal lattice. Such a mechanism gives rise to the magnetic anisotropy energy, which is comparatively small because of the indirect coupling of the spins with the lattice.

On the other hand, in rare-earth metals and alloys, the $4f$ electrons, which are the carriers of the magnetic moments, are situated at the inner part of the electron core and are protected by the outer electron shell from the influence of neighboring atoms. In this case, the effect of the crystalline field is small and the orbital angular momenta of $4f$ electrons remain unquenched. Thus the magnetic moments are associated with the total angular momentum J consisting of the sum of the spin angular momentum S and the orbital angular momentum L. Since this unquenched L is coupled directly with the crystal lattice, the magneto-crystalline anisotropy energy of rare-earth metals and alloys is usually larger than that of the iron-group transition metals and alloys.

Now, how can we measure such a large anisotropy energy? Let us look at some interesting experiments carried out by Chikazumi et al. [6.24]. Since the Gd^{3+} ion has no orbital magnetic moment, the anisotropy energy of metallic Gd is expected to be exceptionally small among the rare-earth metals, and in fact, this has been shown by experiment. Chikazumi et al. measured the anisotropy energy by means of a torque magnetometer and obtained successfully a dramatic change in the anisotropy for Gd alloys containing one species of a rare earth metal, R, in dilute form, such as a light rare-earth metal (R = Ce, Pr, Nd and Sm) or a heavy rare-earth metal (R = Tb, Dy, Ho, Er and Tm). According to Chikazumi et al., "since the exchange field produced by the Gd matrix is estimated to be 2000 kOe, we can utilize this field to rotate the magnetic moments of the impurity rare-earth atoms which have strong magnetic anisotropies". Incidentally, we can see from Chap. 2 how much work is required to produce a field of 2000 kOe artificially. "Nature doesn't favour a strong magnetic field." This is a saying by Chikazumi, but the interesting thing is that his idea was "to utilize a strong field produced by nature".

One example of the results of torque measurements, at 4.2 K in the plane containing the c-axis, is shown in Fig. 6.16, where a dotted line is the torque curve for pure Gd and a solid line is the torque curve for the 1.8% Tb–Gd alloy. We can see a surprising difference between both the torque curves. Such a remarkable change in torque curves was also found for other Gd alloys with other rare-earth impurity metals. These Gd-based dilute alloys form a hexagonal close-packed lattice and the uniaxial symmetry. For uniaxial crystals, the magnetocrystalline anisotropy energy of the impurity single atom is expressed by

$$\varepsilon_a = DP_2(\cos \theta_i) + EP_4(\cos \theta_i) + FP_6(\cos \theta_i) + \ldots , \qquad (6.8)$$

where D, E and F are the anisotropy constants per single atom, P_2, P_4 and

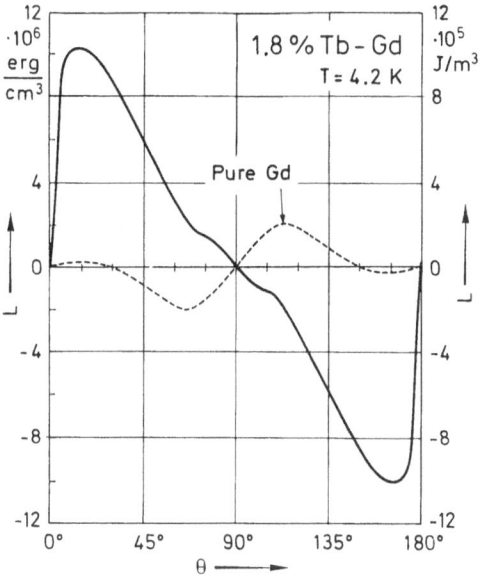

Fig. 6.16. Torque curves of Gd and the 1.8% Tb–Gd alloy. The angle θ of the external field was measured from the c-axis and the curves were measured at 4.2 K in a field of 31 kOe [6.24]

P_6 are Legendre polynomials, and θ_i is the angle of the impurity moment measured from the c-axis. From analyses of the measured torque curves, the anisotropy constants for various rare-earth impurities doped in Gd were obtained. Figure 6.17 shows values of D as a function of the number of $4f$ electrons of the doped atoms.

Now, let us consider the mechanism of the magnetocrystalline anisotropy of a rare-earth impurity atom in Gd. We can see that this anisotropy arises from the interaction between the $4f$ orbits and the crystalline field produced by the surrounding Gd ions. Charge clouds of $4f$ electrons are schematically shown in Fig. 6.18. For example, for Tb with eight $4f$ electrons, the first seven electrons occupy the orbits with the magnetic quantum number m_l from $+3$ to -3 and form a spherical charge distribution. The eighth electron enters the orbit with $m_l = +3$ and exhibits a pancake-like charge distribution spreading perpendicular to the z-axis, which corresponds to the direction of the total angular momentum J. On the other hand, the axial ratio c/a of the crystal lattice for Gd is 1.599 which is less than the ideal value 1.633 for a hexagonal close-packed lattice. Therefore, in Gd containing Tb impurity atoms, the crystalline field produced by Gd ions has a field gradient which acts on the pancake-like charge distribution of Tb so as to make its z-axis, or the direction of J, perpendicular to the c-axis. As a result, the direction of easy magnetization is in the c-plane and

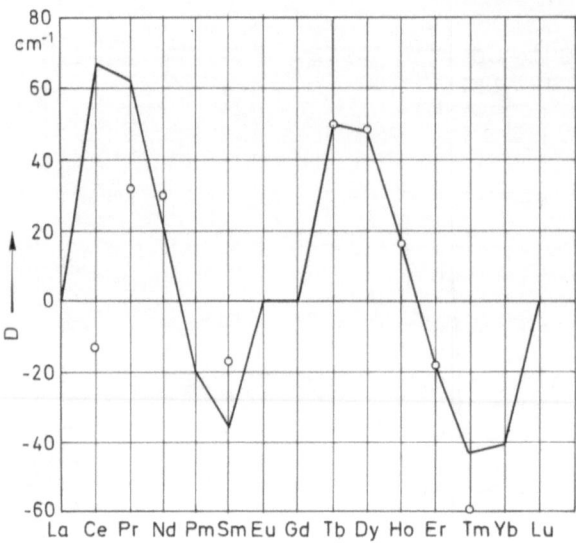

Fig. 6.17. Values of the anisotropy constant D of various rare-earth impurity atoms doped in Gd as a function of the number of $4f$ electrons of the impurity atoms. Solid lines show the theoretical values [6.24]

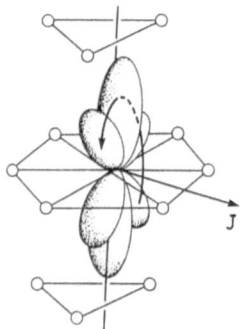

Fig. 6.18. $4f$ electron cloud of Tb in Gd [6.25]

thus D becomes positive. In this way, the calculation of D was carried out for various rare-earth impurities doped in Gd metal. In Fig. 6.17, the solid lines show the theoretical values, which agree with the experimental results reasonably well for heavy rare-earth atoms. However, there is no agreement between calculation and experiment for most of the light rare-earth impurities. This is thought to be due to the effect of excited states and a contribution of an anisotropic part of the s–f exchange interaction (Appendix A.28).

b) Forced Magnetostriction of Gd

Several investigations have been made on the magnetostriction of rare-earth metals. Let us look at some interesting results obtained from the forced magnetostriction of Gd. Forced magnetostriction arises from an increase in the alignment of the thermally agitated spin system when a fairly strong field is applied to a ferromagnet. For rare-earth metals, Bozorth and Wakiyama [6.26] measured the forced magnetostriction of single crystals of Gd for the first time. Their experimental results are shown in Fig. 6.19, where we see that the forced magnetostriction $(\partial\lambda/\partial H)$ is positive and unusually large in the c direction, while it is negative and normal in magnitude in the a direction. There exists the well-known thermodynamic relation

$$\frac{\partial\lambda}{\partial H} = -\frac{\partial M}{\partial P}, \tag{6.9}$$

in which M is the magnetization and P the compressive stress. From this relation and the experimental results, we find that when P is applied parallel to

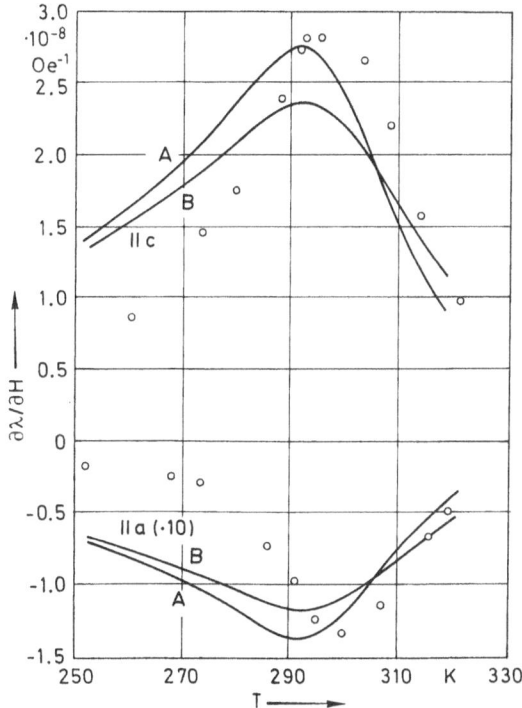

Fig. 6.19. Temperature dependence of the forced magnetostriction, measured parallel to the c-axis and a-axis, of Gd single crystals. Experiment (open circles) [6.26]; theory (curves A and B) [6.27]

the c-axis there is a decrease in the molecular field, as in Fe, and when applied parallel to the a-axis there is an increase in the molecular field, as in Ni. Bozorth and Wakiyama [6.26] also measured the anomalous thermal expansion of single crystals of Gd, the results of which are in good agreement with those of $\partial\lambda/\partial H$.

In rare-earth metals, the exchange interaction between $4f$ spins is mainly attributable to indirect exchange interaction via the conduction electrons. Tonegawa [6.27] investigated theoretically the forced magnetostriction and anomalous thermal expansion of the single crystal of Gd by calculating the strain dependence of this exchange interaction. In Fig. 6.19, the solid curves show the results of this calculation, which give a fairly good temperature dependence of the forced magnetostriction. A part of the magnetic moment of Gd is thought to be due to polarization of the conduction electrons. Also, it is pointed out that the conduction electrons play an important role in the magnetocrystalline anisotropy of Gd [6.28]. Thus, such a situation should be considered in discussing the magnetostriction of Gd.

6.4 Realization of High Magnetic Permeability. "The Focus of Zero" in Magnetic Anisotropy and Magnetostriction

"The focus of zero", a detective story, is one of the well-known novels written by Seichō Matsumoto. Magnetic anisotropy and magnetostriction are the important factors which control easy or hard magnetization of magnetic materials. The guiding principle for realizing a high magnetic permeability is to decrease magnetic anisotropy and magnetostriction, or to search for the conditions where both quantities are zero, if possible. We may say that this principle is to look for "the focus of zero". For alloys, values of magnetic anisotropy and magnetostriction can be controlled by changing constituent elements or compositions. Here we see this fact for Sendust.

Sendust was discovered by Masumoto and Yamamoto [6.29] at the Institute for Iron, Steel and Other Metals of Tohoku University. In order to obtain cheap materials with high magnetic permeabilities, many attempts were made and measurements of permeabilities were carried out for hundreds of samples of Fe–Si–Al alloys with various compositions. Finally, the initial permeability $\mu_i = 35,100$ and the maximum permeability $\mu_m = 120,000$, which are higher than those for Permalloy (78.5% Ni–21.5% Fe), were found for the alloy containing 85% Fe, 9.5% Si and 5.5% Al. This material is brittle, so it is reduced to a powder for practical applications as high frequency inductors in compressed powder core form. Honda named it "Sendust" after the name of Sendai city where this material was discovered. Recently, much attention has been given to this old magnetic material because of its superior characteristics as a magnetic head material, which requires mechanical wear resistance and high saturation magnetization as well as soft magnetic properties.

The magnetocrystalline anisotropy constant K_1 and the magnetostriction constants λ_{100} and λ_{111} were measured as a function of composition for single crystals of Fe–Si–Al alloys lying near the Sendust composition [6.30]. The saturation magnetostriction λ_s of polycrystalline material was calculated from the single crystal constants. Both lines of $K_1 = 0$ and $\lambda_s = 0$ obtained by analyses of the experimental results are shown in the concentration diagram of the Fe–Si–Al system of Fig. 6.20, together with the data for initial permeabilities of polycrystalline specimens measured by Masumoto and Yamamoto [6.29]. The high initial permeability is seen to be realized near the composition for which both K_1 and λ_s are zero, namely at "the focus of zero" of K_1 and λ_s. It is interesting to note that both values of λ_{100} and λ_{111} were found to be almost zero near the Sendust composition.

6.5 Induced Magnetic Anisotropy.
How to Control the Shape of Magnetization Curves

It is very useful for applications of magnetic materials that we can control the shape of magnetization curves by some specific treatment. Most ferro- and ferrimagnetic materials exhibit an extra magnetic anisotropy and the shape of their magnetization curves changes when they are treated by "some treatment

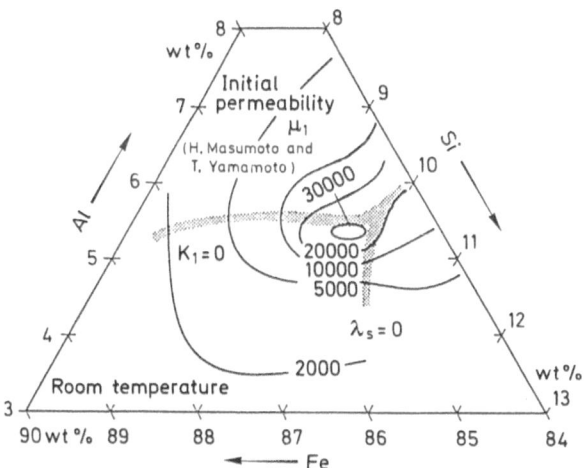

Fig. 6.20. The relationship between the permeability [6.29] and the locus of $K_1 = 0$ and $\lambda_s = 0$ [6.30] in the Fe–Si–Al alloy system

with directional properties". We call this anisotropy the induced magnetic anisotropy, and several such treatments are known. For example, in many ferromagnetic alloys and ferrites, a uniaxial magnetic anisotropy can be induced with relation to the direction of the applied field in which they are heat-treated. This is known as the magnetic annealing effect. Most ferromagnetic alloys exhibit a uniaxial anisotropy associated with the roll direction when they are cold-rolled. This is called the roll magnetic anisotropy. Furthermore, induced magnetic anisotropies can be produced by various methods, such as heat-treatment in a magnetic field through a phase transition, magnetic annealing under neutron-irradiation and heat-treatment in an applied stress. In this section we treat the magnetic annealing effect and the roll magnetic anisotropy which are important for practical applications and also are interesting in a physical sense [6.31]. Many excellent investigations have been carried out in this field in Japan by Chikazumi.

6.5.1 Magnetic Annealing Effect

The magnetic annealing effect of ferromagnetic alloys can be explained in terms of the directional order, which was first proposed by Chikazumi. In Fig. 6.21 [6.32] the directional order is shown schematically together with the random solid solution and the perfect order. Directional order is produced by magnetic annealing and results in uniaxial magnetic anisotropy. This idea was proposed by Néel, Taniguchi and Yamamoto, and Chikazumi and Oomura independently.

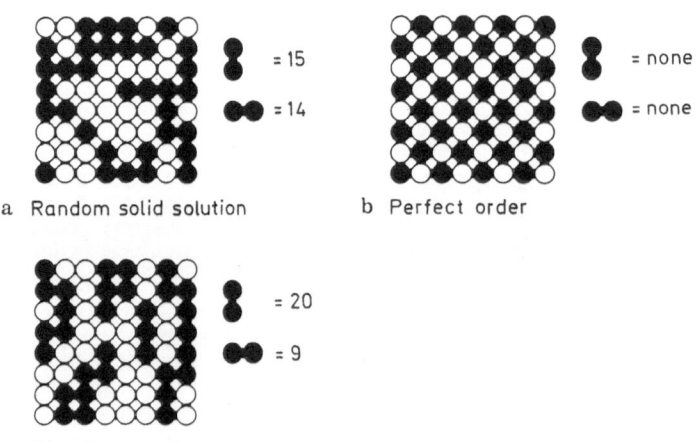

a Random solid solution   b Perfect order

c Directional order

Fig. 6.21a–c. Schematic representation of various atomic arrangements in a 50–50 alloy [6.32]

Let us consider the ferromagnetic alloy of a solid solution type which consists of two kinds of atoms A and B. In this alloy, there exist three sorts of atomic pairs such as $A-A$, $A-B$ and $B-B$. The pseudo-dipole interaction (Appendix A.25) of the atomic pairs depends on the direction of the spontaneous magnetization and also on the kind of atomic pair. When this alloy is annealed at a temperature below the Curie temperature but high enough for diffusion to occur, and at the same time an external magnetic field is applied to align the magnetization throughout the sample, the like-atom pairs tend to be parallel to the direction of magnetization. Then the equilibrium arrangement of pairs becomes anisotropic. When this alloy is quenched, the anisotropic distribution of like-atom pairs is frozen and the directional order is formed as shown in Fig. 6.21c. Such a mechanism gives rise to a uniaxial magnetic anisotropy in the sample after magnetic annealing.

Experiments of magnetic annealing on single crystal samples were first carried out by Chikazumi [6.33]. Figure 6.22 shows the experimental result on Ni_3Fe. This experiment was made before the Néel–Taniguchi theory and a part of the results was published in a paper by Kaya in 1953. The beautiful experiment by Chikazumi was referred to in Néel's paper. As shown in Fig. 6.22, the theoretical curve A explains the behavior of the experimental curve B qualitatively, but there is a quantitative difference between them. When Chikazumi talked about his experiment at the Physical Society of Japan, his words were impressed on the mind of the present author, who was one of Professor Chikazumi's students in those days. "Néel said that the experiment is in agreement with the theory considering the experimental difficulty. But I

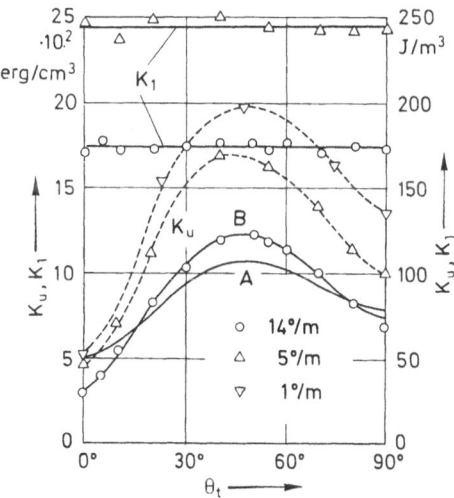

Fig. 6.22. Induced uniaxial anisotropy constant K_u as a function of the direction of the magnetic annealing field in the (110) plane of Ni_3Fe [6.33]

would rather say that the theory is in agreement with the experiment considering the theoretical difficulty." His words were something like these and were
filled with his confidence in experimentation. Afterwards, studies of magnetic
annealing effects were carried out at Tohoku University and Grenoble
University.

6.5.2 Roll Magnetic Anisotropy

The origin of roll magnetic anisotropy was thought to lie in the directional order
produced by rolling. In studying the mechanism for the formation of this
ordering, Néel and Taniguchi–Yamamoto considered the stress-induced directional order, while Chikazumi–Suzuki–Iwata considered the slip-induced directional order [6.31]. However, the experimental results obtained by Chikazumi et
al. for fcc Ni_3Fe single crystals and bcc Fe_3Al single crystals were found to be
explained successfully in terms of the slip-induced directional order [6.34].

During the process of plastic deformation of a crystal, the slip of atomic
planes occurs along specific crystallographic planes. In a fcc crystal, such as
Ni_3Fe, the slip deformation takes place on $\{111\}$ planes and in $\langle110\rangle$ directions.
Figure 6.23 shows an A_3B type fcc superlattice after a one-step displacement in
the $[01\bar{1}]$ direction along the (111) plane. No like-atom pairs exist before slip,
while A–A and B–B pairs are produced across the slipped plane after slip, as
shown in Fig. 6.23. Such an unbalanced distribution of like-atom pairs results in
the directional order in the crystal.

Chikazumi et al. [6.34] derived the uniaxial magnetic anisotropy energy E_a
which results from this mechanism. Since the like-atom pairs are not induced by

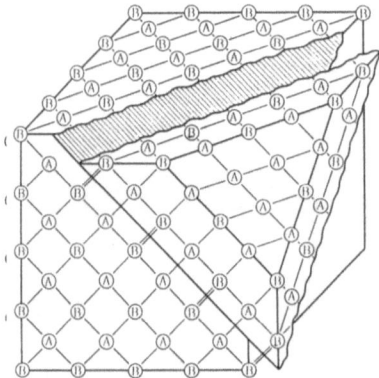

Fig. 6.23. Schematic representation of the production of like-atom pairs by a single-step
slip on the (111) plane in the $[01\bar{1}]$ direction in an A_3B-type superlattice [6.34]

the even-number slip step, the expression for E_a contains the probability p_0 of having an isolated dislocation. Also, the expression should have the probability p' of creating dislocations on a new atom plane and the long-range order parameter S to take into account the initial state of atomic arrangement in the crystal. Detailed calculations give E_a as

$$E_a = \tfrac{1}{8} N l_0 \, p \, S^2 \sum |s_i| \, f_i(\alpha_1 \alpha_2 \alpha_3) \; , \tag{6.10}$$

where N is the number of atoms per unit volume, $l_0 = l_{AA} + l_{BB} - 2l_{AB}$, $p = p_0 p'$, s_i is the slip density which is defined as the average number of dislocations that run through a given plane in the crystal and the suffix i specifies the sort of slip systems. $f_i(\alpha_1, \alpha_2, \alpha_3)$ is a function of direction cosines $(\alpha_1, \alpha_2, \alpha_3)$ of the spontaneous magnetization, which is determined by specifying the slip system. This type of roll magnetic anisotropy is called the long-range order fine-slip type. Chikazumi et al. [6.34] also derived an anisotropy called the short-range order coarse-slip type. On the basis of these two types of anisotropy, the experimental results of roll magnetic anisotropy for Ni_3Fe, Fe_3Al, Ni_3Mn and Ni–Co alloys were successfully explained.

A paper entitled "Slip-Induced Directional Order in Fe–Ni Alloys. I. Extension of the Chikazumi–Suzuki–Iwata Theory" was published in the Journal of Applied Physics [6.35]. This is the paper written by Chin at Bell Laboratories. Chin extended the theory of Chikazumi et al. to various cases and to technological applications. Slip-induced anisotropy has been utilized for controlling the magnetic characteristics of "Twistor" materials, which are used as memory devices. The most extensive follow-ups to the investigations by Chikazumi et al. have been carried out by Chin et al. [6.36] at Bell Laboratories. Roll magnetic anisotropy has also been investigated by Takahashi [6.37] at Iwate University who developed the theory of slip-induced directional order using the theory of dislocations.

As mentioned at the beginning of this chapter, the investigation of magnetic anisotropy and magnetostriction has a long history and has been carried out extensively for various materials. However, even for iron-group transition metals and alloys, which seem to have been investigated fully, interesting phenomena have been found and new aspects have been developed. Although the subject of this chapter is an old research problem, it will continue to take on new features for as long as the study of magnetism continues. To know the old is to know the new.

References

6.1. K. Honda, S. Kaya: Sci. Rep. Tohoku Imp. Univ. **15**, 721 (1926)
      S. Kaya: Sci. Rep. Tohoku Imp. Univ. **17**, 639, 1157 (1928)
6.2. L. Néel: J. Phys. Radium **15**, 225 (1954)

6.3.   T. Yamada, S. Chikazumi: *Measurement of Magnetic Anisotropy* (in Japanese) *Magnetism*, Experimental Physics, Vol. 17 (Kyoritsu Publishing Co., Tokyo 1968) 293–325
6.4.   K. Abe, S. Chikazumi: Jpn. J. Appl. Phys. **15**, 623 (1976)
6.5.   T. Wakiyama, S. Chikazumi: *Measurement of Magnetostriction* (in Japanese) *Magnetism*, Experimental Physics, Vol. 17 (Kyoritsu Publishing Co., Tokyo 1968) 327–347
6.6.   R.M. Bozorth: J. Appl. Phys. **8**, 575 (1937)
6.7.   K. Honda, H. Masumoto, S. Kaya: Sci. Rep. Tohoku Imp. Univ. **17**, 111 (1928)
6.8.   H. Gengnagel, U. Hofmann: Phys. Status Solidi **29**, 91 (1968)
6.9.   H.P. Klein, E. Kneller: Phys. Rev. **144**, 372 (1966)
6.10. M. Asdente, M. Delitala: Phys. Rev. **163**, 497 (1967)
6.11. N. Mori: J. Phys. Soc. Jpn. **27**, 307, 1374 (1969)
6.12. J.J.M. Franse, G. De Vries: Physica **39**, 477 (1968)
6.13. J.J.M. Franse, J. de Phys. Coll. C1 **32**, 186 (1971)
6.14. G. Aubert, Y. Ayant, E. Belorizky, R. Casalegno: Phys. Rev. **B14**, 5314 (1976)
6.15. W.N. Furey: Thesis, Harvard Univ. (1967); Bull. Am. Phys. Soc. **12**, 311 (1967)
6.16. R. Gersdorf: Phys. Rev. Lett. **40**, 344 (1978)
6.17. R. Pauthenet, Y. Barnier, G. Rimet: J. Phys. Soc. Jpn. **17**, Suppl. B-I, 309 (1962)
6.18. D. Watanabe, T. Sekiguchi, T. Tanaka, T. Wakiyama, M. Takahashi: Jpn. J. Appl. Phys. **21**, L179 (1982)
6.19. S. Chikazumi, T. Wakiyama, K. Yosida: Proc. Int. Conf. Magnetism, Nottingham, 756 (1964)
6.20. T. Wakiyama: Magnetism, Magnetic Materials-1972, AIP Conf. Proc. **10**, 921 (1972)
6.21. M. Takahashi, S. Kadowaki: J. Phys. Soc. Jpn. **48**, 1391 (1980)
6.22. T. Tanaka, M. Takahashi, S. Kadowaki, T. Wakiyama, D. Watanabe, M. Takahashi: J. Magnetism and Magnetic Materials **31–34**, 843 (1983)
6.23. T. Wakiyama: Solid State Phys. (in Japanese) **13**, 733 (1978)
6.24. S. Chikazumi, K. Tajima, K. Tōyama: J. de Phys. Coll C1 **32**, 179 (1971)
6.25. K. Tajima, S. Chikazumi, K. Tōyama: Solid State Phys. in (Japanese) **5**, 121 (1970)
6.26. R.M. Bozorth, T. Wakiyama: J. Phys. Soc. Jpn. **18**, 97 (1963)
6.27. T. Tonegawa: J. Phys. Soc. Jpn. **19**, 1168 (1964)
6.28. J.J.M. Franse, R. Gersdorf: Phys. Rev. Lett. **45**, 50 (1980)
6.29. H. Masumoto: Sci. Rep. Tohoku Imp. Univ. **25**, 338 (1936)
       H. Masumoto, T. Yamamoto: J. Inst. Metal (in Japanese) **1**, 127 (1937)
6.30. T. Wakiyama, M. Takahashi, S. Nishimaki, J. Shimoda: IEEE Trans. Magn. **17**, 3147 (1981)
6.31. S. Chikazumi: *Physics of Ferromagnetism*, Vol. II (in Japanese) (Syōkabō Publishing Co., Tokyo 1984)
6.32. C.D. Graham, Jr.: Chap. 13 *Magnetic Annealing, in Magnetic Properties of Metals and Alloys* (Am. Soc. Metals, Cleveland, Ohio 1959)
6.33. S. Chikazumi: J. Phys. Soc. Jpn. **11**, 551 (1956)
6.34. S. Chikazumi, K. Suzuki, H. Iwata: J. Phys. Soc. Jpn. **12**, 1259 (1957)
6.35. G.Y. Chin: J. Appl. Phys. **36**, 2915 (1965)
6.36. G.Y. Chin, E.A. Nesbitt, J.H. Wernick, L.L. Vanskike: J. Appl. Phys. **38**, 2623 (1967)
6.37. S. Takahashi: Phys. Status Solidi **A42**, 201, 529 (1977)

7. The Intermediate Field Between Pure and Applied Magnetism. Importance of Accurate Measurements of Magnetization Curves

Minoru Takahashi

Around 1945, most of the researchers who were concerned with magnetism, including magnetic physics and the development of magnetic materials such as iron core or permanent magnets, studied the alloys of the ferromagnetic $3d$ transition metals Fe, Co, Ni, and elements near them in the periodic table such as Cr, Mn, Cu or the cheaper Si and Al. At that time, they investigated the initial susceptibility μ_0, maximum susceptibility μ_m, the behavior of magnetization within a region of rotation and the shape of hysteresis loops (squareness, hysteresis loss). These parameters were obtained from the magnetization curves measured by a ballistic current method or by an abstraction method. They also studied dc behavior such as the magnetoresistance or magnetostriction which accompanies magnetization. Physicists studied the mechanism of the magnetization process or the origin of magnetism and metallurgists developed practical magnetic materials using metallurgical methods. On the other hand, electronic engineers worked so independently of physicists or metallurgists that they even used different terms such as "permeability" and "coercivity" instead of "susceptibility" and "coercive force", and they were interested in magnetic properties in alternating fields. Although there were very clear boundaries between the two groups, an "intermediate field" did not seem to exist.

Among those same physicists, however, there were two groups which were slightly different from each other; one consisted of the pure physicists who were interested in magnetic moments and susceptibility in ferro- or paramagnets of pure metals and their temperature dependence. The other consisted of the applied physicists who tried to enlarge the susceptibility or coercive force of a permanent magnet by adding a third or fourth element to alloys of Fe, Ni and Co. So something like a vague boundary was formed between the pure and applied physicists.

After about 1950, i.e. postwar, techniques using high frequency circuits (microwave techniques) for the measurement of magnetization were brought from the U.S.A. Pure physicists gradually changed their techniques from static methods using a dc magnetic field to dynamical magnetization measurement methods, using FMR, ESR and Mossbauer effect etc. They started to look at the intrinsic magnetization of magnets, and at the same time, looked for new objects for study: ferrimagnets, antiferromagnets and metamagnets, such as the compounds $M^{+2}O$, $M^{+2}F_2$, the rare earth metals Gd $\rightarrow$ Tm and low-dimensional magnets such as $CoCl_2 \cdot 2H_2O$. In addition, they studied magnetism under limiting conditions such as extremely low temperatures or high pressures.

On the other hand, the applied physicists were still interested in the mechanisms of static magnetization, and engineers were studying materials for high frequency use such as magnetic thin films of oxide, YIG and so forth, as well as permanent magnets of Alnico metal or oxides such as barium ferrite. In other words, pure physicists were interested in materials and experiments for investigating the origin of magnetism which were not necessarily of much practical use. Then applied physicists and some engineers began to be concerned with saturation magnetization, magnetostriction, magnetic anisotropy and resistance, i.e. the properties which characterize magnets as practical materials. Following this trend, it was the applied physicists who studied magnetization characteristics, magnetostriction or magnetic anisotropy to the extent that they could communicate more easily with the engineers. This applied group gradually formed an "intermediate field" and acted as an intermediary between the pure physicists and the engineers.

About 1950, the origin of the field cooling effect and the magnetic after effect (Appendix A.22) was studied actively in Kaya's laboratory at Tokyo University. In particular, in order to shed light on the origin of the field cooling effect, Chikazumi measured the magnetization curves and the magnetic anisotropy of Fe–Ni permalloy which was an important practical material for engineering. He obtained many fruitful results in the field of basic magnetism, which made him a pioneer of connectors. But at that time, as I remember, many pure physicists doubted whether it was worth studying binary or ternary alloy magnets before the magnetism of simple single element metals was well understood.

After about 1965, topics familiar to metallurgists, electrical engineers and applied physicists appeared one after another in the main magnetism section of the Meeting of the Physical Society of Japan: reverse of magnetic anisotropy in low-concentration Fe–Co alloys, Fe–Ni Invar problems, anomaly in the low-temperature magnetization of Fe_3O_4 and so forth. Here, it must be noted that it was Chikazumi, who studied the magnetization characteristics of the most popular magnetic materials, Fe–Ni alloys, under various thermal treatments. Furthermore, he abstracted topics for magnetism in those practical materials and made them the main theme in the magnetism section. Engineers rather respected the researchers in this "intermediate field" in those days, because they took account of saturation magnetization, magnetic anisotropy and magnetostriction which govern the magnetic characteristics, in order to make components for computers or permanent magnets smaller and more sophisticated.

In the 1970s, pure physicists became interested in magnetism in the disordered state. An apparatus for neutron diffraction was improved to be more precise, various types of equipment for measuring magnetism became large, and the word "physics under extreme conditions" took the place of "physics of limited states". On the other hand, components for electronic devices became smaller and smaller, and Japanese devices reached world level, some of them becoming praised as model ones.

In the 1980s, researchers in every field aimed at "creative science" and were concerned with new materials (magnets). The pure physicists tended to study the microscopic origin of magnetism, such as explaining the magnetization of various magnets in terms of their band structures. On the other hand, engineers began to try making magnets smaller, effectively using the small changes in accompanied magnetization or minute phenomena. So communication between pure physicists and engineers again became more difficult. Then the researchers in the "intermediate field" between pure and applied magnetism became very important persons. They studied the characteristics of magnetization necessary for applications, such as saturation magnetization (M_s), magnetic anisotropy (K) or magnetostriction (λ), and made some efforts to understand the problem of complicated magnetization processes (Fig. 7.1).

These are the researchers who attempted to explain various phenomena in magnetization in terms of macroscopic structures and intuitive models, who were interested in both magnetic engineering and pure magnetism, and who studied books such as Becker and Doring: Ferromagnetismus, Bozorth: Ferromagnetism, Kaya: Ferromagnetism, Chikazumi: Physics of Ferromagnetism. In other words, they were the researchers who made efforts to produce new magnetic phenomena, new materials and new techniques, discussing in simple terms the difficult concepts in both pure and applied magnetism. Now, however,

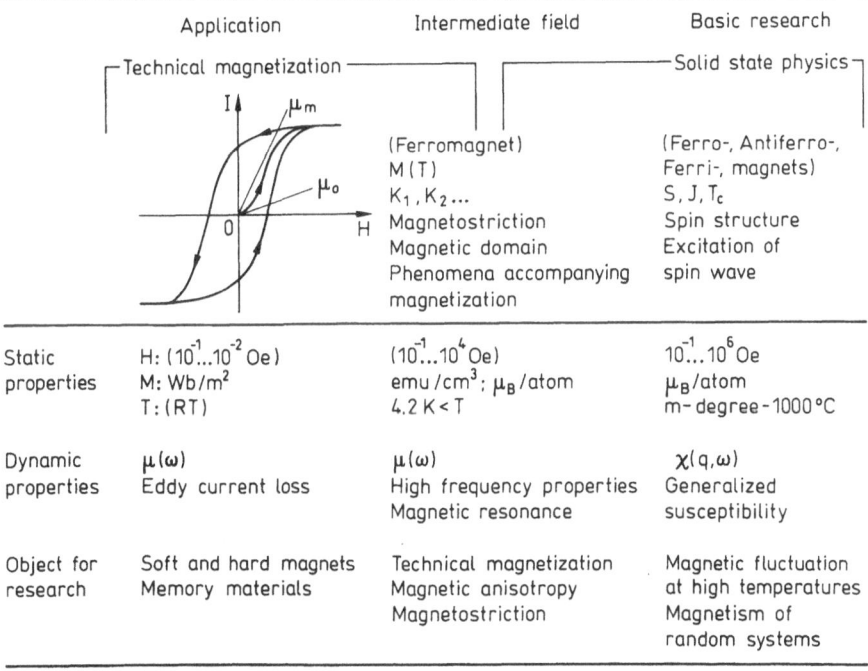

	Application	Intermediate field	Basic research
	Technical magnetization	(Ferromagnet) $M(T)$ $K_1, K_2 ...$ Magnetostriction Magnetic domain Phenomena accompanying magnetization	Solid state physics (Ferro-, Antiferro-, Ferri-, magnets) S, J, T_c Spin structure Excitation of spin wave
Static properties	H: ($10^{-1}...10^{-2}$ Oe) M: Wb/m^2 T: (RT)	($10^{-1}...10^4$ Oe) emu/cm^3; μ_B/atom 4.2 K < T	$10^{-1}...10^6$ Oe μ_B/atom m−degree−1000°C
Dynamic properties	$\mu(\omega)$ Eddy current loss	$\mu(\omega)$ High frequency properties Magnetic resonance	$\chi(q,\omega)$ Generalized susceptibility
Object for research	Soft and hard magnets Memory materials	Technical magnetization Magnetic anisotropy Magnetostriction	Magnetic fluctuation at high temperatures Magnetism of random systems

Fig. 7.1. Areas studied by researchers concerned with magnetism

these researchers in this "intermediate field" are spoken ill of by engineers as "they propose many grand ideas, but do not make practical, technical products", and by some pure physicists who say that the "magnetism of materials cannot be clarified these days by magnetization curves, magnetic anisotropy and magnetostriction alone". But it is very important to work in this "intermediate field", for example to measure magnetization curves accurately, for the development of both pure magnetism and of applied magnetic engineering. And I will show why in the following.

7.1 Technical Terms and Figures

A single technical term is often used with different meanings by pure physicists and engineers, while different terms are often used for the same magnetic phenomenon. This has made discussion very difficult between physicists and engineers to the point sometimes that they cannot communicate with each other at all. For example, the "impurity effect" in pure research is often used in the range from a few at.% to more than 5 at.%. This is very strange for engineers, because they consider "impurities" as unwanted atoms or molecules in samples, so the number is around a few ppm ($\sim 0.0001\%$), at least nowadays. And it is not clear whether the words "short range" and "long range", which are often used in the explanation of characteristics of materials, mean a scale of structural analysis in X-rays, i.e. tens or hundreds Å, or a more macroscopic scale in the range of a few microns. Furthermore, the order of length is not indicated definitively by "long wavelength" or "short wavelength" for spin waves in the theory of spin fluctuations, making it difficult to understand the intrinsic state of a magnet expressed by such waves. Also in discussions of magnetism using band structures, the same unclear expressions are used. It would be quite helpful if pure physicists could explain the intrinsic magnetic state using an intuitive model based on atoms at lattice points in real space.

Recently, the magnetism of $3d$ transition metals at finite temperatures is reported to have been made clear. This would imply that all the metals and alloys of Fe, Co and Ni can be explained. But in fact it means only that weak ferromagnetism has been interpreted. To add to the confusion the terms "weakly ferromagnetic" or "nearly ferromagnetic" are not clearly defined. Thus there are many terms or expressions which are very familiar to pure physicists but not to the man in the "intermediate field" even after much study.

Researchers in the "intermediate field", who are usually concerned with ferromagnets, take paramagnets to have quite a weak magnetization which increases linearly with magnetic field, considering the character of magnetization in paramagnets above the Curie point. They also consider that paramagnets can be distinguished from ferromagnets by the temperature dependence of the susceptibility χ in a constant magnetic field. However, χ in a constant field has no meaning near the Curie temperature, in constrast to the case at temperatures far above the Curie point. This is because, as the temperature is decreased, the curve of χ versus H of paramagnets resembles that of ferromagnets with a large magnetic anisotropy. The shape of the magnetization curve becomes Langevin type and χ increases non-linearly with increase of magnetic field. Furthermore, the text books say that in paramagnets χ, not the initial susceptibility χ_0, increases exponentially on cooling at low temperatures. It would be better for the text books to use χ_0 instead of χ.

Accurate measurements of magnetization curves at low temperatures make it possible to distinguish paramagnets from ferromagnets and a high magnetic field (few tens of thousands Oe) is not always necessary. So the experiment becomes very easy, as can be judged roughly from the temperature dependence of magnetization in a low magnetic field (few mOe). Recently, magnets have been divided into many classes and frequently have been characterized by a χ or by an anomaly in the temperature dependence of the magnetization in a constant field. In these experiments, the magnetization curve should be measured also and the definition of χ on the magnetization curve should be described, else it may not be clear what kind of magnets they are.

Next, as an example of confusion with figures, I will consider the presentation of the temperature dependence of magnetization. Experimentalists take the abscissa as K or °C and the ordinate as magnetization per unit volume or unit mass. On the other hand, pure theorists usually take the abscissa as kT/J (k: Boltzmann constant, J: exchange integral) and the ordinate as a product of μ_B and various kinds of coefficients. So, it is not straightforward to compare experimental results with theory, as the value of J is usually unknown. Normalization by T_C makes it easier to compare, but doesn't bring enough agreement between experiment and theory, as in the case of Fe or Ni, where the experimental and theoretical lines actually cross each other.

Sometimes, pure physicists say that the fine structure of intrinsic magnetization can be explained well by such and such a theory and such and such a problem is solved. It may sound to young researchers as well as those in the "intermediate field" that all the problems are solved and they might lose their desire to do research. So it is better to say that such and such a specific character of such and such limited materials in a limited region of temperatures can be explained.

Although pure physicists always consider the magnetism of materials in an ideal state, applied physicists study the character of magnetization in relation to the macroscopic structure of materials. All researchers in pure, "intermediate", and applied magnetism attempt to explain magnetic phenomena which can be seen or experienced, so it is desirable that pure physicists attempt to clarify macroscopic magnetic phenomena on the basis of their studies on microscopic intrinsic magnetism.

On the other hand, engineers use many terms which cannot be understood even by applied physicists let alone pure physicists. Often they seem to describe magnetic behavior, using the terms of pure magnetism thoughtlessly, without sufficient understanding or even with misunderstanding. This makes it difficult for them to communicate with each other, and means that pure and applied magnetism become even more separate.

Sufficient understanding of the terms used by pure physicists and engineers and explanations from both sides will contribute to the development of both pure magnetism and engineering.

7.2 Estimation of Saturation Magnetization and Curie Temperatures

Metallurgists and electricians who deal with soft magnetic materials often estimate saturation magnetization M_s from the measurement of the magnetization curve in a magnetic field of 1–100 Oe. On the other hand, researchers in pure magnetism usually obtain the saturation magnetization by extrapolating the linear part of the magnetization curve in magnetic fields higher than the technical saturation point to zero field, or take the value of magnetization in the maximum applied magnetic field of tens of thousand Oe as the saturation value.

The shape of a magnetization curve changes remarkably with the scale of units of the abscissa (magnetic field) or the ordinate (magnetization). Figure 7.2 shows an example where the magnetization seems to be saturated at both 1 Oe and 1 kOe, if one takes the same point of the abscissa as the maximum applied field and uses the same ordinate. As a result, the saturation values for the two curves are quite different. And it is also seen that the values of M_s depend on the upper limit of the applied field, 1, 10, 10^3 Oe, which are obtained by extrapolating the magnetization curves at higher fields to zero. According to Chikazumi, the magnetization curves are still increasing steeply even in a magnetic field of tens of thousand Oe, on the scale of maximum field of $5-10 \times 10^5$ Oe. So a serious error may be caused if the saturation magnetization, which is a basic magnetic quantity and material constant, is estimated from the magnetization curve. It is also necessary to take care when measuring the temperature dependence of the saturation magnetization. For example, if the saturation magnetization is measured in a field lower than tens of Oe in spite of a saturation field of hundreds of Oe, it will show an anomalous temperature dependence, when the sign of the magnetic anisotropy or magnetostriction changes with temperature.

Generally, the temperature dependence of saturation magnetization is measured by increasing the temperature in a constant high magnetic field or by extrapolating the magnetization curve in a high field to zero at each temperature. An example of this is shown in Fig. 7.3. However, these methods produce very different curves when the magnetization decreases gradually with field. In this case the Arrot plot method for weak magnets is used, i.e. the magnetization

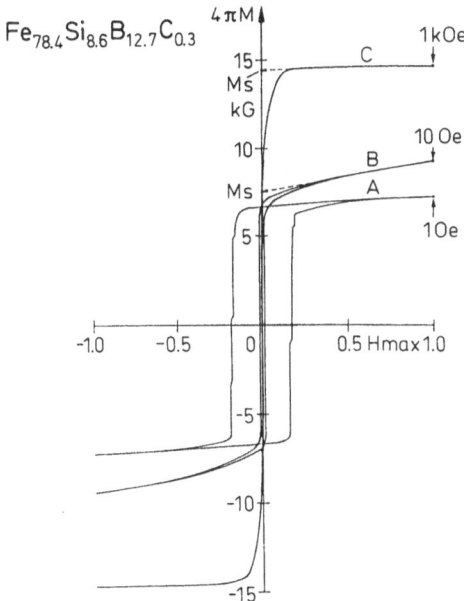

Fig. 7.2. An example of the difference of saturation magnetization M_s determined by the extrapolation $H \to 0$ by changing the scale of unit of abscissa. Magnetization curves, A and B, measured at a weak field can be seen as saturate. The value of saturation magnetization differs by the magnitude of the measuring magnetic field

curve is plotted on a scale of σ^2 for the ordinate and H/σ for the abscissa and the plot is extrapolated to $H/\sigma = 0$ (curve (c)). The result is different from the above two curves so it cannot be decided which is the true temperature dependence of the saturation magnetization.

The methods of determining the Curie temperature are as follows:

(1) The point at which the temperature axis is crossed by extrapolating the σ vs. T line around T_C.
(2) The point at which the temperature axis is crossed by extrapolating the straight line of σ^2 vs. T.
(3) The temperature at which the straight line of σ^2 vs. H/σ crosses the origin (Arrott plot method).
(4) The temperature at which the magnetization measured in a weak field (few mOe–few Oe) shows a peak (Hopkinson effect).

The value of T_C determined by method (2) varies by more than $10°$ depending whether one uses σ_s^2 vs. T or σ^2 vs. T (at $H = 15$ kOe) plots [plots (A) and (B) in Fig. 7.3]. Furthermore, T_C determined by the Arrott plot method is lower by over $20°$ than that obtained by method (2). Such differences are occasionally

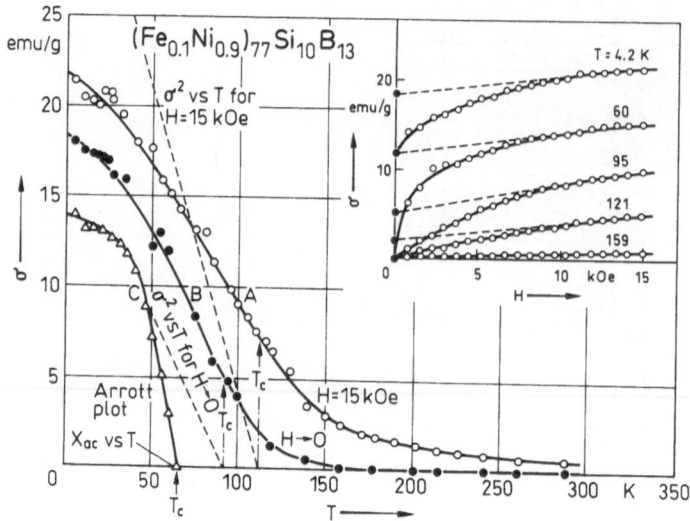

Fig. 7.3. Temperature dependence of magnetization and an example of the difference of Curie temperature determined by various methods. Inset shows the magnetization curves at various temperatures: *A* Magnetization at $H = 15$ kOe ($T_c = 110$ K; determined by σ^2 vs T plot). *B* Magnetization determined by the extrapolation $H \rightarrow 0$ ($T_c = 90$ K; σ^2 vs T plot). *C* Arrott plot method ($T_C = 66$ K). Susceptibility at weak magnetic fields ($T_C = 63$ K)

experienced in disordered system (including amorphous alloys), reentrant spin-glass such as 35% Ni–Fe Invar alloy and so forth. The saturation magnetization and the Curie temperature are important constants in a material, because they are necessary for developing iron core materials for transformers etc., as well as being magnetic quantities which are important clues for the investigation of intrinsic magnetization.

Nowadays, one of the most interesting problems in magnetic engineering is how to increase M_s or how to control T_C. So the method of measuring and controlling M_s and T_C is a serious problem for consideration and discussion between basic researchers and engineers.

7.3 Magnetic Anisotropy

The problem of magnetocrystalline anisotropy and induced magnetic aniso-tropy by magnetic annealing or cold rolling is an old-fashioned one for pure physicists. On the other hand, it is one of the most interesting for researchers studying applications such as bubble memory or vertical recording.

Magnetic anisotropy can be obtained by the measurement of magnetization curves or by magnetic torque measurements. In the following, I will point out the problems in the methods of obtaining magnetic anisotropy from magnetization curves and some of the problems concerned with magnetic anisotropy itself.

7.3.1 Problems in the Methods of Obtaining Magnetic Anisotropy from Magnetization Curves

In the process of obtaining magnetic anisotropy from magnetization curves, it is difficult to ascertain the magnetic field at which both the magnetization curves for easy and hard axes coincide. For example, in the case of the inset (a) in Fig. 7.4, there exist one or two orders in the magnetic energies obtained in the following cases; in one case, magnetization curves for two axes are considered to coincide at $H_s \simeq 10$ Oe, and in the other case, $\Delta M \simeq 10$ G (almost the experimental error) is thought to still remain at H_s and the two lines are considered to coincide at about 15 kOe. That is, approximating the figure enclosed by two curves above 10 Oe to be a triangle, the magnetic anisotropy energy for this part is evaluated as $(1/2)\Delta M \times H \simeq (1/2)10 \times 15 \times 10^3 = 7.5 \times 10^4$ erg/cm^3. When the corresponding point is taken to be 10 Oe, this large energy is neglected and the magnetic anisotropy is overlooked.

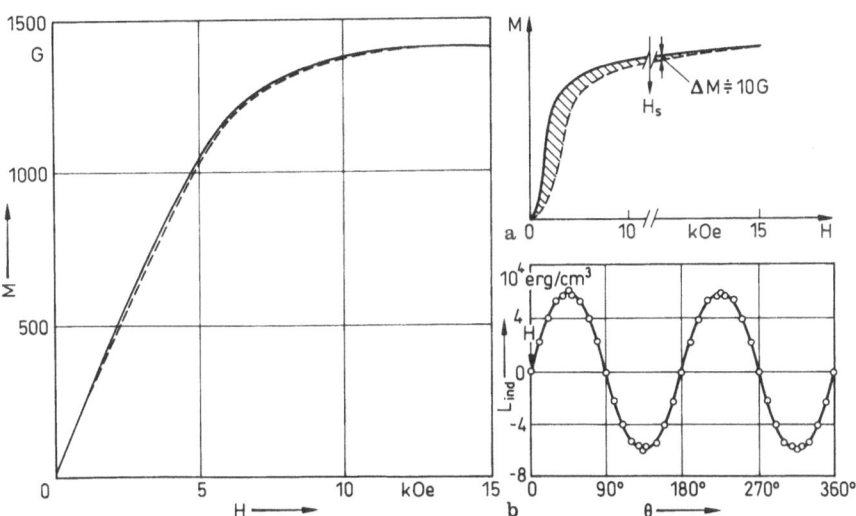

Fig. 7.4. Magnetization and torque curves for the magnetic annealed Co disk sample, solid line: $H \parallel$ magnetic annealed direction, broken line: $H \perp$ magnetic annealed direction. Torque curve clearly exhibits the uniaxial induced magnetic anisotropy

Further, Fig. 7.4 shows the magnetization curves for a disk sample of polycrystalline Co for the direction of applied magnetic field during heat-treatment (solid line) and for the direction perpendicular to it (broken line). Both lines coincide nearly exactly within experimental error and magnetic anisotropy is considered not to be induced. However, the torque curve for this sample obtained by a torque magnetometer shows a large uniaxial character, as shown in inset (b) of Fig. 7.4, and it is found that a large magnetic anisotropy of 10^4–10^5 erg/cm^3 is induced.

Engineers often estimate the magnetic anisotropy from magnetization curves and discuss various phenomena of magnetization, but it is necessary to be extremely careful using this method.

For alloys possessing a small crystalline magnetic anisotropy and magneto-striction, such as permalloys, the magnetic anisotropy energies estimated from both magnetization curves and torque curves are nearly equal.

7.3.2 Sign Reversal in Magnetocrystalline Anisotropy

The magnetocrystalline anisotropy of the transition metals, Fe, Co and Ni, or alloys of these metals does not attract much interest from pure physicists. But magnetocrystalline anisotropy is not only one of the basic physical quantities in magnetism but is also one of the most important properties for the development of magnetic materials.

It is strange that a systematic study of magnetocrystalline anisotropy in the binary alloys, Fe–Ni and Co–Ni, has not yet been reported, in spite of the very famous work by Kaya [7.1] on single crystals of Fe and Co. In particular, there are scarcely any studies on Co-rich alloys, perhaps because of the difficulty of making single crystals. In the following, some of the problems concerned with the magnetocrystalline anisotropy of alloys of Fe, Co and Ni are mentioned.

a) Sign Reversal of Magnetocrystalline Anisotropy
  with Temperature or Concentration

The magnetic phase diagram [7.2] of magnetocrystalline anisotropy in Ni–Co alloys is shown in Fig. 7.5. As seen in this figure, the sign of the magnetocrystal-line anisotropy constant at room temperatures changes with Co concentration from negative → positive → negative, within the γ(fcc) phase. And the sign reverses on warming in alloys of a certain region of concentration. The change of sign on warming is also observed in single metal Ni or Co, and the change of magnetic anisotropy constant versus concentration is known to exist also in the alloys Ni–Fe [7.4] and Fe–Co [7.5]. On the other hand the saturation magnetiz-ation, Curie temperature, crystal structure and lattice constants all vary mono-tonically with temperature or concentration. This sign reversal anomaly cannot be explained at present, even taking account of the magneto-elastic effect caused

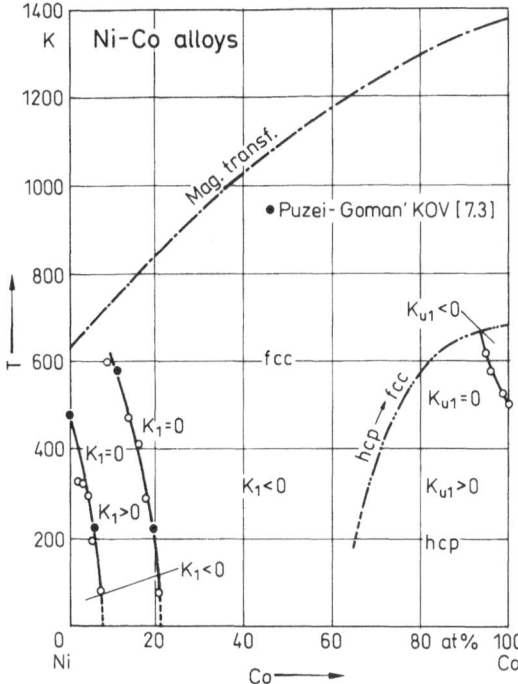

Fig. 7.5. Phase diagram of magnetocrystalline anisotropy for Ni–Co alloys. The sign of the magnetocrystalline magnetic anistropy constant changes even in the same solid solution

by the magnetostriction. Thus clarification of the origin of magnetic anisotropy is important for both pure and applied fields.

b) Sign Reversal of Magnetocrystalline Anisotropy
   and the Development of Magnetic Materials

In 1963, Chikazumi et al. [7.6] discovered that the uniaxial magnetocrystalline anisotropy in Co changes its sign suddenly from positive to negative by adding only about 1% Fe. This is a very remarkable discovery in the study of magnetic anisotropy in the ferromagnets Fe, Co and Ni, and I am certain that it has given an important impetus to the development of magnetic materials, especially permanent magnetic materials. Here, the influence of this discovery on the development of magnetic materials is briefly mentioned.

If the arrangement of the atoms in amorphous alloys caused by highly rapid quenching of molten alloys of 3d transition metals and metalloid atoms such as B, P, C, Si and Al etc. is completely random, the magnetism must be homogeneous and magnetic anisotropy shouldn't be observed. In almost all amorphous alloys however, uniaxial anisotropy of the order of $10^3 \sim 10^4$ erg/cm^3 is observed. Takahashi et al. [7.7] made clear that the temperature dependence of the uniaxial magnetic anisotropy is due to a large uniaxial magnetocrystalline anisotropy in the metallic compounds Fe_2P, Fe_3P, $Fe_2B(Fe_3B)$ and $Fe_2C(Fe_3C)$ which appear after the crystallization of these amorphous alloys.

Table 7.1 lists the physical characteristics of these metallic compounds. It is seen that the magnitude of the magnetocrystalline anisotropy is of the same order as that of a Sm_2Co_{17} permanent magnet which has a maximum energy product of 20–30 MG·Oe. However, it is found that the easy axis in some of these compounds is in the c-plane and the Curie temperature in some of them is lower than room temperature. Now, recalling the above phenomenon in the Co–Fe alloy discovered by Chikazumi et al., it would seem highly likely that it would be possible to reverse the sign of the easy axis by adding a small amount of a third element to these metallic compounds. Furthermore, if the saturation magnetization and Curie temperature could be enlarged by some means like the case of Fe-nitride, it would be expected that a cheap Fe-based permanent magnet comparable with or better than the Sm–Co permanent magnets could be developed. Indeed, Sagawa et al. [7.20] have recently devised an Fe–Nd permanent magnet which is much superior to Sm_2Co_{17} and which has attracted attention from the world as the great discovery of a magnetic material of this century. However, it is not so surprising that this alloy, which contains B to the amount of about 10%, has been found, considering the above indications by Takahashi [7.21].

The discovery of the sign reversal of magnetic anisotropy in ε phase Co–Fe alloys and Fe–Nd permanent magnets has been a great success, reminding us that an important discovery can be made from a simple idea, a challenging spirit and a great deal of effort.

7.4 Magnetostriction and Magneto-Elastic Energy

Magnetostriction, like magnetocrystalline anisotropy, is one of the basic characteristics of a magnetic material and is an important property to consider when looking for materials for producing ultrasonic or surface elastic wave devices and stress or strain sensors.

It should be noted also that the value of the saturation magnetostriction λ_s used by engineers, which is obtained from the linear (longitudinal) magnetostriction, depends on the distribution of magnetic domains. The values of the saturation magnetostriction found in tables in text books are obtained for samples which have been annealed in order to remove the stress. When the samples are heat-treated in a special way or dealt with mechanically, the value of the saturation magnetostriction should be changed because of the change in the

Table 7.1. Magnetism and crystal structure for intermetallic compounds exhibiting a larger uniaxial magnetocrystalline magnetic anisotropy

	SmCo$_5$	Sm$_2$Co$_{17}$	Fe$_2$P	Fe$_3$P	Fe$_2$B	Fe$_3$B	Fe$_2$C	Fe$_3$C
M$_s$ [G] (at 20°C)	770 [7.8]	955 [7.8]	710 [7.12]	950 [7.13]	1270 [7.9]	1240 [7.15]	(139 emu/g) [7.18]	890 [7.13]
T$_C$ [°C]	724 [7.8]	930 [7.8]	− 7 [7.12] (at 4.2 K)	420 [7.13]	742 [7.9]	540 [7.15]	1. 380 [7.9] 2. 247	210 [7.9]
K$_1$ [erg/cm^3] (at 20°C)	1.3 × 10^8 [7.8]	3.2 × 10^7 [7.11]	2.3 × 10^7 [7.12] (at 4.2 K)	− 5.5 × 10^6 [7.13]	− 4.3 × 10^6 [7.14]		1.2 × 10^6 [7.19]	
crystal structure	hexagonal [7.9]	hexagonal [7.9]	hexagonal [7.9]	tetragonal [7.9]	tetragonal [7.9]	1. tetragonal 2. orthorhombic	1. hexagonal [7.9] 2. orthorhombic	orthorhombic [7.9]
lattice constant [Å]	a = 5.002 [7.10] c = 3.964	a = 8.395 [7.10] c = 12.216	a = 5.864 [7.9] c = 3.46	a = 9.107 [7.9] c = 4.46	a = 5.109 [7.9] c = 4.249	1. a = 8.63 [7.16, 17] b = 4.29 2. a = 4.45 b = 5.43 c = 6.66	1. a = 6.27 [7.9] b = 21.4 2. a = 3.82 b = 4.72 c = 12.5	a = 4.525 [7.9] b = 5.087 c = 6.743

[a] Magnetic anisotropy constant obtained for a pseudo-single crystal

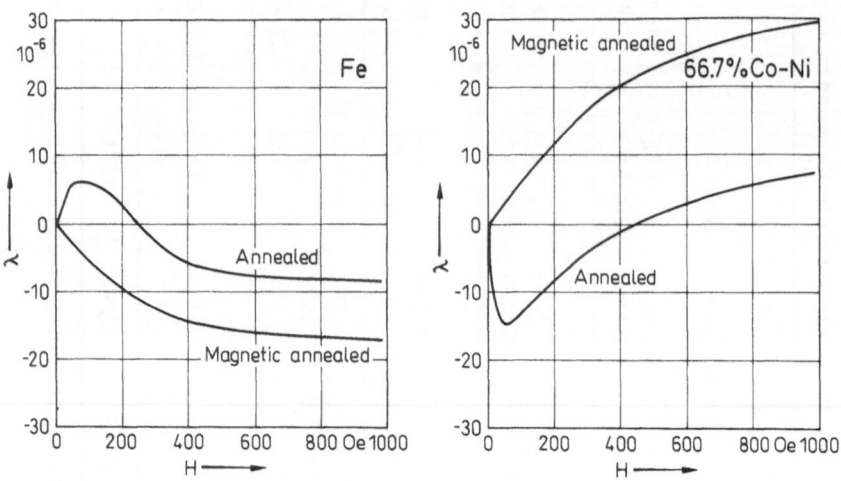

Fig. 7.6. An example of linear magnetostriction. The dependence of λ on magnetic field and its saturation value differ by the distribution of magnetic domains

distribution of magnetic domains (Fig. 7.6). Researchers working on applications today often use the word "saturation magnetostriction constant" as the value of the linear magnetostriction at the maximum applied magnetic field used in the measurement, but the expression "constant" should be avoided, since the value depends strongly upon the history of the material (distribution of magnetic domains). This is a good example where the usage of such terms can cause misunderstanding between pure, "intermediate field" and applied physicists.

A systematic study of magnetostriction constants, $\lambda_{100}, \lambda_{111}$ for alloys of Fe, Ni and Co has not been reported, although the constants have been known for long time to show complicated behavior, similar to their magnetocrystalline anisotropy, such as reversing sign or exhibiting a maximum or a minimum as the concentration or temperature is varied. Now that excellent methods have been developed for measuring magnetostriction, this is an area of research should be studied further.

It is also necessary to measure Young's modulus and the elastic modulus, because magnetostriction is a type of elastic modification. Therefore, the elasticity must also be taken into account. Furthermore, magnetostriction should be treated in terms of a magneto-elastic interaction energy in order to be considered in the same way as saturation magnetization or magnetic anisotropy. If magnetostriction is studied further in order to investigate the origins of magnetism which accompany a very hard effort, it will certainly lead to new topics in magnetism and new applications.

7.5 Spin Glasses, Hopkinson Effect and the Invar Problem

The problem in Invar or spin glasses is one of the topics which needs to be understood well by researchers in the "intermediate field" in order for them to keep up with the pure and applied physicists. In the following, the problem of the Fe–Ni Invar alloy which shows the character of a spin glass, as well as Invar itself, will be considered briefly from the pure, "intermediate" and application points of view.

a) The Invar Effect

Researchers in the applied fields consider Invar alloy as a "practical alloy which has quite a low thermal expansion coefficient ($\alpha = 1.2 \times 10^{-6}/°C$) at room temperature." On the other hand, for the basic researcher, an Invar alloy is not only an alloy with a small α, but also includes alloys with negative thermal expansion coefficients, like 50% Fe–Pt alloy, alloys which reverse the sign of α at a certain temperature (i.e., with no regard to the temperature range), alloys with a clear negative magnetic contribution (spontaneous volume magnetostriction) to the thermal expansion curve and even Elinvar type alloys. Furthermore, some use the name "Invar" for an alloy which exhibits at least one anomalous magnetic characteristic. The expressions "Invar in a narrow sense" or "Invar in a general sense" are occasionally used.

Metallurgists will direct their attention to the equilibrium phase diagram of the alloy to consider the mechanism in terms of the appearance of 35% Ni–Fe Invar characteristics. They will therefore look for the coexistence of two phases α and γ, since the 35% Ni–Fe alloy is on the border of a phase transition from the phase α to γ.

On the other hand, pure physicists investigate the spin states of the material by means of neutron diffraction and try to explain the characteristics of Invar alloys in terms of their electronic states. Sometimes however, it is not clear whether they are trying to solve the Invar problem itself or to explain the anomalous physical phenomenon in Invar alloys. In any case researchers in the "intermediate" and applied fields still hope that the pure physicists will be able to explain the origin of the simple and naive problem of Invar phenomena, as a result of their experimental and theoretical investigations of basic magnetism. This example highlights one of the reasons, not limited to Invar, why there is a gap between the pure and applied physicists.

b) Spin Glasses, Hopkinson Effect

Fe–Ni Invar alloy has been one of the main subjects in physics conferences over the last ten years. In this material we observe the phenomenon of a spin glass (Appendix A.32), which is a current topic in pure magnetism, and the Hopkinson effect, which was a topic in the 1890s. The Fe–Ni Invar characteristics have been discussed from various points of view; the coexistence of the α and γ phases as mentioned above, the concentration for the transition from a ferromagnet to an antiferromagnet and the coexistence of high spin and low spin states. From any point of view, it is easy to suppose that the alloy is in a non-equilibrium state, both structurally and magnetically, on a microscopic scale. Since there may be a tendency of transform to the lower energy more stable state from the non-equilibrium state, it is very natural to expect some anomaly in the magnetization, when it is investigated in detail. With this idea in mind, we measured the magnetization in a weak field and obtained the result shown in Fig. 7.7. As seen in this figure, the magnetization in a high magnetic field decreases monotonically with temperature, but the magnetization in an alternating magnetic field of $\tilde{h} = 0.1$ Oe shows an anomaly at about 30 K similar to that found in reentrant spin glasses and a Hopkinson peak slightly below the Curie point. As is well known soft magnetic properties are much influenced by impurities, vacancies, magnetocrystalline anisotropy and magnetostriction. In this case we confirmed that these effects can be neglected.

On the other hand, although it is scarcely studied, the Hopkinson effect can be easily explained as the phenomenon caused by the vanishing of the magnetic anisotropy and the magnetostriction just below the Curie temperature. Its dependence on the magnetic field however, is quite similar to that of spin glasses. So these two phenomena should be considered to have some common relationship with regard to the influence of an external magnetic field on the spins. This is supposing that the Hopkinson effect originates from the singular behavior of intrinsic magnetization when there is a transition from the ferromagnetic to the paramagnetic state around T_C.

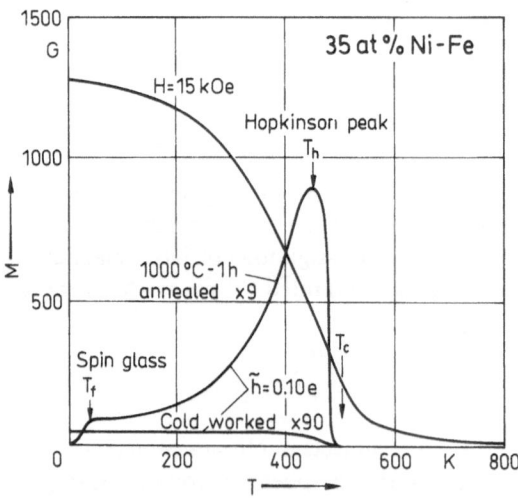

Fig. 7.7. Anomalous temperature dependence of magnetization measured at a weak magnetic field and Hopkinson peak for 35 at.% Ni–Fe Invar alloy

Nothing was thought to be left to study about the magnetic anisotropy, magnetostriction and magnetization in a weak magnetic field in the alloys of the ferromagnetic $3d$ transition metals, Fe, Ni and Co, because these magnetic phenomena had been well known for so long. However, they can now be regarded as objects for further magnetic study, in which many new phenomena may be hidden, as evidenced above. Considering that researchers interested in basic magnetic effects regard magnetization curves and the study of technical magnetization saturation to be in the "intermediate field", we cannot help recognizing again that the existence of this "intermediate field" is very important for pure magnetism.

7.6   The Problem of Communication Between Applied Researchers and Those in the "Intermediate Fields" or in Basic Research

a) Materials for Magnetic Memory

Materials for bubble memory and vertical magnetic recording both have their easy axis perpendicular to the film surface. Bubbles are not observed in crystalline Co–Cr films used for magnetic memory, but in the case of rare metallic garnets and amorphous Gd–Co films bubbles are observed, yet they all possess the same perpendicular magnetic anisotropy. Bubbles with a diameter of about 0.3 μm should be observed easily in Co–18% Cr films, because its saturation magnetization is 380 G, its perpendicular magnetic anisotropy constant is 1.2×10^6 erg/cm^3 and the thickness of the film is 4000 Å, but they have not been found so far. It may be that the state of the microscopic magnetization in the sample is different even in films with the same perpendicular magnetic anisotropy. In addition, the origin of the large coercive force in fine Fe_2O_3 particles coated with Co or Fe_2O_3 magnetic memory materials doped with Co, remains an unsolved problem for researchers in the "intermediate field".

b) Materials for Permanent Magnets

Typical permanent magnets are Sm_2Co_{17} and $SmCo_5$. The equilibrium phase diagram of this alloy system is complicated, and the magnetic properties change remarkably according to the method of sample preparation, and the heat and surface treatment. The origin of large coercive forces in materials which are sensitive to these treatments is a difficult problem that cannot be explained simply by high resolution structural analysis, magnetic anisotropy or magnetostriction etc. Researchers in basic fields of study tend not to look at magnetic materials which have large coercive forces such as magnetic memory materials or permanent magnetic materials. However the behavior of the local spins in these materials, which is of interest to those in basic research, may well contribute to the origin of the large coercive force. So the problem may be left to be solved step by step from the engineer, to the "intermediate field" to basic research. Therefore, on this point at least, the "intermediate field" is necessary and can justify its existence.

7.7 Conclusion

In order to make progress in magnetism, including pure magnetism, applied magnetism and magnetic engineering, it is necessary to be interested in and be concerned with the essence and the application of magnetism, considering the phenomenon of technical magnetization and the basic mechanisms of magnetization. The researchers in the "intermediate field" should make efforts to explain all phenomena of technical magnetization simply and schematically. I believe that these efforts can create a communication bridge between the pure physicists and the engineers concerned with applications and development of magnetic materials. To put it more concretely, researchers in this field should learn and understand pure magnetism well and find possibilities for application from the basics of magnetism and bring them to the engineers. In the other direction, they should extract subjects for the study of pure magnetism from the investigations of the engineers, i.e., various magnetization phenomena. The role of the researchers in the "intermediate field" is extremely necessary to develop both pure magnetism and its application, since magnetism nowadays has become specialized into narrow areas and is studied intensively within these narrow boundaries. I believe that it is the ability and leadership of Chikazumi himself that have made the study of practical materials made from permalloys and Invar alloys one of the main topics in pure magnetism, from the simple measurements of magnetic anisotropy, magnetostriction and magnetic domain patterns, which decide the shape of the magnetization curves. Basically, he introduced the ideas of materials physics into materials engineering. Finally, I would like to point out that this is the reason why the "Physics of Ferromagnetism" [7.20] by Chikazumi is famous all over the world as the basic text on magnetism and is used daily by people from every area of magnetism.

The author wishes to sincerely thank Dr. I. Matsubara, Professor K. Hisatake and Associate Professors T. Miyazaki and Migaku Takahashi for their co-operation in the English translation and in preparing the manuscript.

References

7.1. S. Kaya: Z. Phys. **84**, 705 (1933)
7.2. S. Kadowaki, M. Takahashi: J. Phys. Soc. Jpn. **38**, 1612 (1975)

7.3. I.M. Puzei, V.I. Goman'kov: Fiz. Metal. Metalloved **23**, 636 (1967)
7.4. I.M. Puzei: Fiz. Metal. Metalloved **11**, 686 (1961)
7.5. R.C. Hall: J. Appl. Phys. **31**, 1575 (1960)
7.6. S. Chikazumi, T. Wakiyama, K. Yoshida: Proc. Int. Conf. on Magnetism, Nottingham, 1964 (Institute of physics and the Physical Society, London, 1965) p. 756
7.7. M. Takahashi, T. Miyazaki, K. Takakura, N. Ikeda: Abstract of the Japan Phys. Society (1976, April) p. 49
7.8. Edited by Non-traditional Technology: Magnetic Materials in Recent Years (Kougyouchousakai, 1981) p. 35
7.9. S. Chikazumi: Handbook of Magnetism (Asakura, 1975)
7.10. K.H.J. Buschow, A.S. van der Goot: J. Less-common Met. **14**, 323 (1968)
7.11. R.S. Perkins, S. Stassler: Phys. Rev. **B15**, 477 (1977)
7.12. H. Fujii, T. Hokabe, T. Kamigaichi, T. Okamoto: J. Phys. Soc. Japan **43**, 41 (1977)
7.13. M. Takahashi, T. Miyazaki: Jpn. J. Appl. Phys. **20**, 1821 (1981)
7.14. A. Iga, Y. Tawara, A. Yanase: J. Phys. Soc. Japan **21**, 404 (1966)
7.15. M. Takahashi, M. Koshimura, T. Abuzuka: Jpn. J. Appl. Phys. **20**, 1821 (1981)
7.16. J.L. Walter, S.F. Batram, R.R. Russell: Meta. Trans. A **9A**, 803 (1978)
7.17. U. Herold, U. Koster: Z. Metallkunde Bd **69**, H.5 327 (1978)
7.18. L.J.E. Hofer, E.M. Cohn, W.C. Pecbles: J. Am. Chem. Soc. **71**, 189 (1949)
7.19. P. Blum, R. Pauthenet: Acad. Sci. **30**, 1501 (1953)
7.20. M. Sagawa, T. Fujimura, M. Togawa, H. Yamamoto, Y. Matsuura: Abstract of the Conference of the Japan Institute of Metals (Oct. 1983) p. 551
7.21. M. Takahashi: Abstract of 14th research meeting for new materials (Oct. 1980)
7.22. S. Chikazumi: Physics of Ferromagnetism (Shōkabō, Tokyo, 1959)

8. Amorphous Magnetic Materials

Tadashi Mizoguchi

A long time ago, in prehistoric times, people were already using natural glass produced by volcanic activity as arrowheads. In museums one can see glassware manufactured and used in ancient Egypt and Mesopotamia. But although amorphous solids have been used by people from ancient time it is relatively recently that amorphous metallic alloys have been available. The first report which appeared about 20 years ago in "Nature" was on an amorphous Au–Si alloy by Professor Paul Duwez at the California Institute of Technology. The first ferromagnetic amorphous alloy $Fe_{75}P_{15}C_{10}$ was also produced by his group with a piston anvil method. In 1973 Dr. Chaudhari et al. at IBM reported that amorphous GdCo alloys prepared by sputtering showed ferrimagnetism.

When we started the study of amorphous ferromagnetic alloys at Gakushuin University in 1971 few people had a perspective on the basic study and application of amorphous alloys. Since then great progress has been made all over the world. As it is almost impossible to cover all the progress here in this short article the author would like to deal mainly with topics which he is interested in or to which he has made some contribution himself.

First the magnetization and temperature dependence of typical ferromagnetic and ferrimagnetic amorphous alloys are described. We then go on to deal with magnetic anisotropy on both the atomic and macroscopic scale in amorphous alloys, which were thought to have an isotropic structure on average. Finally the effects of structural relaxation and preparation conditions of amorphous alloys on their magnetic properties are discussed.

8.1 Magnetization and Temperature Dependence of Amorphous Magnetic Materials

Typical well-known amorphous magnetic alloys are classified in the following groups: 1) Fe- and/or Co-based $3d$ transition metal ferromagnetic alloys with about 10–30% of metalloid elements such as B, C, Si and P; 2) Fe- and/or Co-based ferromagnetic alloys with early transition metals, e.g. Zr, Nb, Ta etc.; 3) Fe- and/or Co-based ferrimagnetic alloys with rare earth metals, e.g. Gd, Tb

etc. A concise description of the basic magnetic properties of these amorphous
magnetic alloys will be given below.

8.1.1 Amorphous Ferromagnetic Alloys

Fe, Co, Ni and Gd are ferromagnetic in their pure elemental state; however, they
are not stable in the amorphous state at room temperature. Addition of 15–25%
metalloid elements to Fe or Co makes it possible to make an amorphous state
which shows ferromagnetism. The saturation magnetic moment of quasibinary
$3d$ transition metal alloys, $(T_{1-x}M_x)_{80}B_{10}P_{10}$ with 10% of B and P as added
metalloid elements is shown in Fig. 8.1 as a function of average valence electron
number of $3d$ transition elements. The magnetization of corresponding crystalline alloys without the metalloid elements, which is well known as the
Slater–Pauling curve (Appendix A.27), is also shown in the same figure. The
crystalline $3d$ alloys in this range form a bcc phase for $N < 8.7$ and an fcc phase
for $N > 8.7$. In these quasibinary amorphous alloys the average magnetic
moments of the transition metals are reduced compared to those of crystalline
alloys of corresponding N, probably because of the effect of the additional 20%
of metalloid elements, B and P. They behave, however, similarly to those in
crystalline alloys, that is, decreasing with increasing N with a slope of

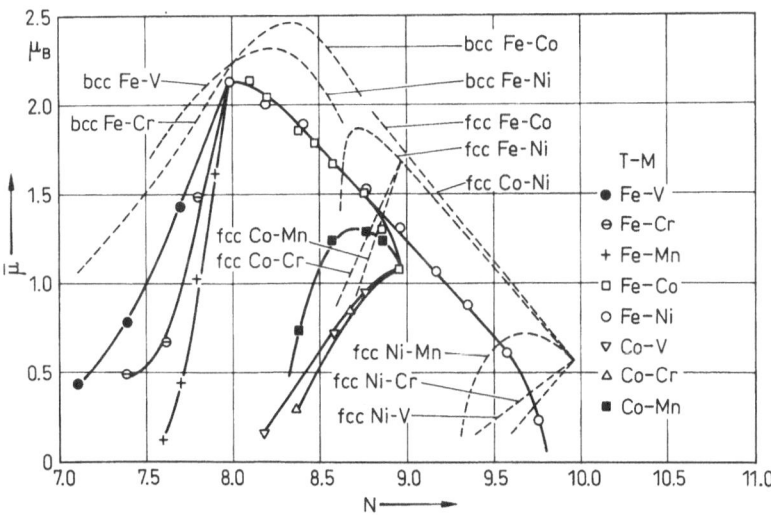

Fig. 8.1. The average saturation magnetic moment per metal atom in amorphous
quasibinary transition metal alloys, $(T_{1-x}M_x)_{80}B_{10}P_{10}$, versus average valence electron
number, N (solid lines). Broken lines in the figure represent Slater–Pauling curves for
crystalline alloys [8.8]

$- 1(\mu_B/N)$ parallel to the Slater Pauling curve as shown in Fig. 8.1. In these amorphous alloys Fe, Co and Ni may be considered to have 2, 1 and $0\mu_B$, respectively, and to couple ferromagnetically. The behavior of V, Cr and Mn which replace Fe in amorphous $Fe_{80}B_{10}P_{10}$ alloys is interesting. The total magnetization of the alloys drastically drops with the addition of these impurities, which may affect the electronic states of neighboring Fe atoms to reduce their magnetic moment. If we assume that the Fe moment is not changed it follows that V, Cr or Mn has a magnetic moment of $- 5$, $- 4$ or $- 3\mu_B$, respectively (negative sign means antiferromagnetic coupling to the host Fe moments).

The reduction rates of the Fe moment in amorphous alloys by the addition of metalloid elements were studied in detail by Mitera et al. If the reduction is assumed to be due to transfer of electrons from the metalloid atoms to an unfilled d-band of Fe, the number of transferred electrons from B, C, Si, Ge and P is 1.07, 1.32, 0.94, 0.64 and 1.64, respectively. Higher valence metalloid elements, and lighter ones of the same valence, have the greater effect on reducing the Fe moment in these amorphous alloys.

The positive exchange interactions between the magnetic atoms induce the long range magnetic ordering, that is, the ferromagnetism. It is clear that this is the case even in the amorphous phase where there is no long range atomic ordering. According to the molecular field approximation the Curie temperature, T_C, is given as follows:

$$T_C = \frac{2S(S + 1)}{3k} \left\langle \sum_j J_{ij} \right\rangle_i \tag{8.1}$$

where S is atomic spin, k, the Boltzmann constant and J_{ij} represents the exchange interaction between atoms i and j. The sum should be taken for all neighboring atomic pairs and averaged.

The exchange interaction in the amorphous Fe alloys, $Fe_{80}G_{10}E_{10}$ ($G = B$, P; $E = B$, C, Si, Ge and P), was estimated from the observed T_C. It is interesting to find that the exchange interaction between Fe moments in these alloys is a monotonous function of the magnetic moment independent of the metalloid species. It ranged from 470 K to 510 K with an average Fe moment of 1.98 to $2.13\mu_B$.

In Fig. 8.2, the composition dependence of T_C in the quasibinary amorphous $3d$ transition metal alloys is plotted versus the average valence electron number of the metal atoms. The T_C of crystalline alloys, which is also shown in the figure in broken lines for comparison, shows a drastic discontinuity between the bcc phase and the fcc phase indicating that the exchange interaction is quite sensitive to the crystal structure. In amorphous alloys it changes continuously with composition because their is a homogeneous single phase of dense random close-packed atoms. The exchange interaction between Fe and Co is strongest, giving the highest T_C in the Fe–Co quasibinary system.

Ferromagnetism in amorphous alloys of transition metals without metalloids, (Fe, Co, Ni)–Zr was reported by Onuma [8.6] in 1980. About 10%

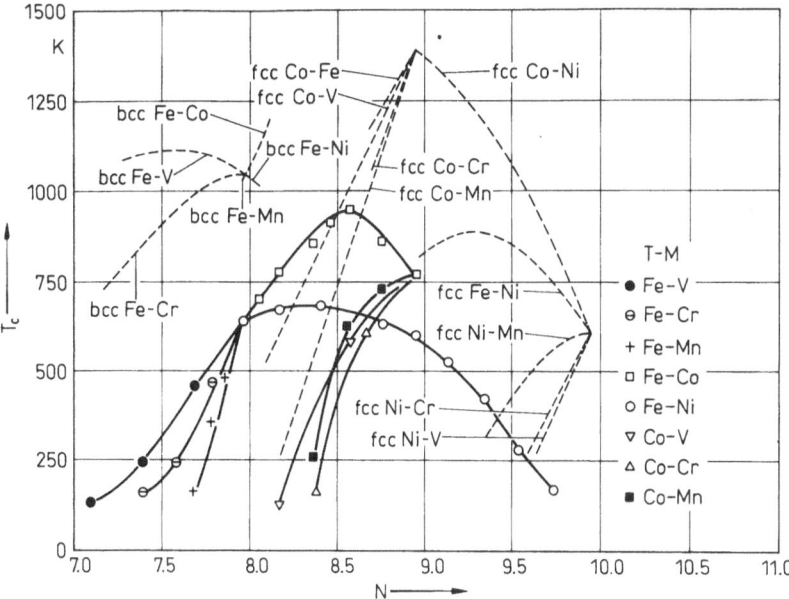

Fig. 8.2. The Curie temperature, T_C of amorphous quasibinary transition metal alloys, $(T_{1-x}M_x)_{80}B_{10}P_{10}$ versus average valence electron number, N, of metallic atoms (solid lines). The broken lines represent the Curie temperatures of the crystalline alloys [8.8]

addition of early transition elements, e.g. Zr, Hf, stabilizes the amorphous phase. The saturation magnetic moment of Co increases, while that of Fe decreases with decreasing Zr content in these amorphous alloys. The extrapolated moment of Co and Fe to 0% Zr is 1.75 and $0\mu_B$, respectively. The average magnetic moment of $3d$ transition metals in the amorphous $(Fe, Co, Ni)_{90}Zr_{10}$ alloys attains a maximum at $N = 8.4$, the same as in the bcc Fe–Co alloys, and decreases with decreasing N. The Fe moment in amorphous $Fe_{90}Zr_{10}$ is $1.3\mu_B$.

What are the characteristics of ferromagnetism in amorphous materials. Let us examine the temperature dependence of the magnetization. In general it is roughly given by molecular field theory, but the magnetization of the amorphous $Co_{70}B_{20}P_{10}$ alloy turns out to fall off more rapidly than predicted by this theory. The microscopic local environments in the amorphous phase differ from site to site and so there must be a distribution of exchange interactions. Handrich has interpreted the rapid temperature dependence by taking into account the fluctuation of the exchange interaction.

Now we have the interesting question of whether or not there is a clear magnetic transition in amorphous alloys where the exchange interactions and even the magnetic moments have different distributions because of the variety of microscopic local environments. Detailed measurements of the magnetization of an amorphous $Co_{70}B_{20}P_{10}$ alloy around the Curie temperature has confirmed that there is a definite second order critical transition in this alloy, and accurate critical exponents have been obtained. The thermal fluctuations of the magnetization near the second order transition have long wavelengths compared to the atomic scale fluctuation due to the amorphous structure. Thus amorphous ferromagnets seem to behave as quite homogeneous systems.

Fluctuations of the magnetization in a ferromagnet at low temperatures are well described by the excitation of spin waves, giving the following temperature dependence for the magnetization,

$$M(T) = M_0(1 - BT^{3/2} - CT^{5/2} - \cdots) . \tag{8.2}$$

The second term which gives a $T^{3/2}$ temperature dependence comes from the first term of the spin wave dispersion relation in a ferromagnet,

$$E(q) = \hbar\omega_q = Dq^2 + Eq^4 + \cdots . \tag{8.3}$$

The Eq^4 term in the above equation adds the $-CT^{5/2}$ term to the magnetization. There is also a relationship between the coefficients in the above equations,

$$B = \frac{2.612g\mu_B}{M_0}\left(\frac{k}{4\pi D}\right)^{3/2} . \tag{8.4}$$

Accurate magnetization measurements at low temperatures and inelastic neutron scattering experiments for the amorphous $(Fe_{93}Mo_7)_{80}B_{10}P_{10}$ alloy have been carried out [8.11]. The temperature dependence of the magnetization was well expressed by only a $-BT^{3/2}$ term for $T < T_C/3$. While the neutron scattering experiment showed clear excitation of spin waves, their energy deviated appreciably from the Dq^2 dependence. Besides, the calculated value of B from the D obtained by neutron scattering was only 76% of that directly observed in the magnetization measurement. The wavelength of spin waves observed by inelastic neutron scattering is relatively long (wave number $q < 0.25A^{-1}$), but there is a possibility that the spin waves have a low energy tail with much shorter wavelengths and becoming diffusive, which contributes to the magnetization fluctuation and reduces it more rapidly. There are reports, however, that the above relation between B and D holds in other amorphous alloys, while in crystalline Invar alloys the relation does not hold. The discrepancy may be related to the instability of occurrence of magnetic moments in both amorphous and crystalline phases.

So far we have dealt with the magnetization and temperature dependence of most simple amorphous ferromagnets. The situation for magnetic ordering becomes even more complex where negative exchange interactions are present. In the amorphous phase one does not expect that antiferromagnetic ordering with regular sublattices or a coherent screw structure will occur as in the crystalline phase. In general a spin glass state is realized at low temperatures in systems which contain a certain fraction of negative exchange interactions. This is a state in which the magnetic moments of the atoms are frozen in random directions and there is no spontaneous magnetization. The transition from the

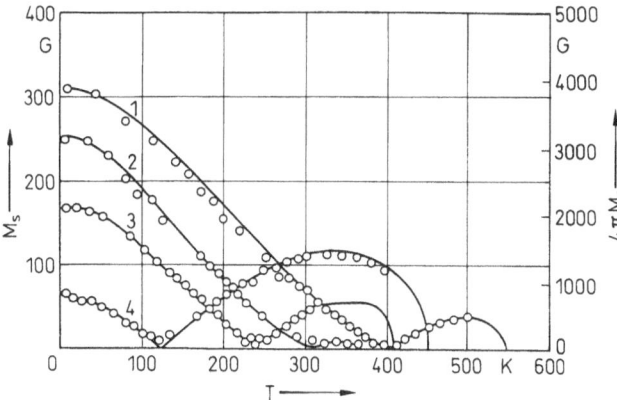

Fig. 8.3. The temperature dependence of the saturation magnetization of an amorphous $(Gd_{1-x}Co_x)_{1-y}Mo_y$ alloy. The values for x, y are as follows: *1*) 0.789, 0.078; *2*) 0.835, 0.154; *3*) 0.850, 0.154; *4*) 0.875, 0.153 [8.9]

paramagnetic state to the spin glass state was clearly observed in an amorphous $Gd_{37}Al_{63}$ alloy at low fields by using a sensitive squid magnetometer [8.12]. The magnetic susceptibility of this alloy at fields ~ 0.1 Oe showed a cusp at the transition temperature. Below the transition appreciable thermal hysteresis and relaxation phenomena were observed.

The magnetic ordering in amorphous rare earth (RE) and transition metal (TM) alloys is determined by the exchange interaction between RE moments and TM moments via conduction electrons. The magnetic moments of light rare earth atoms (La–Eu) couple parallel to those of transition metal atoms while moments of heavy rare earths (Gd–Lu) couple antiparallel to those of transition metals. Amorphous thin films of Gd–Co and Tb–Fe are especially interesting because they show perpendicular magnetic anisotropy. They have been studied intensively as possible candidates for bubble magnetic materials or magneto-optical recording materials.

The temperature dependence of the magnetization of amorphous $(Gd_{1-x}Co_x)_{1-y}Mo_y$ alloys is shown in Fig. 8.3. As seen in the figure the magnetization of this ferrimagnetic alloy behaves differently from an ordinary ferromagnet in that the magnetization decreases to zero at a temperature below the Curie temperature as the compensation of subnetwork magnetization occurs. This peculiar temperature dependence of the total magnetization can be interpreted as follows. In these alloys the exchange interaction between like atoms, that is, the interactions between transition metal atoms and those between rare earth metal atoms are positive while the exchange interaction between transition metal atoms and rare earth metal atoms are negative. (Mo affects the magnetic moment of Co, but it is nonmagnetic and causes no essential change in the sign of the exchange interactions.) Then the lowest energy state is realized in the ferrimagnetic configuration in which the transition metal moments are coupled antiparallel to those of the rare earth metal atoms. In this manner a long range magnetic ordering is realized in an amorphous phase in which each subnetwork of ferrimagnetism contains both TM and RE elements.

In the TM-RE amorphous alloys the exchange interaction between the TM moments is the strongest, the next is that between TM and RE moments which is about an order of magnitude weaker and the weakest, by two orders of magnitude, is those between RE moments. Let us consider a case where at low temperatures the subnetwork moments of RE atoms are greater than that of the TM atoms. The RE subnetwork moments decrease faster with increasing temperature than the TM ones because coupling between the former is weaker than between the latter. At the compensation temperature both subnetwork moments become equal in quantity but with opposite sign so that the net moment is zero. Above that temperature the TM moments exceed those of the RE ones. The exchange interactions of these ferrimagnetic alloys can be estimated from the temperature dependence of the total magnetization using the molecular field approximation. The following values were obtained by Hasegawa et al. for $(Gd_{1-x}Co_x)_{1-y}Mo_y$ amorphous alloys [8.9]:

$$J_{Co-Co} = (5.1 - 2.8) \times 10^{-14} \text{ erg} ,$$

$$J_{Co-Gd} = -2.3 \times 10^{-15} \text{ erg} , \tag{8.5}$$

$$J_{Gd-Gd} = 2.0 \times 10^{-16} \text{ erg} .$$

The total magnetization of heavy RE and TM amorphous alloys and their temperature dependence is quite sensitive to the composition of the alloys because it depends on the difference between the subnetwork moments. On the other hand it becomes easier to get perpendicular magnetic thin films since low magnetization makes the magnetostatic energy low. Applications in thermomagnetic recording are also possible by utilizing the abrupt increase of coercive force at the compensation temperature.

8.2 Magnetic Anisotropy

Amorphous materials in which there is no long range atomic ordering are usually considered to be isotropic over the whole system. We were surprised, then, to see the report by Chaudhari, Cuomo and Gambino [8.13] that amorphous Gd–Co thin films produced by sputtering had a strong perpendicular magnetic anisotropy which exceeded the shape anisotropy of the thin films. Real nature sometimes presents us with unexpected variety. For the origin of this magnetic anisotropy we can consider the following:

(1) Spin-orbit interactions
   (a) Single-ion anisotropy which is effective for non-S state ions in a local ligand field. In order to get a macroscopic overall anisotropy there must be an anisotropic distribution of the axial directions of the ligand field.
   (b) Anisotropic exchange interactions (including pseudo-dipole interactions) (Appendix A.25).
(2) Classical magnetic dipolar interactions.
   (a) Internal shape effect due to inhomogeneity or microstructures of thin films.
   (b) Anisotropic distribution of atomic moment pairs.
(3) Inverse effect of magnetostriction with anisotropic distribution of stress.

The mechanisms cited above are not exclusive, that is, they may coexist. It is difficult, however, to decide qualitatively the ratio of their contributions. Here we discuss a mechanism of magnetic anisotropy due to the classical dipolar interaction so as to understand the relation between structure and magnetism, since magnetic anisotropy is quite sensitive to structure anisotropy. Even a slight deviation from structural isotropy which cannot be detected by direct diffraction methods may induce appreciable magnetic anisotropy.

8.2.1 Macroscopic Magnetic Anisotropy

First we discuss macroscopic magnetic anisotropy in amorphous alloys caused by mechanism (2.a), the internal shape effect which is most easy to understand. The magnetostatic energy in a homogeneous ferromagnetic thin film becomes $2\pi M^2$ due to the demagnetization effect when the magnetization is directed perpendicular to the film plane. This is the result of the integral of all the dipole-dipole atomic pair interactions in the films.

Sometimes thin films have complex internal columnar structures depending on their preparation conditions even if they look homogeneous. In this case it is easy to expect a perpendicular magnetic anisotropy due to these internal structural heterogeneities. It should be noted, however, whether this internal shape effect can cause sufficient perpendicular magnetic anisotropy to exceed the magnetostatic energy $2\pi M^2$ in order to make a perpendicularly magnetized film.

For simplicity we deal with columnar structure films of two phases whose magnetizations are M_1 and M_2. The uniaxial anisotropy constant K_u becomes in this case,

$$K_u = \pi V(1 - V)|M_1 - M_2|^2 ,$$ (8.6)

where V or $1 - V$ is the volume fraction of each phase. The average macroscopic magnetization is

$$M = VM_1 + (1 - V)M_2 .$$ (8.7)

Then the ratio of the anisotropy energy to the magnetostatic energy is

$$Q = \frac{K_u}{2\pi M^2} = \frac{1}{2} \frac{V(1-V)(1-x^2)}{[V+(1-V)x]^2} ,$$
(8.8)

where $x = M_2/M_1$. Let us consider the simplest case where one phase is non-magnetic ($M_2 = 0$) and the other phase is precipitated in it as thin needle-like columns with their long axis perpendicular to the film plane. Even in this extreme case the volume fraction of the ferromagnetic phase must be less than one third ($V < 1/3$) to get a perpendicularly magnetized film ($Q > 1$). If $x > 0.1010$ we can not expect perpendicular magnetic films based on this internal shape mechanism.

The internal shape effect, that is, the magnetic anisotropy due to anisotropic heterogeneity of thin films is easy to explain, however, quantitatively speaking it may be insufficient to explain perpendicular magnetic films. For crystalline ferromagnetic films there must be a contribution from the crystalline anisotropy as well as from the internal shape effect to make perpendicular magnetic films because the crystalline axis aligns along the columnar axis.

How can we understand the origin of the magnetic anisotropy of amorphous perpendicular magnetic films. Especially why do amorphous films of non-S state Gd with $3d$ transition metals have such a strong anisotropy energy which exceeds the magneto-static energy? It is indeed difficult to specify the mechanism responsible for this magnetic anisotropy. Here we confine ourselves to discussing mechanism (2.b), that is, the classical dipolar interactions between microscopic atomic moments in an amorphous phase.

Let us consider an amorphous ferrimagnet which consists of two species of magnetic atoms A and B, in which there are positive exchange interactions between like atoms and negative ones between unlike atomic pairs. In this system atomic moments of A and B atoms couple antiparallel to make a ferrimagnetic arrangement. This is the case, in amorphous Gd–Co alloys, that is, molecular field theory gives a relatively good approximation with negative exchange interactions between Co atoms and Gd atoms and positive ones between like atoms. The basic magnetic structure is determined by the exchange interactions. We now consider the energy due to the classical dipolar interactions which are much less weaker than the exchange interactions in this amorphous ferrimagnet.

In order to get the total dipolar energy we have to sum up for all atomic pairs in the amorphous structure. Let the probability of finding an A or B atom at the point r from an atom A be $P_{AA}(r)$ or $P_{AB}(r)$. Using these atomic distribution functions the averaged dipolar interaction energy between A–A pairs is written as follows,

$$U_{AA} = \int_0^{2\pi} \int_0^{\pi} \int_0^{\infty} u_{AA}(r) P_{AA}(r) r^2 \sin\theta \, dr \, d\theta \, d\varphi .$$
(8.9)

The same expressions are obtained for B–B and A–B atomic pairs.

The magnetic dipolar energy per unit volume is written as

$$E(\alpha, \beta, \gamma) = (1/2)[N_A(U_{AA} + U_{AB}) + N_B(U_{BB} + U_{BA})] \ , \tag{8.10}$$

where α, β and γ are direction cosines of the magnetic moment. If $P_{IJ}(r)$ are isotropic, E becomes independent of the direction of the magnetic moments. We assume anisotropy in the radial distribution functions and expand them using spherical harmonics as follows

$$P_{IJ} = (N_J/N)R_{IJ}(r)[1 + p_{IJ}(r)(3\cos^2\theta - 1) + \cdots] \ , \tag{8.11}$$

where $R_{IJ}(r)$ is the isotropic radial distribution function, and θ is the angle between the atomic pairs and the normal to the film plane.

The parameters p_{IJ}, which express the order of anisotropy of the atomic distribution, can be important only for nearest neighbor pairs in the amorphous phase where A and B atoms are mixed together on a microscopic scale. Let us put $p = -p_{AB} = -p_{BA}$ as an anisotropy parameter for the atomic distribution of different kinds of atoms, and Z as a number of nearest neighbor atoms. If $p > 0$, atomic pairs of unlike atoms dominate in the direction of the film plane over the normal to the plane. $p_{AA} = (x_B/x_A)p$ and $p_{BB} = (x_A/x_B)p$ can be derived when the ratio of number of A and B atoms among nearest neighbor atoms is equal to x_A/x_B.

The magnetic anisotropy energy K_u can be obtained as a coefficient in a term containing $(1 - \cos^2\theta)$. Finally we get an expression of the anisotropy energy as follows,

$$K_u = (3/5ND^3)pZx_Ax_B(|M_A/x_A| + |M_B/x_B|)^2 \ , \tag{8.12}$$

where D is the distance between nearest neighbor atomic pairs, and N is the number of atoms per unit volume.

In ferrimagnetic materials the total magnetization is the difference between the subnetwork magnetizations, $M = |M_A - M_B|$, while the anisotropy energy K_u is proportional to $(|M_A/x_A| + |M_B/x_B|)^2$ which is larger than M^2. Therefore $Q = K_u/2\pi M^2 > 0$ can be realized even for a small anisotropy of atomic distributions $(p > 0)$. This must be one of the origins of the large magnetic anisotropy in amorphous ferrimagnetic films making them perpendicular magnetic films.

The situation is easy to understand if one looks at the characteristics of the dipolar interaction. Let us think about a microscopic atomic configuration, for example, as shown in Fig. 8.4. Surrounding a central atom, like atoms prefer to sit at up and down sites and unlike atoms sit inside by side positions in the film plane, which correspond to the anisotropic atomic distribution of $p > 0$. If the magnetic moments are aligned perpendicular to the film plane as shown in (a), the directions of the magnetic moments of surrounding atoms coincide with the component of the dipolar field produced by the central atom in this atomic configuration. On the other hand, if the magnetization lies in the plane of the film as shown in Fig. 8.4b the directions of the magnetic moments of the surrounding atoms are opposite to the component of the dipolar field produced

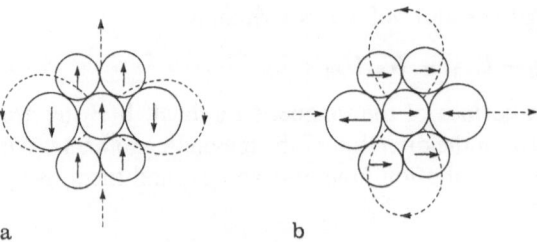

a   b

Fig. 8.4a,b. Local atomic arrangements and the magnetic dipolar interaction in an amorphous ferrimagnet. **a** Magnetic moments perpendicular to the film plane. **b** Magnetic moments in the film plane

by the central atoms. Therefore, it is clear that the dipolar interaction energy in this system is lower when the magnetization is perpendicular to the film plane than when it is in the plane. The ferrimagnetic arrangement promotes this situation. Perpendicular anisotropy can be expected from this dipolar mechanism if there are more like-atom pairs than unlike-atom pairs in the perpendicular direction and the opposite pair distribution holds in the parallel direction to the film plane.

Now how much anisotropy of atomic pair distribution is necessary in order to explain the actually observed perpendicular magnetic anisotropy? For example, amorphous $Gd_{14}Co_{12}Mo_{11}Ar_3$ alloy films produced by bias sputtering have a perpendicular magnetic anisotropy energy of 8×10^4 erg/cm^3 at 0 K. If this observed magnetic anisotropy is entirely attributed to the dipolar mechanism there must be an anisotropy of $p = 0.01$ in atomic pair distribution, that is, only 1% of the anisotropy of nearest neighbor pairs of unlike-atoms causes the appreciable amount of magnetic anisotropy. It is impossible to detect this slight structural anisotropy directly by diffraction methods, but the magnetic anisotropy is so sensitive to the anisotropy of atomic pair distributions that sufficient magnetic anisotropy arises by this mechanism.

8.2.2 Microscopic Local Magnetic Anisotropy

In this section we consider the microscopic local magnetic anisotropy which each magnetic moment feels in an amorphous structure. This microscopic anisotropy is appreciable especially for non-S-state rare earth metal ions in amorphous alloys.

In a single crystal the local magnetic anisotropy for each atomic site appears directly as a macroscopic crystalline anisotropy, while in an amorphous phase the directions of the local anisotropy axes can differ from site to site so that it is hard to detect them by macroscopic measurements as they are averaged out. There are, however, some phenomena which are caused by the microscopic local anisotropy. For example, the coercive force in these systems becomes quite high at low temperatures, and the magnetization varies gradually with time.

For a more qualitative discussion, let us consider the response of the magnetization to an external field, that is, the magnetization process. According to phenomenological theories, the magnetization, M, in an external field, H, is written in the following form,

$$M^2 = b(H/M) + c \ . \tag{8.13}$$

Using appropriate exchange constants it is possible to calculate the magnetization in the external magnetic field at finite temperatures by the molecular field theory. For amorphous $Tb_x Fe_{1-x}$ ($x = 0.64$ and 0.74), the calculated value of the coefficient b was about 2–3 times larger than the observed value in experiments. This means that the observed change of magnetization in the external field is much less than the value calculated by molecular field theory with appropriate exchange constants. This discrepancy may be attributed to the pinning effect of Tb magnetic moments by the local magnetic anisotropy.

In order to deal with this mechanism quantitatively we introduce a simple uniaxial local anisotropy energy for rare earth metal atoms, $-D(n_i \cdot J_i)^2$, where J_i is the angular momentum and n_i is a unit vector in the easy direction of the local anisotropy axis for the i-th atom in an amorphous alloy. For simplicity we take D as a constant and only the directions of n_i are randomly distributed from site to site. It should be noted that the alignment of the J_i either parallel or antiparallel to n_i will serve to minimize the local anisotropy energy. In amorphous alloys with $3d$ metals, Fe or Co, and heavy rare earth metals the exchange energy is lowered when the magnetic moments of $3d$ atoms and those of rare earth metals couple antiparallel since the exchange constant between them is negative. In the system in which both the exchange interaction and microscopic local anisotropy coexist, the magnetic moments of rare earth metal atoms cannot align parallel to each other due to the local microscopic anisotropy, however, their directions fall in a hemisphere whose axis is antiparallel to the subnetwork magnetization of $3d$ atoms.

The energy for the i-th rare earth atom in this system may be written as follows,

$$E_i = -(B_{RE} - B_{TM} + B_0)g\mu_B J_i - D(n_i \cdot J_i)^2 \ , \tag{8.14}$$

where B_{RE} and B_{TM} are the molecular fields from the rare earth metal and $3d$ transition metal subnetworks, respectively, and B_0 is the external field. Taking the z axis as the direction of the external field or the net magnetization we define angles θ_J and θ_n as those between J_i and z, and n_i and z, respectively. The second term in the above equation can be written as $-DJ^2 \cos^2 \phi$, where ϕ is the angle between J_i and n_i. The thermal average of the z component of J_i, $\langle J_i \cos \theta_J \rangle$, at temperature T is written as follows,

$$\langle J\cos\theta\rangle = \frac{\int\limits_0^{2\pi}\int\limits_0^{\pi/2} J\cos\theta_J\exp[-E_i(\theta_J,\phi,\theta_n)/kT]\,d\Omega_J}{\int\limits_0^{2\pi}\int\limits_0^{\pi/2}\exp[-E_i(\theta_J,\phi,\theta_n)/kT]\,d\Omega_J}\,, \tag{8.15}$$

where the integral over the solid angle of the J_i directions should be performed for a hemisphere of $+z$. This thermal average of the z component of the J_i differs from site to site depending on the directions of the n_i. For the molecular field approximation we need the averaged subnetwork magnetization of rare earth metals, which is obtained by the integration over solid angles of the direction of the n_i, that is,

$$\langle J_R\rangle = (J/4\pi)\int\langle\cos\theta_J\rangle_i\,d\Omega_n\,. \tag{8.16}$$

In practice the calculation can be performed by successive approximations. We can get a self-consistent solution when the procedure becomes convergent. The values of D which result in the closest agreement with experiment are 4.0×10^{-15} and 9.6×10^{-15} erg for $x=0.64$ and 0.72 in the Tb_xFe_{1-x} amorphous alloys [8.5]. These magnitudes seem to be reasonable compared to the crystalline anisotropy energy of Tb, 1.8×10^{-14} erg, in a single crystal of Tb. In this way we can estimate the microscopic local magnetic anisotropy which is hidden by cancellation in amorphous alloys.

8.3 Magnetism, Preparation Conditions and Structural Relaxation of Amorphous Alloys

An important point about amorphous solids is that they are in a nonequilibrium phase. They are possibly even in different states at the same temperature and pressure with the same composition depending on their preparation conditions and history because they are not in a uniquely determined equilibrium state which has the lowest free energy. By thermal annealing at relatively low temperatures, even at room temperature in some cases, structural relaxation can take place which causes changes in certain kinds of properties with time. The disaccommodation effect of initial permeability in amorphous magnetic materials is an example which is important in practical applications. Here we introduce the results of an experimental study on structural relaxation characteristic in amorphous alloys by measuring the Curie temperature, which is an intrinsic quantity of magnetic materials.

In order to get an amorphous phase a liquid alloy must be cooled with a quenching rate which is fast enough to suppress the nucleation or growth of the crystalline phase. Through this process a quasistable supercooled liquid state is frozen at a certain temperature into a fixed amorphous solid state. This freezing temperature depends on the cooling rate of the quenching process. Thus amorphous alloys at ambient temperature are in one of various possible nonequilibrium states, which correspond to the quasiequilibrium states at the different freezing temperatures. This is a reason why amorphous alloys can differ slightly depending on the preparation conditions.

The physical properties of amorphous alloys are then affected by the preparation conditions. For example, the Curie temperature of amorphous alloys prepared from an identical ingot differs according to the quenching rate. It also changes appreciably on thermal annealing. The experimental results for amorphous $(Fe_{1-x}Ni_x)_{80}B_{10}Si_{10}$ are given in the following [8.10].

There have been few direct measurements of the real cooling rate of the rapid quenching process. We have measured the temperature of a ribbon specimen on a disk rotating at high speed by using an infrared thermometer and a small mirror which was tilted by a pulse motor in order to measure three different points on the disk during a single rapid quenching process. We then knew the temperature drop on the rotating disk as a function of distance from a nozzle which ejected the molten alloy. The faster the disk rotated the thinner was ribbon and the higher the cooling rate attained. It was concluded from the experimental results that an amorphous phase of this alloy was obtained for faster cooling rates than 2.4×10^6 K/s at 500 °C and an average rate of 5×10^5 K/s from the melting point to 500 °C.

The Curie temperature of these amorphous alloys which had the same composition but were prepared with different cooling rates was measured carefully. It turned out that the thinner ribbon which was fixed from the supercooled liquid state at higher temperatures with a faster cooling rate had a lower Curie temperature. The difference between those of 20 μm and 70 μm thick ribbons was as much as 3 degrees.

What happens when the amorphous alloys are annealed at lower temperatures than the freezing temperature. In this case they approach a quasiequilibrium state with a higher T_C because it corresponds to a state at a lower temperature than the freezing temperature. The Curie temperature of this amorphous alloy indeed increased over 30 °C after annealing at 250 °C for about one month. This is clear evidence of variation of the state during annealing, which is different from the onset of crystallization. Once it begins to crystallize a phase separation occurs and T_C decreases drastically.

The Curie temperature changes linearly with the logarithm of isothermal annealing time as follows,

$$T = p \log(t_a/t_0) , \tag{8.17}$$

where t_0 is a kind of incubation time which is longer for lower temperatures. It can be expressed with an activation energy E as follows,

$$t_0 = t_1 \exp(E/k_B T) . \tag{8.18}$$

For this amorphous alloy t_1 is 3.0×10^{-12} s, and $E = 1.25$ eV.

The slope p in the above expression and the saturation value of T_C decrease with increasing annealing temperature. In Fig. 8.5 the variation of T_C after

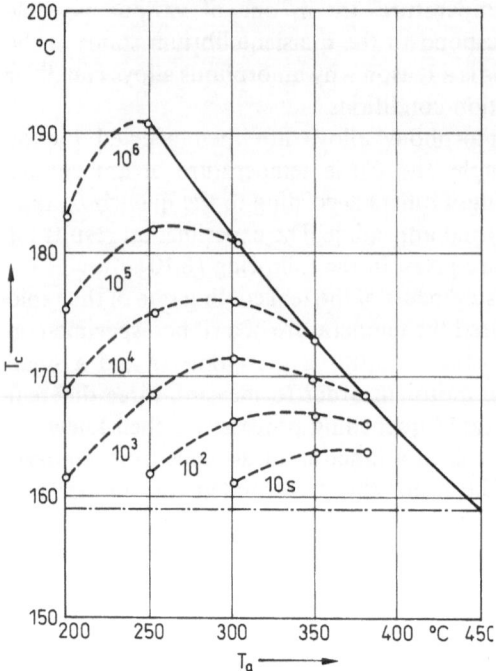

Fig. 8.5. The dependence of the Curie temperature, T_C, of an amorphous $Fe_{25}Ni_{55}Si_{10}B_{10}$ alloy on annealing at constant temperature, T_a. The horizontal dash-dotted line represents T_C of the as-quenched sample from the liquid state. Broken lines connect data points for the same annealing time at different annealing temperatures. The solid line represents T_C for samples in which structural relaxation is considered to be completed after long enough annealing at T_a [8.10]

annealing for 10 s to 10^6 s at various temperatures is shown. A solid line in the figure shows the saturation value after sufficiently long annealing at various temperatures, which corresponds to that in a fully relaxed state at each annealing temperature.

The extrapolation of this solid line approaches to the T_C of the as-quenched alloy at 460 °C. This means that the as-quenched amorphous phase corresponds to the fully relaxed state at 460 °C, that is, the quasi-equilibrium super cooled liquid state at 460 °C is fixed into the nonequilibrium amorphous state by rapid cooling.

The question now is what is the structural relaxation of the amorphous alloys which induces the change of the Curie temperature. The first possibility is that microscopic free volume is eliminated by the thermal annealing. Careful measurement of the amorphous alloy before and after the heat treatment showed little change in the density and poor correlation with the Curie temperaure. Egami interpreted the structural relaxation in terms of atomc level stresses which are relaxed by a small displacement of the atoms.

The Curie temperature is mainly determined by the exchange interactions between nearest neighbor Fe–Ni, Fe–Fe and Ni–Ni pairs. The composition dependence of T_C suggests that the Fe–Ni pair interaction gives the predominant positive contribution among them. The elevation of T_C is, therefore, interpreted as an increase in the relative number of Fe–Ni nearest neighbor pairs, that is, a kind of chemical short range ordering in the amorphous alloy. In the amorphous phase, the structural relaxation takes place appreciably even at relatively low temperatures since there is considerable freedom of atomic positions thanks to the lack of long range periodicity.

As shown in Fig. 8.5, T_C attains a higher value by thermal annealing at lower temperatures. It is an interesting question whether an intermediate state can be uniquely specified by the single parameter T_C. For example a state after annealing for 7×10^4 s at 200 °C has the same T_C as the fully relaxed state at 350 °C after long annealing. Can we assume that both states are identical? If this is the case then the former state should show no change on further annealing at 350 °C. Experiment shows that T_C decreases at first and then recovers its initial value.

Greer [8.7] explained this behavior as follows. There is a distribution of processes for the structural relaxation in amorphous alloys from fast processes to slow ones. First an as-quenched alloy shows an intermediate unsaturated T_C by annealing at T_1 for a time in which the fast processes are completed while the slower ones are still not. Then when the annealing temperature is raised to T_2, the T_C decreases first because only the fast processes bring the local configuration to the state of T_2 which has a lower T_C. Later the whole state approaches the fully relaxed state of T_2 includdng the contribution by the slower processes, and T_C is recovered to its saturated value.

As shown in this case magnetic measurements can be effective for investigating nonequilibrium states. The logarithmic time dependence of the overall relaxation process is considered to be a characteristic behavior of the system with a wide distribution of relaxation times, which comes from the many degrees of freedom present in the amorphous phase.

We have considered the Curie temperature of amorphous alloys in relation to structural relaxation, but magnetic anisotropy is also quite sensitive to the relaxation process. The disaccommodation effect in amorphous magnetic materials is induced by the stabilization of magnetic domain walls through self-induced local magnetic anisotropies.

References

Detailed references can be found in the review articles [8.1–5]

8.1. T. Mizoguchi, K. Yamauchi, H. Miyazima: *Amorphous Magnetism of Transition Metal Alloys*, Int. Conf. on Magnetism (1973) TOM 2, 54

8.2.   T. Mizoguchi: *Magnetism in Amorphous Alloys*, AIP Conf. Proc. (1976) 286–291.

8.3.   T. Mizoguchi, G.S. Cargill III: *"Magnetic Anisotropy from Dipolar Interactions in Amorphous Ferrimagnetic Alloys"*, J. Appl. Phys. **50**, 3570–3582 (1979)

8.4.   *Rapidly Quenched Metals*, ed. by T. Masumoto, K. Suzuki (Japan Inst. of Metals, 1982)

8.5.   S. Hatta, T. Mizoguchi, N. Watanabe, Proc. 5th Int. Conf. Rapidly Quenched Metals (Elsevier, 1984)

8.6.   S. Ohnuma, K. Shirakawa, M. Nose, T. Masumoto: IEEE Trans. Mag. MAG-**16**, 1129 (1980)

8.7.   A.L. Gree, J.A. Leake: J. Non-cryst. Solids **33**, 291 (1979)

8.8.   T. Mizoguchi: AIP Conf. Proc. **34**, 268 (1976)

8.9.   R. Hasegawa, B.E. Argyle, L.J. Tao: AIP Conf. Proc. **24**, 110 (1974)

8.10.   T. Mizoguchi, H. Kato, N. Akutau, S. Hatta: Proc. 4th Int. Conf. on Rapidly Quenched Metals (Sendai, 1981) p. 1173

8.11.   J.D. Axe, G. Shirane, T. Mizoguchi and K. Yamauchi: Phys. Rev. B**15**, 2763 (1977)

8.12.   T. Mizoguchi, T.R. McGuire, S. Kirkpatrick and R.J. Gambino: Phys. Rev. Lett. **38**, 89 (1977)

8.13.   P. Chaudhari, J.J. Cuomo and R.J. Gambino: IBM J. Res. Dev. **17**, 66 (1973)

9. Amorphous Magnetic Alloy Ribbons and Their Application

YOSHIMI MAKINO

Amorphous materials are solids such as glass which have a randomly close-packed structure of atoms and/or molecules. The amorphous state is also called the "glass state." All amorphous magnetic materials are artificial: no natural magnetic glass is known to exist.

Amorphous magnetic alloys are composed of $3d$-transition metals and/or $4f$-rare earth metals with usually a metalloid element or elements added to stabilize the amorphous state. Amorphous magnetic alloys are prepared by rapidly solidifying these elements from a gas, ion or molten state which are the ideal random states at the atomic level.

The quenching rate in the solidifying process must be between 10^4 and 10^6 degrees per second. Such a high quenching rate is the major reason no natural magnetic glass exists and the reason amorphous alloys can only be made into powders, slender wires, thin films or thin ribbons.

There are three kinds of amorphous magnetic alloy:

1. T-M system: Consisting of $3d$-transition metals and a metalloid element or elements such as B, Si, C, P, Zr, Nb, Ta, or Ti for stabilizing the amorphous state.
2. R-T system: Consisting of rare earth metals and $3d$-transition metals. These materials can only be made into thin films.
3. Oxide system: The oxides of the T-M system such as a mixture of ferrites and pentaoxiphosphides.

In this chapter, amorphous ribbons of T-M system magnetic alloys will be reviewed and their preparation methods, characteristics, heat treatments and applications, for which these materials are the most advanced presently available, will be discussed.

The R-T system will be treated in the next chapter and the oxide system has not yet reached the stage of application.

9.1 Materials

9.1.1 Preparation Methods

Amorphous alloy ribbons are prepared by the so-called continuous rapid quenching method. In this method the molten alloy is ejected from a narrow nozzle onto the surface of a rotating roller and is solidified rapidly at a quenching rate of 10^5 to 10^6 degrees per second. Figure 9.1a–c shows the three basic methods of preparation.

The fundamental requirement for making amorphous ribbons of good quality is a high quenching rate. The quenching rate depends on the thermal conductivity and rotating speed of the rollers, the ejection rate, and the thermal conductivity and viscosity of the molten alloy in the cases of (a) and (c), plus the contact pressure of the rollers in the case of (b). The critical quenching rate of almost amorphous magnetic alloys in the T-M system is 10^5 to 10^6 degrees per second and the theoretical upper limit of thickness estimated from the thermal conductivity of the alloy is 100 µm. The thickness and width of the ribbon prepared by the single roller method is 20 to 40 µm and 100 to 200 mm, respectively. A photograph of the ribbon is shown in Fig. 9.2.

9.1.2 Classification of Amorphous Magnetic Alloys

The amorphous magnetic alloy compositions for practical use are Fe, Co and Ni with Si and B [9.1]. For example, the magnetic properties of the pseudo-ternary alloys $(Fe–Co–Ni)_{78}Si_8B_{14}$ are shown in Fig. 9.3. These alloys are essentially soft magnetic materials since no magnetic crystalline anisotropy influences their magnetic properties, and the magnetostriction constant λ_s plays an important

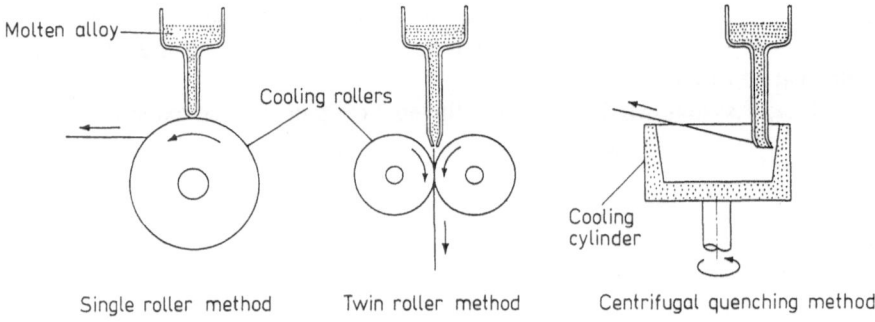

Molten alloy

Cooling rollers

Cooling cylinder

Single roller method   Twin roller method   Centrifugal quenching method

Fig. 9.1. The three basic methods of preparing amorphous alloy ribbons

Fig. 9.2. Photograph of some amorphous alloy ribbons prepared by the single roller method

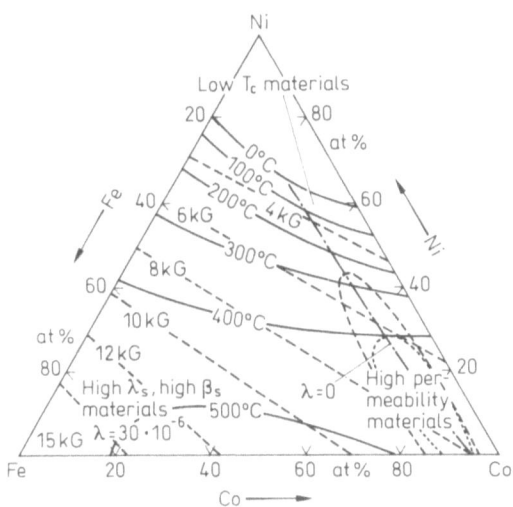

Fig. 9.3. Saturation magnetic induction B_s and Curie temperature T_C of the pseudoternary alloys $(Fe–Co–Ni)_{78}Si_8B_{14}$

role in their magnetic properties. The λ_s changes sign from negative to positive with an increase of Fe content. The saturation induction B_s and the Curie temperature T_C increase with the addition of Fe and/or Co. From these magnetic properties and their fields of application, these pseudo-ternary alloys are classified into three large groups, that is, iron-based high B_s and high λ_s materials; cobalt-based high permeability materials with $\lambda_s = 0$; and nickel-based low B_s and low T_C materials.

9.1.3 Characteristics and Shortcomings

The major characteristics and shortcomings of amorphous magnetic alloys which are directly attributable to the alloys' non-crystalline origin will be described below. A non-crystalline structure is slightly different from a perfectly random structure as a liquid, because it has a short range order of atoms. This short range order gives variety to amorphous structures, and has various influences on the structure-sensitive magnetic properties through the heat-treatment and mechanical processes described later. However, the characteristics and shortcomings of the alloys generally reflect the lack of long range order in their structure.

a) Characteristics

i) Non-magnetocrystalline anisotropy: The magnetocrystalline anisotropy associated with crystal symmetry does not appear in the amorphous state. This unique characteristic is closely related to high permeability μ and small coercive force H_c.

ii) Easy choice of zero λ_s compositions: Crystalline materials with high μ and zero λ_s are limited to compositions having very little magnetocrystalline anisotropy. However, because materials in the amorphous state have no magnetocrystalline anisotropy, it is possible to choose easily zero λ_s compositions without considering the magnetocrystalline anisotropy.

iii) High electrical resistivity: The amorphous state has 4 to 5 times greater electrical resistivity ρ_0 than the crystalline state and this high ρ_0 reduces the eddy current loss.

iv) High hardness and high stiffness: The amorphous state has no slip planes for deformation in contrast to the crystalline state. This property results in amorphous materials with a hardness and stiffness which exceeds ferrite, permalloy and sendust. Thus, the amorphous alloys can be machined with high precision.

v) No grain boundaries: The amorphous state does not contain any grains or grain boundaries. This suppresses the tipping and the magnetic noise at the grain boundaries.

vi) Less corrosion: The addition to the amorphous alloys of small amounts of Cr and P, for example, increases the anti-corrosion property without suppressing the useful magnetic properties.

b) Shortcomings

i) Crystallization: All of the forming processes for device applications must be confined to below the crystallization temperature T_x in order to avoid recrystallization of the amorphous material. This is the most fundamental shortcoming for the application of these materials.

ii) Induced magnetic anisotropy: The permeability of amorphous magnetic materials, especially Co-based alloys, is degraded by an induced magnetic anisotropy generated by heat and/or mechanical treatment. Moreover, the casting process induces magnetic anisotropy so that as-cast amorphous alloys of zero λ_s show low permeability. Though induced magnetic anisotropy has also been reported in crystalline materials, that of amorphous materials is more severe because the short range order in the amorphous state allows the structure to be easily relaxed. Structural relaxation appears in the temperature range of 100 to 300 °C below T_x.

9.1.4   Various Heat Treatments

Generally, magnetic alloys require a suitable heat treatment in order to acquire the desirable magnetic properties required for their application. Particularly for amorphous magnetic alloys, heat treatment is indispensable to stabilize the magnetic properties and control the induced magnetic anisotropy which originates from structural relaxation.

a) Heat treatment for iron-based amorphous magnetic alloys

Iron-based amorphous alloys are mainly applied to magnetostrictive devices and the magnetic cores of power transformers, because of their high λ_s and high B_s.

For materials to be applied to magnetostrictive devices, the heat treatment is carried out considering the residual internal stress and the structural relaxation. The temperature of the heat treatment is chosen over a wide range from just below the crystallization temperature to emphasize the ΔE effect (Appendix A.6) to just above 200 °C for the purpose of structural relaxation. These heat treatments are generally carried out in inert gases such as argon or nitrogen, excepting the special case of a stress-detector application which is annealed in air at 200 °C.

For the cores of power transformers, the heat treatment is given to the cores after wiring in order to reduce the residual inner stress and core loss. The atmosphere of annealing is an inert gas to avoid oxidization and hydrogenation. Magnetic-field annealing is performed by supplying dc or ac currents to the wire of the core at a temperature between 350 and 450 °C. As an example, the result of the heat treatment of two different magnetic cores is shown in Fig. 9.4. As can be seen in this figure, low supplied power means low core loss.

b) Heat treatment for cobalt-based amorphous magnetic alloys

Cobalt-based amorphous magnetic alloys are high permeability materials with zero magnetostriction and are used for electronic devices such as magnetic recording heads and high-frequency transformers. This alloy system is subject to heat treatment mainly to control induced magnetic anisotropy.

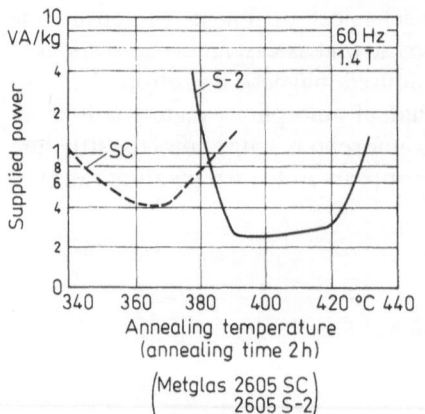

Fig. 9.4. The annealing temperature dependence of supplied power for typical amorphous magnetic cores for pole transformer use

Two methods of eliminating induced magnetic anisotropy have been reported. One is called the rapid quenching method, and is carried out by rapidly quenching from the annealing temperature T_a between the Curie temperature T_C and the crystallization temperature T_x. The other is the rotating magnetic-field annealing method, and is carried out by annealing the sample in a rotating magnetic field and/or rotating the sample in a static magnetic field at a temperature below T_x.

In cobalt-based amorphous alloys with zero magnetostriction, an increase of (Co–Fe) content increases B_s. On the other hand, an increase in the transition metal content raises T_C and lowers T_x. Rapidly quenching from $T_a > T_C$ is useful but is not possible for high B_s materials when $T_C > T_x$. Rotating magnetic field annealing when $T_a < T_x$ is more complex than the rapid quenching method, but the relation between T_C and T_x is not critical and it can be applied to samples which are difficult to quench rapidly such as thin films on a substrate and wired cores. Typical examples of each method are discussed below.

c) Rapidly quenching method

The quenching temperature dependence of the permeability for the amorphous alloy $Fe_{4.7}Co_{70.3}Si_{15}B_{10}$ is shown in Fig. 9.5. The frequency dependence is also shown. From this figure, it is clear that the permeability is improved by quenching from a suitable annealing temperature between T_C and T_x. This improvement of permeability is caused by eliminating the induced magnetic anisotropy generated by the heat and/or mechanical treatment [9.2.3]. However, the permeability of the sample degrades with aging at temperatures below T_C, as can be seen in Fig. 9.6. This degradation of permeability is quite marked in

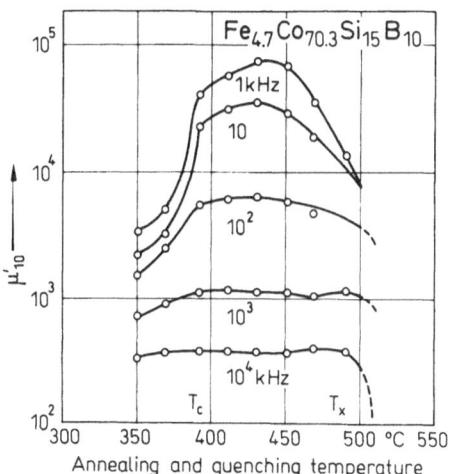

Fig. 9.5. Annealing and quenching temperature dependence of the permeability of $Fe_{4.7}Co_{70.3}Si_{15}B_{10}$

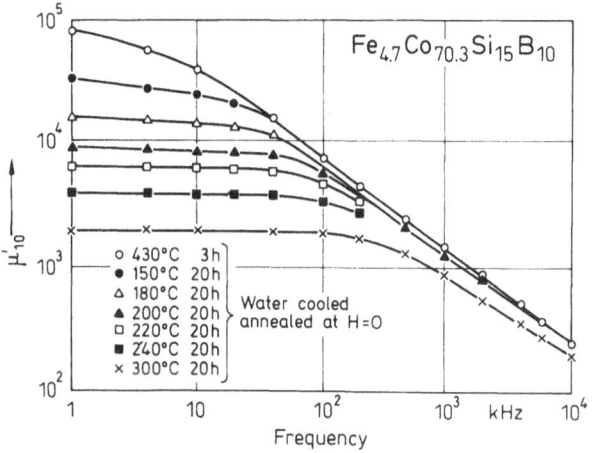

Fig. 9.6. Frequency dependence of the permeability $Fe_{4.7}Co_{70.3}Si_{15}B_{10}$ after annealing at various temperatures

the low frequency range up to 100 kHz when the aging temperature is below 200 °C, and spreads to the high-frequency range above 1 MHz with higher aging temperatures. The degradation is caused by the induced magnetic anisotropy regenerated by the aging, and the previous level of permeability can be recovered by reapplying the rapid quenching method. The regeneration of the induced magnetic anisotropy can also be used to suppress the disaccommodation of the permeability. The induced magnetic anisotropy also flattens the frequency dependence of the permeability, as shown in Fig. 9.6.

d) Rotating magnetic field annealing

Figure 9.7 shows the frequency dependence of the permeability of the amorphous magnetic alloy $Fe_5Co_{75}Si_4B_{16}$ with a T_C of 600 °C and a T_x of 420 °C, where the dotted and solid lines indicate before and after rotating magnetic field annealing [9.4, 5]. The rotating speed dependence of the permeability for the same sample is shown in Fig. 9.8. The permeability increases with increasing rotation speed R, and is proportional to the square root of R with a cusp-shaped period. In order to explain these characteristic curves, Makino et al. proposed a model with the following assumptions: that the magnetic field is applied to the material in a given direction; that the uniaxial anisotropy will be induced along the field direction and the anisotropy energy increases and finally reaches a saturated value K_i^∞ after a certain time τ_0; that both K_i^∞ and τ_0 are determined by the temperature of annealing and the materials; that the growth and decay speed of the induced magnetic anisotropy is constant.

The resultant induced magnetic anisotropy energy K_i is calculated by adding the uniaxial magnetic anisotropy energy generated in units of time of a magnetic field rotation, and is given as:

$$K_i = (\sin(2\pi\tau_0 R/2)\pi\tau_0 R)K_i^\infty \ . \tag{9.1}$$

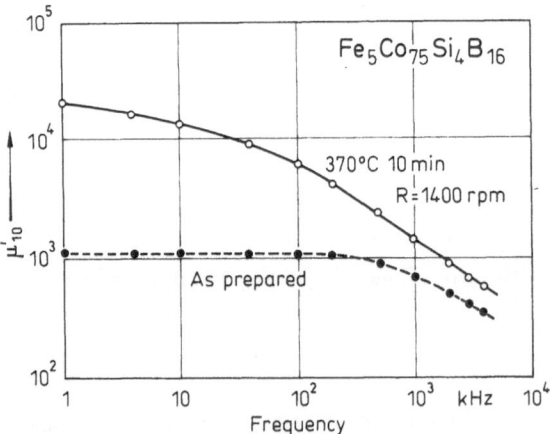

Fig. 9.7. Frequency dependence of the permeability of $Fe_5Co_{75}Si_4B_{16}$ before and after rotating magnetic field annealing

On the other hand, the permeability is proportional to the inverse square root of the anisotropy energy K_i in the magnetic domain wall-displacement magnetization process. In this case, the relation between μ and R is given as

$$\mu \propto (2\pi\tau_0 R/\sin 2\pi\tau_0 R)^{1/2}(K_i^\infty)^{-1/2} . \tag{9.2}$$

This formula suggests that the effect of the rotation magnetic annealing is characterized by the term $(2\pi\tau_0 R/\sin 2\pi\tau_0 R)^{1/2}$.

Figure 9.9 shows the plot of this characteristic term. This plot strongly resembles the plot of the experimental result in Fig. 9.8.

From the experimental results and the theoretical analysis described above, it is clear that magnetic induced anisotropy is uniformly dispersed in the rotating plane by the rotating magnetic field annealing, and as a result eliminated. A variation of this heat treatment in which a static magnetic field is superimposed perpendicular to the rotating plane has been reported to be a useful method for treating magnetic recording head cores during the assembly process.

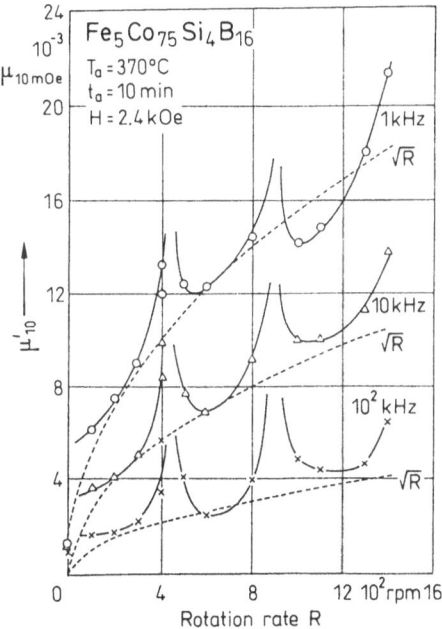

Fig. 9.8. Dependence of permeability on rotation rate for $Fe_5Co_{75}Si_4B_{16}$ after rotating magnetic field annealing

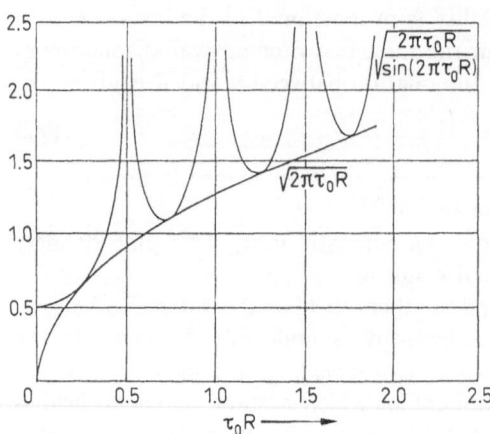

Fig. 9.9. Plot of equation (9.2)

9.2 Applications

Various applications for amorphous magnetic alloy ribbons were proposed in the 1970s when the continuous rapidly quenching method was developed. By the early 1980s, amorphous magnetic alloy ribbons were being used in a number of electronic devices. In addition, the ribbons were finding applications in electronics and measuring instruments.

Table 9.1 lists some appliances and devices using amorphous magnetic materials which are already on the market or in trial production. Photographs of these articles are shown at the end of this chapter. In this section, some of these articles will be described in some detail.

The MM cartridge with amorphous magnetic cores was the first product. Figure 9.10 shows the construction of the cartridge. The pick-up needle traces the groves of a disk, and the moving magnet is vibrated by the phonographic signals recorded in the grooves through the cantilever. The magnetic core transforms the mechanical signals of the magnet into electrical signals. Permalloy has often been used as a core, but this alloy has several problems, such as its relatively poor output and phase frequency characteristics, and its high noise level. Figure 9.11 compares the output frequency characteristics of an amorphous core with those of a permalloy one. The characteristics of the amorphous core are flat throughout the audible frequency· range. Figure 9.12 shows the phase-frequency characteristics of the two types of cores. The amorphous core has almost no phase delay over the entire frequency range. Such output and phase-frequency characteristics reduce signal distortion as is well known. Figure 9.13 shows the signal-to-noise ratio in the audible frequency range of the two

Table 9.1. Applications of amorphous magnetic alloy ribbons

1. Power Transformer	(on the Market)
2. Switching Regulator	(on the Market)
3. Magnetic Recording Heads	
for Audio	(on the Market)
for Video	(on the Market)
for Computer	(in trial production)
for Duplicator	(in Use)
for Height Gauge	(on the Market)
4. Audio Devices	
MM-Cartridge	(on the Market)
Dynamic Microphone	(on the Market)
Transformer for MC-Cartrige	(on the Market)
5. Mavica	(on the Market)
6. Data Tablet	(on the Market)
7. Go Traning Machine	(on the Market)
8. Fluorescent Lamp of Bulb Type	(on the Market)
9. Electric Spark Killer	(on the Market)
10. Various Magnetic Cores	(on the Market)

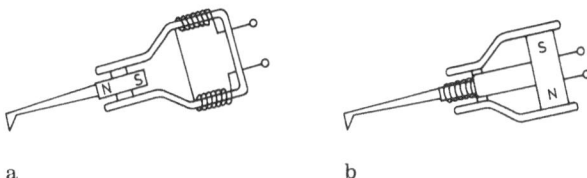

a   b

Fig. 9.10a,b. Construction of cartridges. **a** MM type; **b** MC type

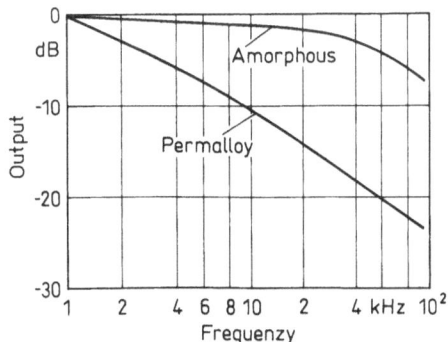

Fig. 9.11. Output-frequency characteristics of the magnetic cores for MM cartridge use

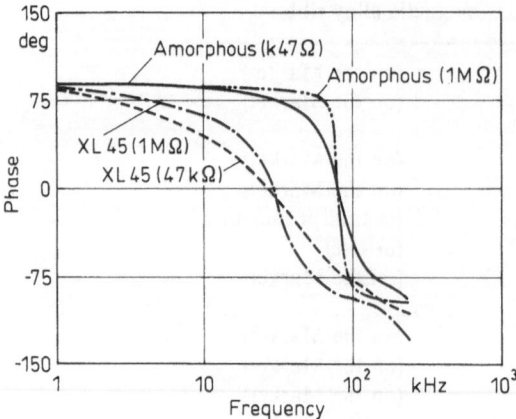

Fig. 9.12. Phase-frequency characteristics of the magnetic cores for MM cartridge use. Permalloy cores are used in XL45.

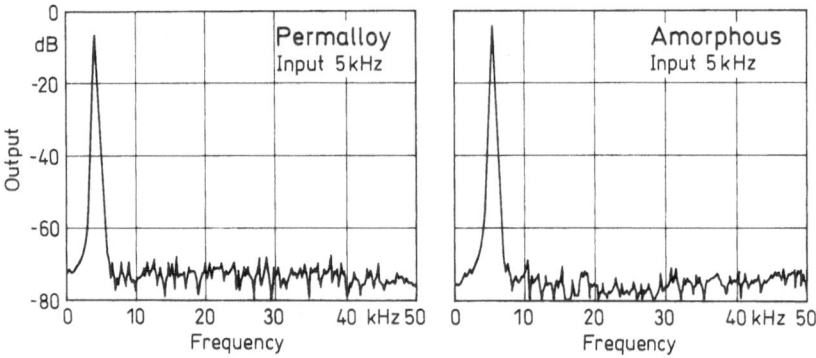

Fig. 9.13. Signal to noise ratio of audio frequency on the magnetic cores for MM cartridge use

types of cores. The noise level of the amorphous core is about 5 dB lower than that of the permalloy core.

Of all the applications of amorphous magnetic materials, the most advanced is in magnetic heads. Amorphous audio heads have been commercially manufactured and used for high and middle-quality tape recorders since 1981, and since 1985 in video heads for 8 mm video recorder use.

Special amorphous heads for duplicators of music tapes and measuring instruments such as height gauges are also now on the market, but they are still relatively few in number. Multi-channel magnetic heads for computer use are currently in trial production. Iron-based amorphous alloys are considered to be the best candidate for use in audio heads of inexpensive tape recorders manufactured in great numbers because of their low material cost. Because these alloys have a high λ_s, however, the strain and stress must be eliminated in the process of head production.

More research on materials of $\lambda_s < 10^{-6}$ with high B_s is required. Cobalt-based amorphous alloys are well balanced as head materials although the

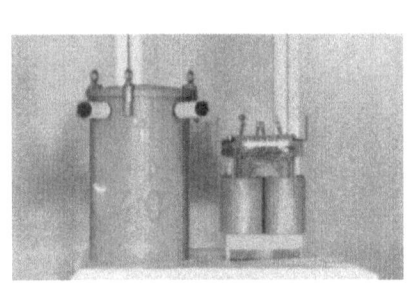

a

b

c

d

e

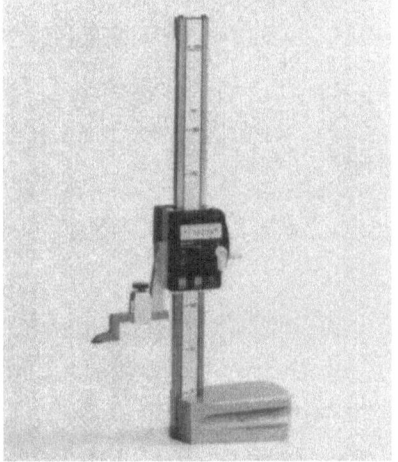

f

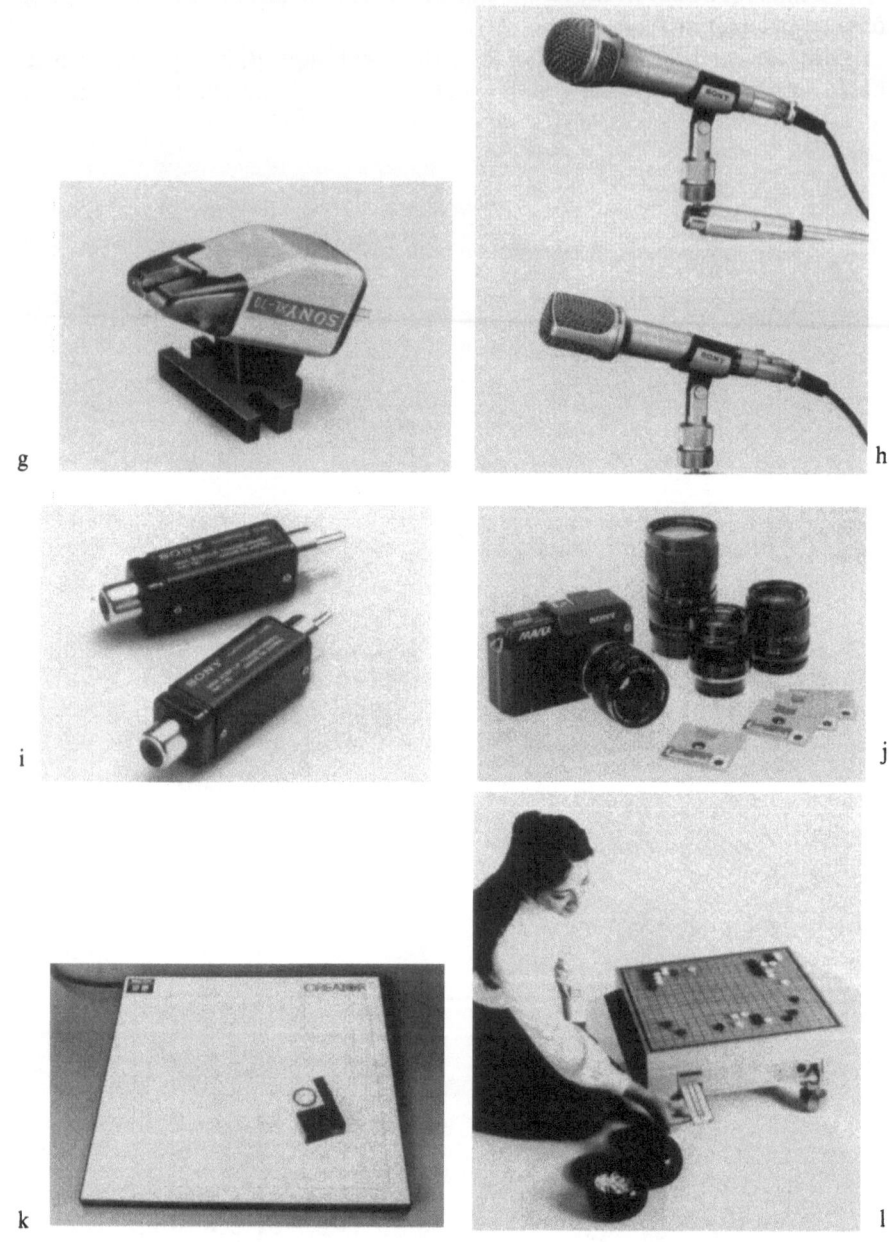

g

h

i

j

k

l

Fig. 9.14(a–l)

m

o

Fig. 9.14a–o. Photographs of the articles listed in Table 9.1. **a** Pole transformer; **b** Switching regulator; **c** Audio-magnetic recording heads; **d** Magnetic heads for computer use; **e** Magnetic heads for duplicator use; **f** Height gauge; **g** MM cartridge; **h** Dynamic microphone; **i** Transformer for MC cartridge use; **j** Mavica; **k** Data tablet; **l** Go-game machine; **m** Fluorescent lamp of bulb type; **n** Electric spark killer; **o** Various magnetic cores for general use

material is not cheap, and audio and/or video heads are already being used in great numbers. However, higher B_s materials with zero λ_s should be developed for high coercivity magnetic recording media [9.6].

The largest scale application of amorphous magnetic materials is as cores for power transformers. The pole transformer is the most suitable application because of its size and operation, and electric power loss in normal operation is expected to be one-third or one-quarter that experienced with silicon steel cores, as estimated by Luborsky in 1987 [9.7]. Table 9.2 shows the performance of various prototype pole transformers with a capacity above 10 kVA in comparison with pole transformers with silicon steel cores. From this table, it is clear that the expected performance has been achieved.

Two big projects for the mass production of amorphous magnetic alloy ribbons and transformers were finished in Japan and the United States, respectively. The results of these projects were reported successfully, and it remains to be seen whether or not these power transformers will be cost effective.

Table 9.2. Performances of amorphous pole transformers (above 10 kVA)

No.	Capacity [kVA]	Voltage [V]	Freq. [Hz]	Core Loss W (%)	Wire Loss W (%)	Noise Phone (%)	Ref. Core	Type
1	15		60	14 (13)	166(79)		Fe–Si	Dry
2	10	210 ~ 105	60	11.8(30)	170(100)	38.5(106)	G-8H	Oil
3	10	210 ~ 105	60	11.8(30)	172(101)	33.6(97)	G-8H	Oil
4	10	210 ~ 105	60	8.6(22)	173(102)	35.3(102)	G-8H	Oil
5	25		60	26 (33)		40.0(87)	M-4	
6	15	210 ~ 105	50	17	292			Dry
7	10	210 ~ 105	50	(28)	(117)	(110)	G-8H	Oil
8	30		60	30 (32)			Fe–Si	Oil

( ) shows comparison with silicon steel core in %

References

9.1. S. Ohnuma, K. Watanabe, T. Masumoto: Phys. Status Solidi A**44**, K151 (1977)
9.2. Y. Makino, K. Aso, S. Uedaira, S. Ito, M. Hayakawa, H. Hotai, Y. Ochiai: Ferrites, Proc. Int. Conf. (1980) p. 699
9.3. I. Kamiya, Y. Makino, Y. Sugiura: Ferrites, Proc. Int. Conf. (1971) p. 483
9.4. O. Kohmoto, H. Fujishima, T. Ojima: IEEE Trans. Mag. MAG-**16**, 440 (1980)
9.5. Y. Makino, K. Aso, S. Uedaira, M. Hayakawa, Y. Ochiai, H. Hotai: J. Appl. Phys. **52**, 2477 (1981)
9.6. Y. Makino: Rapidly Quenched Metals, Proc. Int. Conf. (1985) p. 1699
9.7. F.E. Luborsky: J. Appl. Phys. **49**, 1769 (1978)

10. Magneto-optical Recording

Nobutake Imamura

Any memory that is written and read optically is called an "optical memory". The optical memory is generally characterized by its very high recording density and its capability of processing information en bloc. Laser beams are usually used as the light source. An optical disk is one such information recording device that uses a laser. Here information is recorded by making pits 1 μm or less in diameter in a recording disk with a laser beam and is read by irradiating the beams onto a video or audio disk. Since optical recording enables the light to be squeezed to near its own wavelength, recording densities of 10^8 bits/cm² can be expected in theory.

Magneto-optical memory is a type of optical memory. It records information through a temperature rise in the recording medium produced by absorption of laser beams and is reproduced by the magneto-optical Kerr effect. Since it uses a thin magnetic film as the recording medium, it is possible to erase and re-record. Therefore, the magneto-optical memory combines the two advantages of high density, which is a characteristic of optical memories, and of rewriting capability, which is characteristic of magnetic ones. Magneto-optical memories have been studied since the late 1950s, but problems in the laser source and techniques for its utilization prevented its practical application. However, progress in optical information processing technology in the 1970s has expedited the development of new materials for thin magnetic films [10.1]. Recently, new developments have taken place which enable the practical use of these new materials [10.2–6].

10.1 Principles of Recording, Reproducing and Erasing

As shown in Fig. 10.1a, in magneto-optical recording, a local temperature rise is caused by irradiating light onto the recording medium while simultaneously applying an external magnetic field to direct the magnetization M at the local point along the magnetic field. The magnetic field required to direct the magnetization M of the recording medium, that is the magnetic field H_w required for recording, varies with the temperature of the recording medium. Generally, the higher the temperature, the smaller is H_w. Therefore, even if the

Table 10.1. History of the development of magneto-optical memories

1957	Recording by thermal pen
	Recorded pattern observed by Kerr effect
	Electron beam recording
1960	He–Ne laser available
	Semiconductor laser
	GdIG-Memory,
	CoP-disk memory (Ampex)
1970	MnBi-disk memory (Honeywell)
	EuO-disk memory (IBM)
	MOPS random access memory (Philips)
	MnCuBi-disk memory (NTT)
	CrO_2-disk memory (NHK)
1980	TbFe·GdTbFe-disk memory (KDD), using laser diode, autofocus and autotracking technology
	TbDyFe-disk memory (Sharp)
	MnCuBi·GdTbFe-disk memory (Matsushita)
	GdCo-disk memory (NHK)
	GdTbFe-disk memory (Philips), 2-inch disk
	TbFeCo-disk memory (KDD-SONY)

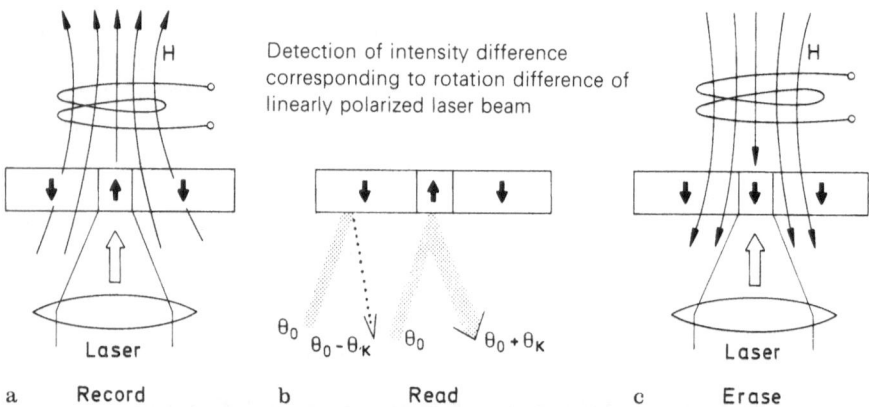

Fig. 10.1a–c Principle of write and read of magneto-optical disks

magnetic field is too small for recording at room temperature, raising the temperature of the recording medium will make recording possible because H_w becomes smaller.

The recording method utilizing the decrease of H_c near the Curie temperature T_C is called Curie temperature recording (T_C-recording). On the other hand, a ferrimagnet has, usually, a compensation temperature T_{comp} in a certain composition area where the magnetization M becomes zero before the Curie temperature is reached. Optical recording is possible because H_c changes rapidly near this temperature. This is called compensation temperature recording (T_{comp}-recording).

The magnetic Kerr and Faraday effects (Appendix A.10), which are both an interaction between light and magnetism, are utilized to optically read the recorded magnetization direction. After laser irradiation onto the recording medium, the record is read by using the reflected light (magnetic Kerr effect), or the transmitted light (Faraday effect). The Kerr effect is generally used because it allows both sides of the disk to be used and is easily incorporated into the optical drive system.

Figure 10.1b illustrates the principle of data reproduction based on the magneto-optical Kerr effect. Light is an electromagnetic wave with an electromagnetic field vector normally emanating in all directions on the plane perpendicular to the light path. When light is converted to linear polarized beams and applied to the recording medium, it is reflected by the surface or passes through the recording medium; at this time the plane of polarization rotates according to the direction of magnetization (M).

For example, if the polarization plane rotates $+ \theta_K$ degrees for upward magnetization, it rotates $- \theta_K$ degrees for downward magnetization. Therefore, if the axis of a polarized light analyzer is set perpendicular to the plane inclined at θ_K degrees to the recording surface, the light reflected by a downward-magnetized surface area cannot pass through the analyzer, whereas the light reflected by upward-magnetized area can be captured by the detector for an amount of $\sin(2\theta_K)$. As a result, upward-magnetized areas appear brighter than downward-magnetized areas. The magneto-optical disk system reproduces the stored data using this principle.

10.2 Requirements for Recording Media

In this section, the desired conditions for the recording medium of the magneto-optical disk will be described and thin film materials and proposed recording media structures will be introduced.

Because of the light-oriented reproduction, the surface of a medium must be smooth enough not to cause irregular reflection. For recording and reproduction, the surface of a thin film, though it varies with manufacturing conditions, is approximately mirror-like provided that the plate underneath it also has a mirror surface.

It is considered that the recording medium for high-density recording must have its magnetization perpendicular to the film surface. The reasons for this are that there is little influence from opposing magnetic fields between neighboring bits allowing high-density recording and also the polarization effect can be used to best advantage in optical reproduction. A condition for the magnetization to be perpendicular to the film surface is that K_u, the anisotropy constant in the direction perpendicular to the film surface, satisfies

$$K_u > 2\pi M_s^2 \quad \text{or} \quad H_k > 4\pi M_s . \tag{10.1}$$

There are many magnetic thin film materials that satisfy (10.1) as shown in Fig. 10.2. However, not all of these materials are suitable for magneto-optical recording as they must also satisfy the following conditions.

Since recording is done through the thermal effect of light, it is desirable that the magnetic field, H_{ex}, used for recording should change greatly in a relatively narrow temperature range and that α, the light absorption coefficient of the medium, should be large. The fraction of light absorbed by the medium is given by

$$(I_0 - I_T)/I_0 = 1 - \exp(-\alpha t) , \tag{10.2}$$

where I_0 is the intensity of incident light and I_T that of transmitted light. It is considered satisfactory if $\alpha t \gg 2$, because about 50% of the light is absorbed. If the thin film is about 1000 Å thick, it is necessary that $\alpha < 10^5$ cm^{-1}. The value of α of thin metal films is usually about 10^5, while that of thin single-crystal films such as magnetic garnet is one or more factors of ten lower, thus requiring higher laser power for recording. T_C should be small for Curie temperature recording, 150–250 °C being a suitable value taking into account the stability at room temperature. For compensation temperature recording, the compensation temperature has to be near room temperature.

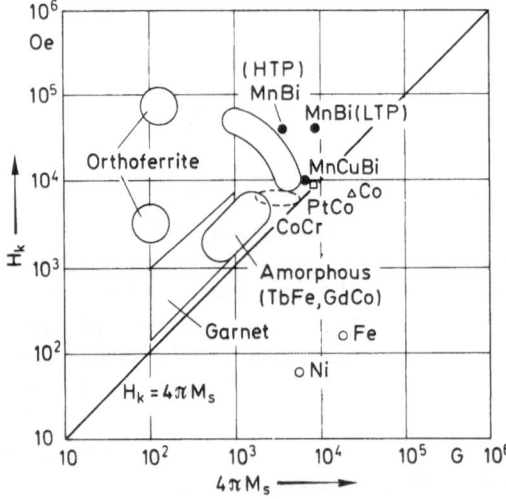

Fig. 10.2. Various perpendicularly magnetized film materials

The S/N ratio for optical reproduction is nearly proportional to the product of the square root of the light intensity and the magneto-optical rotation angle, that is,

$$S/N \sim \sqrt{I}\,\theta \qquad\qquad (10.3)$$

In this equation, I is the transmitted light intensity, I_T, and θ is the Faraday rotation angle, θ_f for the Faraday effect and I is the reflected light intensity, I_R, and θ the Kerr rotation angle, θ_K, for the Kerr effect. If the incident light intensity is I_0, then I_T and I_R are

$$I_T = I_0 \exp(-\alpha t) \qquad\qquad (10.4)$$

$$I_R = R I_0' \,, \qquad\qquad (10.5)$$

where R is the reflectivity of the medium. From these, the performance index is obtained as $\exp(-2\theta_f/\alpha t)$ for the Faraday effect, and $\sqrt{R}\,\theta_K$ for the Kerr effect.

To achieve high-capacity and high-density recording, there must be a stable presence of many small magnetic domains. Where magnetization is perpendicular to the film surface, the diameter of the smallest magnetic domain existing in a stable state is obtained through the principle of bubble magnetic domains. If the diameter of a magnetic-domain d is larger than the film thickness t, d is approximately

$$d = \sigma_w / 2 M_s H_c \,. \qquad\qquad (10.6)$$

Therefore, materials that have a large coercive force H_c are desirable as the recording medium, because small stable magnetic domains can be realized.

10.3 Recording Media

Materials for magneto-optical recording, as shown in Table 2, include thin oxidized magnetic films represented by magnetic garnets, thin polycrystalline films like MnCuBi and MnPtSb, and thin amorphous films like TbFeCo and DyFeCo. Table 10.2 also shows the various production methods and characteristics of these thin films.

In these materials, the most practical ones are thin amorphous rare-earth transition metal alloy films. These amorphous films are usually prepared by sputtering methods, and have a magnetic anisotropy perpendicular to the film plane. This was found by Chaudhari et al. in 1973 for GdCo films which are used as bubble memories. In 1974, they found that GdCo was also suitable for magneto-optical memory, that is, GdCo films can be used as a compensation point writing material. In 1976, it was found that TbFe and DyFe films were also

Table 10.2. Materials and characteristics of the magneto-optical recording media

Material	Recording method	θ_K [degrees]	λ [nm]	Remarks
Amorphous				
DyFe	$T_C \sim 70$	0.13–0.25	(800)	Medium noise is low
TbFe	$T_C \sim 140$	0.24–0.30	(800)	
GdTbFe	$T_C \sim 160$	0.28–0.35	(800)	T_{comp} medium must have
GdFeBi	$T_C \sim 180$	0.30–0.41	(800)	uniform composition
TbFeCo	$T_C \sim 200$	0.27–0.35	(800)	
GdFe	$T_{comp} \sim 220$	0.25–0.35	(800)	Crystallized at 350–400 °C
GdFeCo	$T_C \sim 300$	0.30–0.40	(800)	
GdCo	$T_{comp} \sim 600$	0.28–0.35	(800)	
NdFeCo	$T_C \sim$	0.63	(400)	
Polycrystalline				
MnBi	$T_C \sim 360$	0.7	(633)	MnBi shows a phase transition
MnCuBi	$T_C \sim 200$	0.5	(633)	
CoCr	$T_C \sim 300$	0.07	(633)	θ_K is relatively large, but
MnPtSb	$T_C \sim 209$	1.27	(720)	there is medium noise
PtCo	$T_C \sim 390$	$\theta_F \sim 4.0$	(633)	
CoCrFeO$_4$	$T_C \sim 150$			
BiAlGdIG	$T_C \sim 136$	$\theta_F \sim 18$	(633)	
BiGaYIG	$T_C \sim 177$	$\theta_F \sim 5.8$	(520)	

very suitable for magneto-optical memory, and these can be used as Curie point writing materials.

In these amorphous alloy films, magnetic moments of the heavy rare earth atoms (Gd, Tb, Dy, . . .) and transition metal atoms (Fe, Co) couple antiparallel to each other. Thus the total magnetic moment M_s becomes the difference between the rare earth atom moment M_{RE} and the transition metal moment M_{TM}, that is, $M_s = |M_{RE} - M_{TM}|$. Figure 10.3 shows the temperature dependences of the total magnetic moment M_s and sublattice moments M_{RE} and M_{TM}. The compensation point appears because of the decrease of M_{Gd} which is more rapid than that of M_{Fe} as temperature increases, so that the total magnetic moment M_s becomes zero.

As mentioned before, these amorphous films are prepared by electron beam evaporation or sputtering methods, and have the easy axis of magnetization perpendicular to the film plane near the compensation composition as shown in Fig. 10.4. The origin of the perpendicular easy axis of magnetization is considered to be due to the following factors: magneto-striction caused by the substrate, pair ordering, or columnar structures.

In order to obtain a small written domain, materials that have a large coercive force H_c are desirable as the recording medium as shown in (10.6).

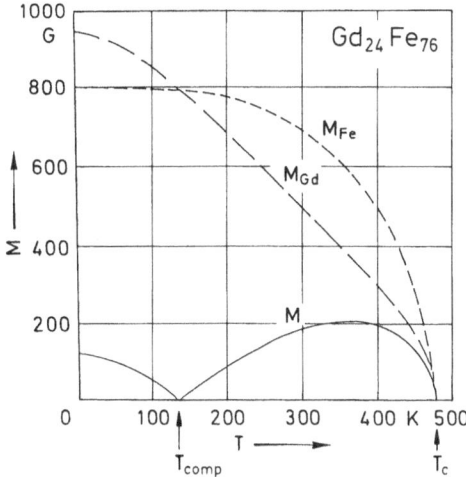

Fig. 10.3. Temperature dependences of sublattice moments and total magnetic moment of a rare-earth-transition alloy GdFe film

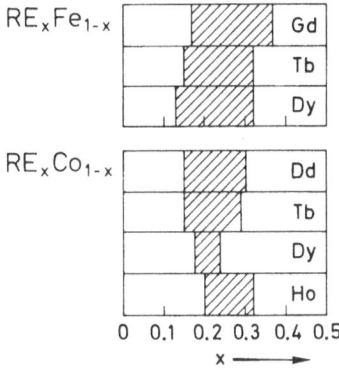

Fig. 10.4. Available composition in which perpendicularly magnetized films were obtained for rare-earth-transition metal alloys

These amorphous films are ferrimagnetic, therefore the coercive force H_c increases near the compensation temperature/or compensation composition, where the value of H_c strongly depends on the materials as shown in Table 10.3 and Fig. 10.5. Alloy films containing Tb have a large coercive force H_c, and the estimated stable minimum domain size is less than 0.1 μm.

The temperature dependences of H_c for GdCo and TbFe are very different from each other, that is, the H_c of GdCo increases only near the compensation

Table 10.3. Coercive force H_c

Material	H_c [Oe]
GdCo	0.5–200
GdFe	10–200
TbFe	150–5500
DyFe	10–800
GdTbFe	800–3000
TbFeCo	2000–10000
MnBi	~ 2500
MnCuBi	~ 1500

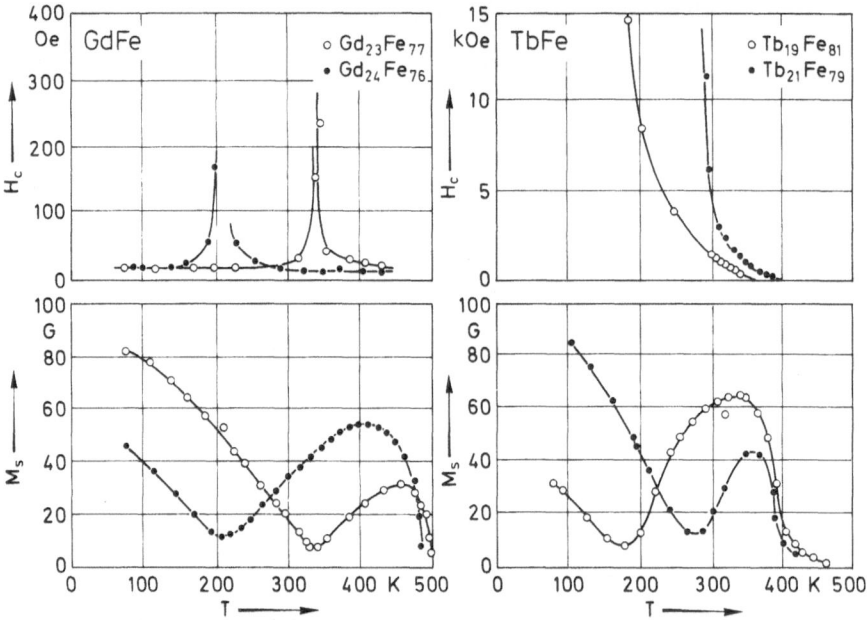

Fig. 10.5. Temperature dependences of coercivity and magnetic moment of TbFe and GdFe alloys

temperature and the H_c of TbFe increases not only near the compensation temperature but also near the Curie temperature T_C. The T_C of TbFe is about 140 °C, therefore. TbFe films can be used as Curie temperature writing material.

In Fig. 10.6, the Curie temperature T_C of iron based alloys such as TbFe and GdFe are plotted together with the compensation temperature T_{comp} of GdCo and GdFe. These T_C and T_{comp} values are less than 200 °C, and so these

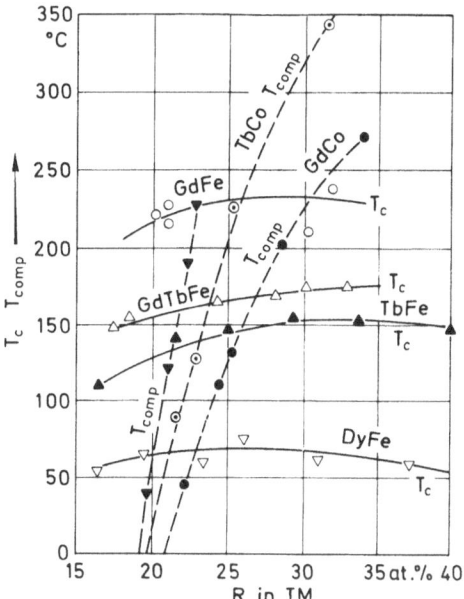

Fig. 10.6. Composition dependence of T_C and T_{comp} for rare-earth-transition metal alloy films

materials can be recorded by a few milliwatts of laser power with less than 1 μs irradiation time. The compositional dependences of T_{comp} of GdCo and GdFe, however, changes more drastically than those of TbFe and GdTbFe as shown in Fig. 10.6. Therefore, these T_{comp}-writing materials are not suitable for making large area uniform films.

The read-out of a written domain is made by the magneto-optical Kerr or Faraday effect, and the S/N value thus obtained is proportional to the Kerr or Faraday rotation angle θ_K or θ_f, respectively. The value of θ_K is about 0.2 degree for TbFe films, and decreases with increase of Tb content as shown in Fig. 10.7. These θ_K values are not sufficient to obtain high C/N values as described later

In order to improve the θ_K value of binary alloys, various kinds of ternary alloy have been investigated. It was found that, in the FeCo based alloys such as TbFeCo and GdFeCo, the θ_K value is increased with increase of the Curie temperature T_C as shown in Fig. 10.8. TbFeCo especially, has a large coercive force, H_c, of more than 5 kOe, and has a Curie temperature suitable for semiconductor laser recording.

The diameter of the laser beam spot D becomes smaller as λ decreases, because D is proportional to λ/NA, where NA is the numerical aperture of the objective lens. The smaller the D value the higher the recording density. Therefore, it is desirable that θ_K becomes large as λ decreases. Figure 10.9 shows

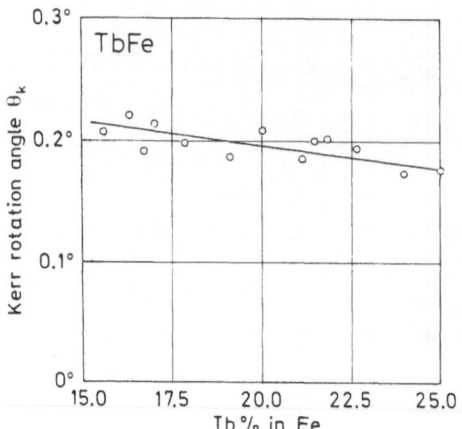

Fig. 10.7. Composition dependence of Kerr rotation angle θ_K of TbFe alloy films

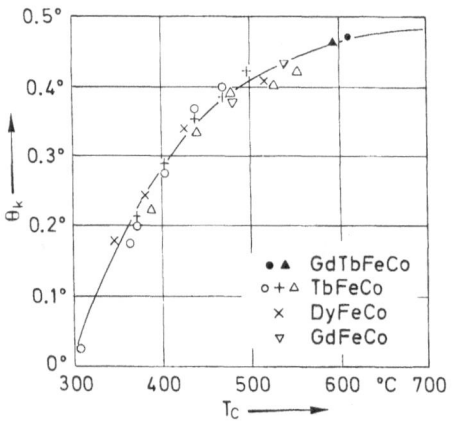

Fig. 10.8. Relation between Kerr rotation angle θ_K and Curie temperature T_C of rare earth–FeCo alloy films (after [10.7])

the wavelength dependence of the Kerr rotation angle θ_K of TbFeCo films. The value of θ_K decreases with decreasing wavelength λ. Recently, NdFeCo was found to be very suitable for high density recording when short wavelength laser beams become available in the future, because the value of θ_K increases with decreasing wavelength as shown in Fig. 10.9. It is important to obtain a sufficiently large S/N for wide application of the magneto-optical memory. There have been various proposals to achieve this, including improvements not only in thin film materials but also in thin film structures. There are a number of representative multi-layer structures as shown in Fig. 10.10.

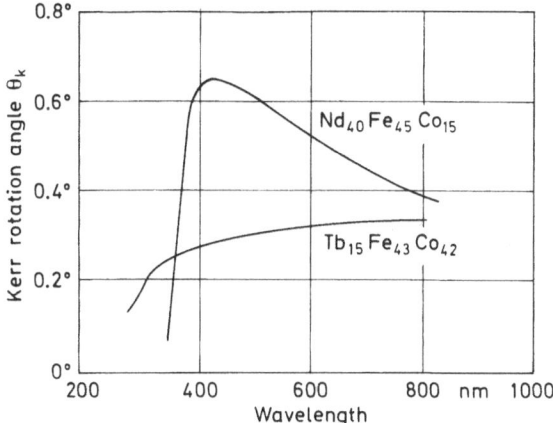

Fig. 10.9. Wavelength dependence of Kerr rotation angle of TbFeCo and NdFeCo (after [10.8])

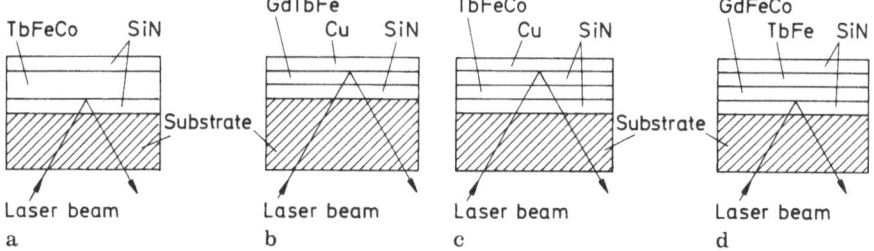

Fig. 10.10a–d. Typical structure of multilayer films for magneto-optical disk memory

A recording medium covered with a protective film is shown in Fig. 10.10a. The reflection coefficient of light changes with the thickness of a SiN film. The Kerr rotation angle θ_K is maximum at the thickness for which the reflection coefficient is minimum, or four times as large as for a film without a protective layer. This phenomenon is called the Kerr effect enhancement, and is considered phenomenologically to be an apparent increase in the Kerr rotation angle by multiple reflection of light. In this case, θ_K increases and the reflection coefficient R decreases; as a result, twice as much $S/N = R$, and θ_K is improved by about a factor of two.

A structural configuration is shown in Fig. 10.10b in which a reflection film is provided on the back of the recording medium so that not only is the light reflected from the recording medium surface but is also reflected after penetrating through the medium. The application of this technique using magnetic

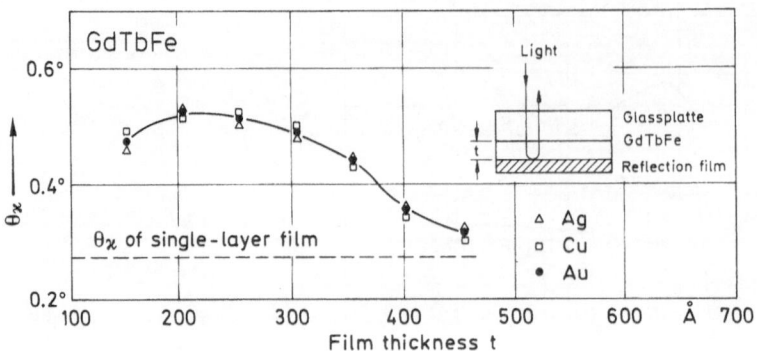

Fig. 10.11. Relationship between θ_K and film thickness in reflection film structures (after [10.9])

garnet, which is nearly transparent in the visible light range and has a small light absorption coefficient, is being studied. Thin amorphous metal films of thickness 500 Å or less also give this same effect as shown in Fig. 10.11.

A two-layer structure consisting of separate recording and reading layers is shown in Fig. 10.10c. This structure is proposed for use where a material with a low Curie temperature is employed in order to improve the recording characteristics. Accordingly, the Kerr rotation angle also becomes smaller. That is, the recording layer should consist of a material with a low Curie temperature and a large H_c, while the reading layer should be of a material with a high Curie temperature and a large θ_K. This kind of material will provide a recording medium capable of recording with low laser power and displaying relatively good optical reproduction characteristics.

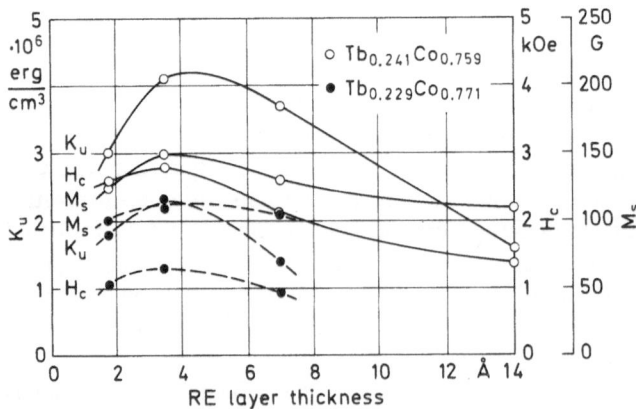

Fig. 10.12. Rare-earth layer thickness dependence of magnetic properties in Tb–Co and Dy–Co films (after [10.10])

A composition-modulated multilayer film also has suitable characteristics for magneto-optical memory, that is, it has the magnetization easy axis normal to the film plane and shows an increase of the Kerr rotation angle for the appropriate modulation length or pitch. These composition-modulated multilayer films are usually deposited by a multi-source dc-sputtering method, in which the rare earth and transition metal targets are placed at opposite positions under the substrate and rotated with various speeds. In the case of Tb–Fe, Dy–Fe, Tb–Co, Dy–Co, the films composed of monoatomic layers of RE and a few atomic layers of TM show superior magnetic properties of large magnetization, coercivity and uniaxial anisotropy energy as shown in Fig. 10.12. All these films show a high carrier-to-noise (C/N) ratio of 50 dB or more at 1 MHz read/write frequencies.

10.4   Dynamic Read/Write Properties

In order to read and write the magneto-optical disk an optical head is used. Figure 10.13 shows the typical arrangement of an optical head. A laser beam, caused by a laser diode, is focused onto the disk surface through the objective lens.

In optical recording, a local temperature rise is caused by irradiating the laser beam onto the magnetic layer while simultaneously applying an external magnetic field to direct the magnetization M at the local heated point along the magnetic field.

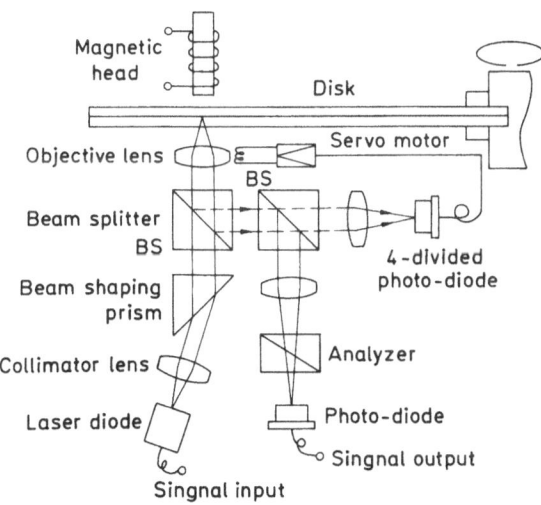

Fig. 10.13. Typical structure of a magneto-optical head

For readout of the recorded domain, the Kerr effect is used. That is, the laser beam is reflected from the magnetic layer, and divided into two beams by a PBS (Polarized Beam Splitter) which is used for differential detection of the Kerr rotation angle.

Figure 10.14 shows a typical readout C/N value as a function of recording laser power. Usually, a few mW of laser power is enough to write the magneto-optical disk. When reading, a laser power of less than 2 mW is used, which is less than half of the recording laser power.

Figure 10.15 shows the C/N value of a magneto-optical disk as a function of recording frequency together with that of another memory. The C/N value shows about 60 dB in the low frequency region, and decreases with increasing frequency. The C/N value is important for choosing the application, that is, more than 45 dB is necessary for digital memory applications, and more than 60 dB at about 6 MHz is desired for analog video applications. From this figure, we can see that magneto-optical disks are available for digital data, audio, and video memories.

Overwrite technology is important for magneto-optical disks, because of present, it is necessary to erase the data before recording new data. Overwrite means that new data is recorded directly onto the old data without an erasing process. Magneto-optical recording using field modulation has been shown to be an effective method for achieving direct overwrite, where the data is recorded using the magnetic field caused by the magnetic head, which is the same as conventional magnetic recording. In this method, the laser beam is used only for heating up the magnetic film, that is, the laser beam is used to restrict the recording position.

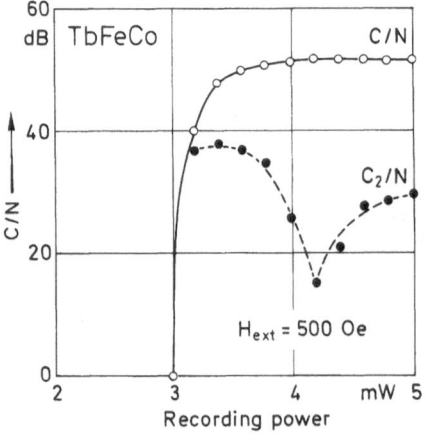

Fig. 10.14. Recording power dependence of carrier-to-noise ratio of TbFeCo films, where C2/N is the C/N value of the second harmonic signal

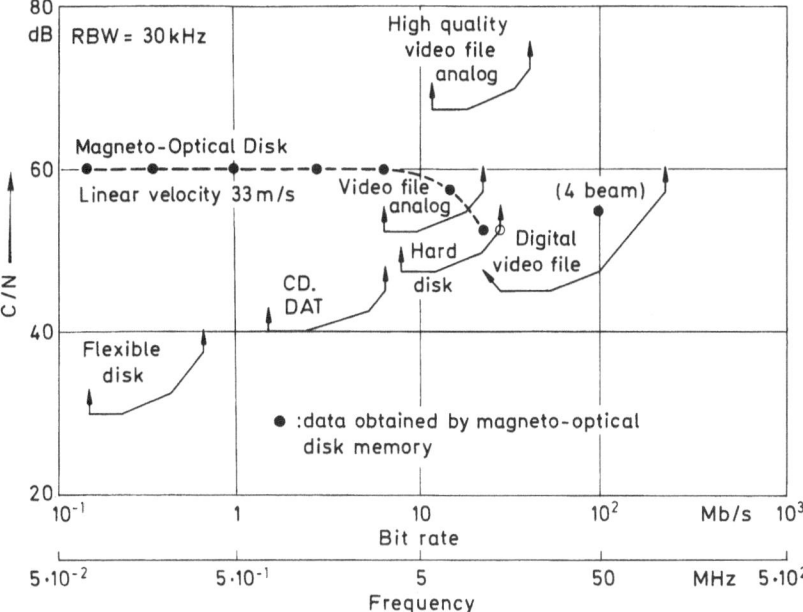

Fig. 10.15. Carrier-to-noise ratio plotted as a function of read/write frequency (bit rate) for a magneto-optical disk together with that for various kinds of memory

One disadvantage of this technique, however, is that a high data rate operation requires the coil diameter and coil-to-media spacing to be small. Recently however, some improvements have been made to this field modulation method. Figure 10.16 is one of these using a floating magnetic head, which is suitable for high speed overwriting. In this method, a focused laser beam continuously irradiates the information track through the disk substrate and a magnetic field supplied from a floating magnetic head is modulated according to the information to be recorded. This magnetic head is a single-pole type, and is capable of producing a field of more than 300 Oe perpendicular to the magnetic film layer. The switching time of the magnetic field is less than 100 ns, therefore, high speed overwriting on a magneto-optical disk at a recording frequency of 5 MHz has been achieved.

The resonant bias coil overwrite technique gives another advantage of the field modulation method, that is, this technique allows high frequency operation with coil-to-media spacings on the order of 1 mm. In the resonant coil overwrite method, the bias coil is part of a resonant circuit, such as that shown in Fig. 10.17. The circuit is driven sinusoidally at a frequency somewhat higher than the desired data rate. Magnetization patterns along the track are thermo-magnetically written by firing short laser pulses at either the positive or negative peaks of the sinusoidal field, as shown in Fig. 10.18. Each laser pulse heats up a

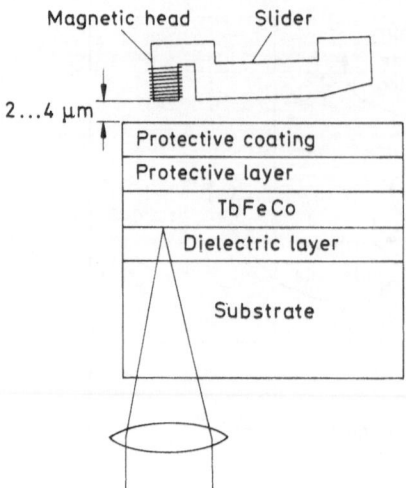

Fig. 10.16. Overwriting method by magnetic field modulation using magnetic head (after [10.11, 12])

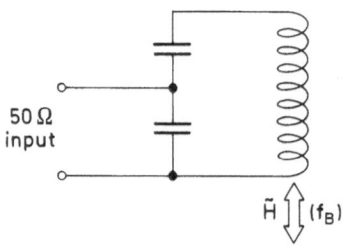

Fig. 10.17. Resonant circuit used to generate oscillation bias field (after [10.13])

micron-sized circular region of the track. Depending on the instantaneous orientation of the field during cooling, magnetic domains of either up or down orientation can be written. The disk velocity is set so that there is substantial overlap between successive circular marks. The overlap ensures that the track will be continuously overwritten and allows variable length marks to be formed.

The other technique is a single beam direct overwrite method, where the information data is recorded by the modulation of the one beam laser power. The recording magnetic layer is an exchange-coupled double-layered thin film which consists of a memory layer (M-layer TbFe) and a reference layer (R-layer: TbFeCo) as shown in Fig. 10.19. The coercivity of each layer depends on the temperature as shown in Fig. 10.16.

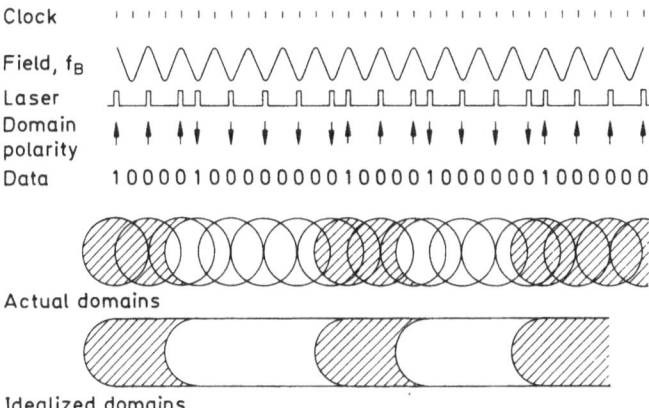

Clock
Field, f_B
Laser
Domain polarity
Data   1 0 0 0 0 1 0 0 0 0 0 0 0 0 1 0 0 0 0 1 0 0 0 0 0 0 1 0 0 0 0 0 0

Actual domains

Idealized domains

Fig. 10.18. Timing diagram and resulting mark shapes for the resonant coil direct overwrite technique. The laser is pulsed at either positive or negative field peaks. Individual marks overlap to form continuous marks of variable length. Binary ones are coded as domain transitions [10.13]

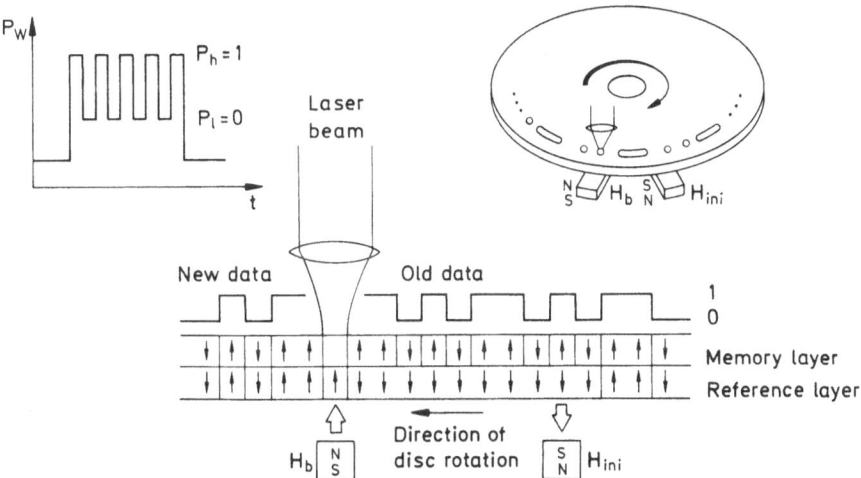

Fig. 10.19. Direct overwrite using exchange coupled layers. When the disk passes over the magnet with the field of H_{ini}, the magnetization of the reference layer switches to the field direction. When it comes to the laser beam, a high power pulse creates domains in both memories (write) and a low power pulse quench domains of the memory layer in the same direction as the reference layer (after [10.14])

When the disk is rotated and subjected to an initializing magnet (H_{ini}), the magnetization direction of the R-layer is initialized. The optical head laser pulse, which has two different power levels P_h and P_l, produces a downward magnetized domain and an upward magnetized domain in the M-layer.

The domain on the disk where the laser is focused is heated up to about T_{C1} by the power of P_l. The M-layer copies the sub-lattice magnetization directions of the R-layer, so that the whole magnetization direction of the M-layers has the same direction of H_{ini}. The domain on the disk where the laser is focused is heated up to about T_{C2} by the power of P_h. The R-layer first reverses just as the ordinary thermomagnetic writing system does. Then the M-layer copies the sub-lattice magnetization direction of the R-layer while it cools down, so that the whole magnetization of the M-layers has the opposite direction of H_{ini}.

10.5   Applications and Other Technologies

As shown in Fig. 10.15, the C/N value of a magneto-optical disk is good enough to use for various kinds of applications, such as digital audio disks, digital data memories, and analog video files. The value of the bit error rate is important for digital memory applications. That is, the bit error rate of the magneto-optical disk is in the range from 10^{-4} to 10^{-6} without any error correction, but it can become less than 10^{-12} which is good enough for data memory applications.

Table 10.4 shows design parameters of some magneto-optical disk drives thus developed. The memory capacity of a disk depends on its size, that is, about 100 MB (Mega Bytes/bits) for a 3.5 inch disk size, 300 MB/surface for a 5.25 inch disk size. These small-size disks are considered to be suitable for personal use memories such as personal computer memories or digital video file memories.

The facsimile is a document transfer device and its data transfer rate is about a few Mb/s. Thus a facsimile is very useful as a documents or picture

Table 10.4. 5.25-inch magneto-optical disk drives

	Capacity [MB]	Data rate [kB/s]	Access time [ms]	Disk-dia. [mm]	Speed [rpm]	Material
Hitachi	275	440	100	120	1200	TbFeCo
Toshiba	268	328	100	130	1200	TbCo
Matsushita	260	980	70	130	3000	TbFeCo
NTT	262	930	65	130	2400	TbFeCo
Sony	325	1000	100	130	1800	TbFeCo
Sharp	190	150	180	130	900	GdTbFe

input/output device for a magneto-optical disk which can record 200 or more A4 size documents per 3.5 inch disk.

As for video file memory applications, a C/N value of more than 55 dB at 6 MHz is necessary for analog video files, and the magneto-optical disk can satisfy this requirement. For real-time digital video recording, a bit rate of more than 100 Mb/s is required, which is very difficult to realize by using one laser beam. One way to achieve such a high bit rate is to divide the source signal into 4 channels and record them on four different tracks. Indeed, video signals were recorded using the following technique, that is, separately PCM coded video and audio signals were fed into a record processor where the sync code, error correction code, etc., were added and then converted into scrambled NRZ recording signals for 4 channels. Fairly high picture quality (S/N > 55 dB, video band width > 4 MHz) was obtained [10.15].

References

10.1. H. Williams, R.C. Sherwood, F.G. Foster, E.M. Kelley: J. Appl. Phys. **28**, 1181 (1957);
       L. Mayer: J. Appl. Phys. **29**, 1003 (1958);
       B. Tsujiyama, S. Yoshii, K. Nishiguchi: IEEE Trans. Magn., MAG-**8**, 603 (1972);
       R. Langlet, B. Caree, J.P. Pivot: IEEE Trans. Magn., MAG-**9**, 401 (1973);
       B.G. Huth: IBM J. Res. Dev. **18**, 100 (1974);
       D. Dhen: Appl. Optics **13**, 767 (1974);
       K. Chida, B. Tsujiyama, A. Katsui, K. Egashira: IEEE Trans. Magn, MAG-**13**, 982 (1977)
10.2. P. Chaudhari, J.J. Gambino: Appl. Phys. Lett **22**, 337 (1973)
       N. Imamura, Y. Mimura and T. Kobayashi: Jpn. J. Appl. Phys. **15**, 179 (1976)
       Y. Mimura, N. Imamura, T. Kobayashi: IEEE Trans. Magn. MAG-**12**, 779 (1976).
10.3. N. Imamura, C. Ota: Jpn. J. Appl. Phys. **19**, L731 (1980)
10.4. K. Ohta, A. Takahashi, H. Yamaoka: Proc. 42nd Conf. on Applied Physics in Japan: (1981) 9pP-8;
       S. Tsunashima, H. Tsuji, M. Kobayashi, S. Uchiyama: IEEE Trans. Magn. MAG-**17**, 2840 (1981);
       Y. Togami, K. Kobayashi, M. Kajiura, K. Sato, T. Teranishi: J. Appl. Phys. **53**, 2335 (1982)
10.5. R.P. Freese, D.H. Davies: SPIE 2nd Symposium on Optical Data Storage, 420–37 (1983);
       F. Tanaka, T. Nagao, N. Imamura: INTERMAG'84 Digest, EB-**10**, 340 (1984)
10.6. N. Imamura, S. Tanaka, F. Tanaka, Y. Nagao: IEEE Trans. Magn., MAG-**21**, 1607 (1985)
10.7. S. Uchiyama: Solid State Physics **20**, 633 (1985) (In Japanese)
10.8. A.E. Bell: Private communication
10.9. K. Ohta, A. Takahashi, H. Yamaoka: Proc. 42nd Conf. on Appl. Phys. in Japan, **20**aA-6 (1981)

10.10. N. Sato, K. Habu, T. Oyama: Digests of the 1987 INTERMAG Conference, CG-07 (1987)
10.11. F. Tanaka, S. Tanaka, N. Imamura: Jpn. J. Appl. Phys. **26**, 231 (1987)
10.12. T. Nakao, M. Ojima, Y. Miyamura, S. Okajima, Y. Takeuchi: Jpn. J. Appl. Phys. Suppl. **26-4**, 149 (1987)
10.13. D. Rugar: IEEE Trans. Mag. MAG-**24**, 666 (1988)
10.14. J. Saito, M. Sato, H. Matsumoto, H. Akasaka: Symp. on Optical Memory (ISOM'87) Technical digest, WA3, p. 9
10.15. T. Nomura, K. Yokoyama, S. Nakagawa, K. Kimoto: Digests of the 1987 INTERMAG Conference, DB-01 (1987)

11. Magnetic Bubble Memories. Solid State File Utilizing Micro Magnetic Domains

Yutaka Sugita

Magnetic bubbles are cylindrical magnetic domains which exist in single crystalline magnetic garnet and orthoferrite films or amorphous Gd–Co films. These domains behave just like bubbles when observed in a microscope, leadng to the name of magnetic bubbles. In 1967, Bobeck [11.1] of Bell Laboratories announced the idea of sequential memory devices using the propagation of magnetic bubbles. In these devices, binary codes '1' and '0' are stored as the presence and absence of magnetic bubbles respectively. In 1969, Bobeck et al. [11.2] and Perneski [11.3] experimentally demonstrated the possibility of these devices by generating, propagating and annihilating magnetic bubbles in a controlled way. The very new idea and high potentiality of these devices attracted many scientists and engineers who were involved in applied magnetics. Since then research on magnetic bubble memories from materials and physics, to devices and systems has been carried out extensively. As a result magnetic memories have been developed as practical solid state files which are now used commercially for electronic switching systems, numerical control machines, robotics, terminals and so on.

The advantages of a magnetic bubble memory are as follows:

(1) Non-volatility.
(2) High reliability and portability because of non-existence of mechanical rotating parts like magnetic disks or floppy disks.
(3) Large-scale integration and small volume.
(4) Maintenance-free

At present 1 Mbit and 4 Mbit devices utilizing 1.5–2.0 μm diameter magnetic bubbles are commercially available and used for the various systems mentioned above because of those advantages. Furthermore, 16 Mbit to 64 Mbit devices are now being developed.

With progress in the technology of magnetic bubble devices, research on materials and physics of magnetic bubbles has advanced dramatically, which in turn contributes much to the progress of devices. Growth-induced magnetic uniaxial anisotropy in garnets, complicated wall structures and dynamics of magnetic bubbles, the effect of ion-implantation into garnets, the magnetization process in micro-shaped patterns, etc. are all important and interesting topics both from basic and from engineering points of view. The physics, materials, devices and systems of magnetic bubbles are described below.

11.1 Physics of Magnetic Bubbles

11.1.1 Stability of Magnetic Bubbles

Let us consider the magnetic domain structure in a magnetic thin film with uniaxial anisotropy whose easy axis is perpendicular to the film plane. Here we treat the case where the uniaxial anisotropy constant K_U is much larger than the shape anisotropy $2\pi M_s^2$, which directs the magnetization perpendicular to the film plane.

In the demagnetized state, the stripe domain structure is energetically stable, where the magnetization is directed upwards and downwards periodically as shown in Fig. 11.1a. The width of the stripe domain W is determined by the balance of magnetostatic energy due to the magnetic poles on the film surface and the domain wall energy.

When a magnetic bias field H_B is applied perpendicular to the film plane, the domain with the magnetization parallel to H_B expands and the domain with the magnetization anti-parallel to H_B shrinks, as shown in Fig. 11.1b. As H_B is increased, the shrinking domain becomes a labyrinth, and when H_B reaches a critical value H_2, it becomes circular as shown in Fig. 11.1c and d. This is called a magnetic bubble. When H_B is increased further, the magnetic bubble becomes smaller and collapses suddenly when H_B reaches a critical value H_0 as shown in Fig. 11.1e and f. The magnetic bubble is stable for $H_2 < H_B < H_0$. The diameter of the bubble is decreased with increasing H_B as shown in Fig. 11.2. The fields H_0 and H_2 are called the bubble collapse field and the run-out field respectively.

A magnetic bubble is kept stable by the balance of wall energy E_W, Zeeman energy E_H due to the bias field and magnetostatic energy E_M due to free poles on

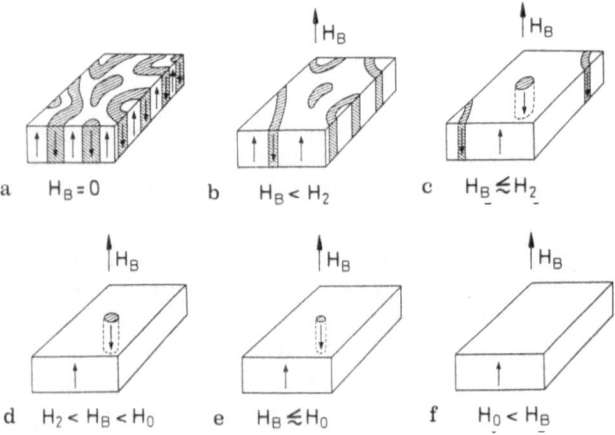

Fig. 11.1a–f. Magnetic domain structure vs bias field

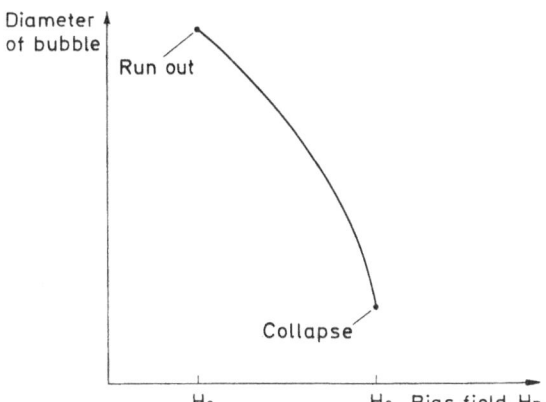

Fig. 11.2. Diameter of a magnetic bubble vs bias field

the surface of the film. Since E_W and E_H make a bubble smaller and E_M makes it larger, a balance is obtained. The situation is expressed by the equations below [11.4]. The total energy E is expressed by

$$E = E_W + E_H + E_M$$

$$= \pi d h \sigma_W + \frac{1}{2}\pi d^2 h M_s H_B - 4\pi M_s^2 I\left(\frac{d}{h}\right)\pi h^3 \ . \tag{11.1}$$

Here d is the diameter of a magnetic bubble, h the height, σ_W the wall energy density, and M_s the saturation magnetization respectively. $I(d/h)$ is a complicated function including an elliptical integral as shown in footnote 1.

From the minimization of E with respect to d, the following equation is obtained:

$$\frac{l}{h} + \frac{d}{h}\frac{H_B}{4\pi M_s} - F\left(\frac{d}{h}\right) = 0 \ . \tag{11.2}$$

Here $l \equiv \sigma_W/4\pi M_s^2$ is a material constant called the characteristic length. $F(d/h)$ is a function including an elliptical integral as shown in the footnote.

Footnote 1.

$$I(x) = \int_0^x F(x)dx$$

$$F(x) = \frac{2}{\pi}x^2\left[\frac{1+x^2}{x}E\left(\frac{x^2}{1+x^2}\right) - 1\right]$$

$$E(x) = \int_0^{\pi/2} (1 - x^2\sin^2\varphi)^{1/2}d\varphi \quad \text{(Perfect elliptic integral of the second kind)}$$

$F(d/h)$ can can be approximated [11.16] by an analytic function $(d/h)[1 + (3/4)(d/h)]^{-1}$. Balance of the three terms in (11.2) is illustrated in Fig. 11.3. The first and second terms are expressed by a line with a y of l/h and a slope of $H_{\mathrm{B}}/4\pi M_{\mathrm{s}}$. The third term $F(d/h)$ is curve 2.

Equation (11.2) holds at the intersections A and B of the line and the curve. At the point A, $\partial^2 E/\partial d^2$ is positive which means that the point A is a stable point. On the other hand the point B is an unstable point since $\partial^2 E/\partial d^2$ is negative at B. As H_{B} is increased, the slope of line l is increased and d for the point A is decreased. Then H_{B} reaches some critical value at which the line touches the curve of $F(d/h)$ at some point corresponding to the collapse of a magnetic bubble. The magnetic bubble collapses at some definite value of d, so H_0 can be obtained as the value satisfying the above condition. Taking into consideration the stability of a magnetic bubble for a change from a circular to an elliptical shape, a critical value of $H_{\mathrm{B}} = H_2$ can be calculated where a magnetic bubble changes to a strip domain [11.4, 5].

As mentioned above, it has been demonstrated that a magnetic bubble can be stable for a bias field H_{B} ranging from H_2 to H_0. The size d changes with various materials from 0.3 μm to 10 μm, and details of these materials will be described in Sect. 11.3.

11.1.2 Domain Wall Structure of a Magnetic Bubble

The diameter of a magnetic bubble for some H_{B} should be determined uniquely according to the theory mentioned above. However, experiments show that diameters of magnetic bubbles in garnet films are not uniquely determined for H_{B} as shown in Fig. 11.4 [11.6-9]. The relationship between d and H_{B} is expressed by many curves. The reason for this is that the domain wall structure

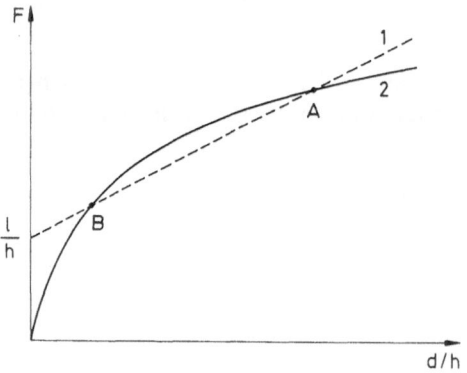

Fig. 11.3. Balance of forces exerted on a magnetic bubble

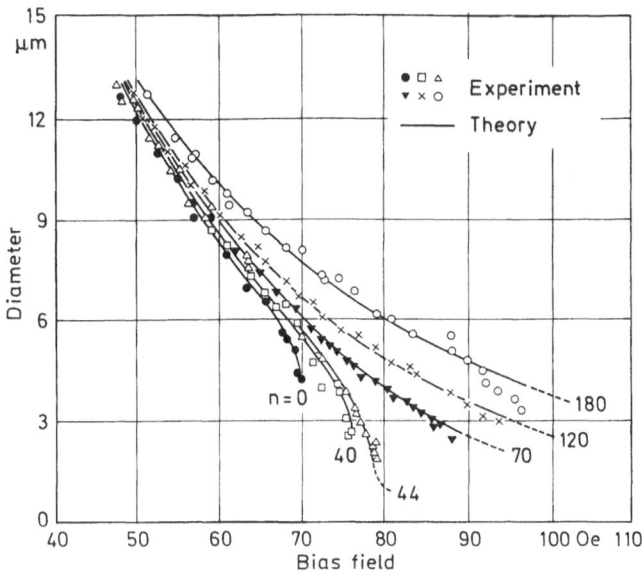

Fig. 11.4. Diameter of a magnetic bubble vs bias field

of magnetic bubbles is not a simple $180°$ Bloch wall (Appendix A.4) but a complex wall having a definite number of Bloch lines in a Bloch wall as shown in Fig. 11.5. As the number of Bloch lines increases, the d vs H_B curve shifts to a higher H_B region and H_0 increases. When a domain wall of a magnetic bubble includes a definite number of Bloch lines, n, the wall energy density σ_W is not constant for changing d. σ_W is expressed as [11.8]

$$\sigma_W = \sigma_{W0}\left[1 + \left(\frac{n\delta}{\pi d}\right)^2\right]^{1/2} . \tag{11.3}$$

Here σ_{W0} is the wall energy density for a $180°$ Bloch wall without any Bloch line

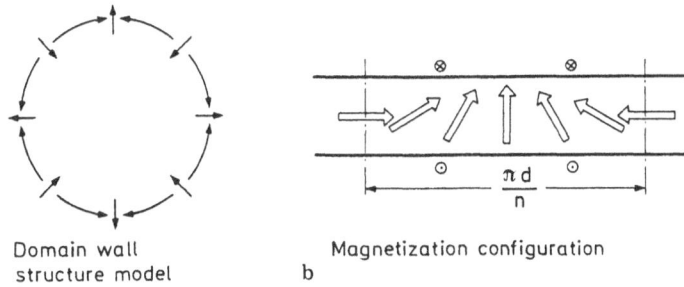

a Domain wall
   structure model

b Magnetization configuration

Fig. 11.5a,b. Domain wall structure of a magnetic bubble

and δ is the wall width. As is well known, σ_{w0} and δ are expressed as $4(AK_u)^{1/2}$ and $\pi(A/K_u)^{1/2}$ respectively, where A is the exchange stiffness constant (Appendix A.8). In this case, the equation for energy balance (11.2) should be modified using the modified l as expressed in the next equation

$$l = l_0\left[1 + \left(\frac{n\delta}{\pi d}\right)^2\right]^{-1/2} . \tag{11.4}$$

Here l_0 is the characteristic length $\sigma_{w0}/4\pi M_s^2$ for a wall without any Bloch line. The d vs H_B curves can be calculated using (11.2) and (11.4), which explain the experimental results well as shown in Fig. 11.4 [11.8].

Magnetic bubbles with more than several tens of Bloch lines are called hard magnetic bubbles, since they hardly collapse even at higher H_B. On the other hand, magnetic bubbles with zero or two Bloch lines are called normal or soft magnetic bubbles. The d vs H_B curve in the lowest H_B region with lowest H_0 corresponds to soft magnetic bubbles.

11.1.3 The Motion of a Magnetic Bubble

A magnetic bubble moves in the direction of lower bias field when it is subjected to the gradient of the bias field. The velocity v of a magnetic bubble in this case is expressed by [11.10]

$$v = \frac{1}{2}\mu_w\left(\Delta H - \frac{8}{\pi}H_c\right) . \tag{11.5}$$

Here μ_w and H_c are the wall mobility and coercive force for a planar wall respectively. ΔH, called the drive field, is the difference between bias fields at both ends of a magnetic bubble. Experiments show that v differs from the value obtained by (11.5) to about one tenth of that value. Some magnetic bubbles move at an angle deviating considerably from the gradient of the bias field. These phenomena come from the complicated domain wall structures of magnetic bubbles [11.9, 11] as mentioned in the previous section. As the collapse field H_0 is increased, that is, the number of Bloch lines in the walls is increased, v is decreased and the angle of deflection is increased as shown in Fig. 11.6. The reason for the decrease in v is that the drive energy due to ΔH is consumed in shifting the Bloch lines along the wall of a magnetic bubble rather than in moving the magnetic bubble itself. As the number of Bloch lines is increased, the integrated value of ϕ, the rotation angle of the magnetization in the center of the wall, along the circumference is increased. The integrated value divided by 2π is called the S number

$$S = \frac{1}{2\mu}\oint d\phi . \tag{11.6}$$

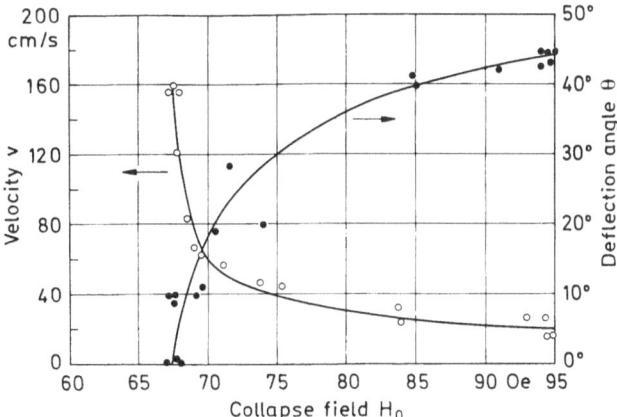

Fig. 11.6. Velocity and deflection angle of a magnetic bubble vs collapse field

A magnetic bubble with a large S moves with a large deflection angle θ from the direction of the bias field gradient due to a strong gyrotropic effect [11.11]. θ is theoretically obtained [11.11] from

$$\tan \theta = \frac{4\mu_w}{\gamma d} \cdot S .\tag{11.7}$$

Here γ is the gyromagnetic ratio. It should be noticed here that the simplest magnetic bubble without any Bloch lines has an S of 1 and a θ with a definite value, and also that the magnetic bubble with two Bloch lines and an S of 0 can move parallel to the bias field gradient [11.11] as shown in Fig. 11.7.

In actual magnetic bubble devices, hard magnetic bubbles with low v and large θ are suppressed using several methods, otherwise they can cause operating errors. Only $S = 1$ and $S = 0$ magnetic bubbles exist in actual devices. Details will be given in Sect. 11.3.

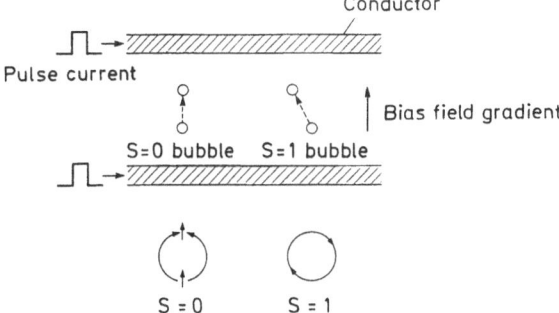

Fig. 11.7. Direction of motion of a magnetic bubble in a bias field gradient

The velocity of magnetic bubbles with $S = 1$ or 0 increases linearly with increasing ΔH according to (11.5) and saturates for larger ΔH as shown in Fig. 11.8. This is caused by the transient generation of reversed domains in the wall during motion at higher v as shown in Fig. 11.9. The driving energy is consumed by the generation and annihilation of these domains, which saturates the velocity of motion of a magnetic bubble [11.12]. The saturation value of v, v_0, is given theoretically by [11.12]:

$$v_0 = C \frac{\gamma A}{hK_u^{1/2}} \ . \tag{11.8}$$

Here C is a constant depending on d/h and ranging from 7.1 to 8.3. This equation only explains the experimental values semi-quantitatively, however. v_0 is of the order of $1.0-3.0 \times 10^3$ cm/s in garnet films in practical use.

The existence of v_0 is one of the factors which limit the operational speed of magnetic bubble devices, but up to now it has not been a serious limitation.

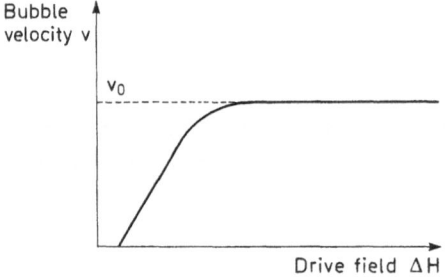

Fig. 11.8. Velocity of a magnetic bubble vs drive field

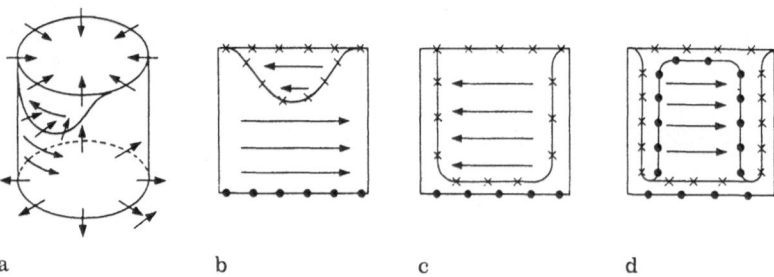

Fig. 11.9a–d. Change of domain wall structures of a magnetic bubble during motion

11.2 Magnetic Bubble Materials

11.2.1 Requirements for Magnetic Bubble Materials

The first requirement for a magnetic bubble material is that the magnetization is directed perpendicular to the film plane. That is, the uniaxial magnetic anisotropy K_u which directs the magnetization perpendicular to the film plane should be larger than the shape anisotropy $2\pi M_s^2$ which directs the magnetization parallel to the film plane. In other words, the uniaxial anisotropy field H_k should be larger than the demagnetizing field $4\pi M_s$.

$$\left.\begin{array}{l} K_u > 2\pi M_s^2 \\ H_k > 4\pi M_s \end{array}\right\} . \tag{11.9}$$

Several materials meet this requirement: Orthoferrites $RFeO_3$ (R: rare-earth ion); garnets $(R_1 R_2)_3(FeX)_5O_{12}$ (R_1, R_2: rare-earth ions, X: non magnetic ion such as Ga, Al and so on); hexagonal ferrites and also rare earth–transition metal amorphous alloys such as Gd–Co. K_u and $4\pi M_s$ for these materials are plotted in Fig. 11.10.

The second requirement to be satisfied is that the diameter of magnetic bubbles d should be small (0.3–5 μm). This is because the storage density of

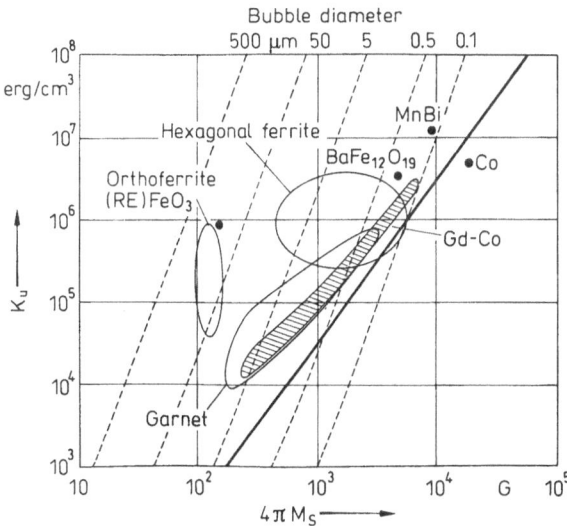

Fig. 11.10. K_u vs $4\pi M_s$ for various materials

magnetic bubble devices increases with decreasing d, since the area of the 1-bit cell is about $4d \times 4d$. d is roughly given by [11.5]

$$d \simeq 8l = \frac{8\sqrt{AK}}{\pi M_s^2} \ . \tag{11.10}$$

In the plot of $K_u - 4\pi M_s$, d is shown in Fig. 11.10. It is shown that in garnet, hexagonal ferrite and Gd–Co with a large $4\pi M_s$, d can be smaller than 1 μm.

The third requirement is that the coercive force H_c should be smaller than 1 Oe. Magnetic bubble devices utilize the propagation of magnetic bubbles in the film, so a low coercive force is necessary to move magnetic bubbles smoothly with a small driving force. In this respect garnets are the best.

In addition to the above three requirements, it is very important that uniform and defect-free films of several μm in thickness can be prepared with good reproducibility. In this respect garnets are also excellent.

Garnet is the only material that can meet all the requirements for a magnetic bubble material. The fabrication and magnetic properties of garnet films will be described below.

11.2.2 Fabrication and Magnetic Properties of Garnet Films

The garnet films considered here are rare-earth iron garnets whose chemical formula is $(R_1 R_2 - -)_3 (\text{Fe}, X)_5 O_{12}$. Here R_1, R_2 are rare-earth ions such as Sm, Eu, Lu or Y, Bi. More than two kinds of rare-earth ions are necessary to induce the uniaxial magnetic anisotropy K_u. X is an ion such as Ga, Ge and Al which replaces some of the Fe ions in order to adjust $4\pi M_s$. Garnet is a cubic crystal as shown in Fig. 11.11. Rare earth ions and Fe ions occupy three kinds of

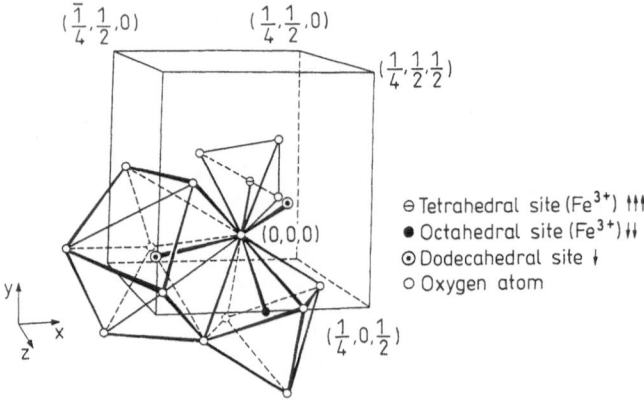

Fig. 11.11. Crystal structure of a garnet

sites in the lattice of oxygen ions. The rare-earth ions occupy dodecahedral sites surrounded by eight oxygen ions. The Fe ions occupy both tetrahedral and octahedral sites surrounded by four and six oxygens respectively. The unit cell of garnet is composed of 96 oxygens, 24 Fe ions partly substituted by Ga or Al ions occupying the tetrahedral sites, 16 Fe ions occupying the octahedral sites and 24 rare-earth ions occupying the dodecahedral sites.

Single crystal magnetic garnet films are epitaxially grown on non-magnetic single crystal substrates by liquid phase epitaxy (LPE) [11.13]. In most cases, about 0.5 mm thick $\{111\}$ plates of $Gd_3Ga_5O_{12}$ (GGG) single crystals are used as substrates. A single crystal GGG boule is grown by the Czochralski method. After being sliced to a wafer the surface of the GGG wafer is polished defect-free by mechano-chemical polishing using colloidal silica. In a Pt crucible, the melt composed of garnet constituents such as oxides of Sm_2O_3, Y_2O_3, Fe_2O_3, Ga_2O_3 and a flux of Pbo and B_2O_3 is kept at a high temperature of $900-1000\,°C$. When the concentration and temperature of the melt are suitably selected, the melt can be in a stable state of super-saturation for a long time. When the GGG substrates are immersed into this supersaturated melt, single crystal garnet films are epitaxially grown on the substrates. When the substrates are held horizontally and rotated at $30-100$ rpm in the melt, the variations in composition and thickness of the films can be smaller than $\pm 1\%$ in one wafer. The defect density can be smaller than $1/cm^2$.

The advantage of this method is that garnet films of any composition composed of many elements can be grown with good reproducibility if the concentration and the temperature of the melt are controlled very precisely. In this respect this method is very practical.

Garnet is ferrimagnetic. As shown in Fig. 11.14 the magnetizations of the Fe ions in the tetrahedral and octahedral sites are directed anti-parallel to each other. The magnetization of rare-earth ions in the dodecahedral sites is anti-parallel to that of the Fe ions in the tetrahedral sites. The total magnetization of garnet is obtained by summing the above three sublattice magnetizations. The magnetization of the garnet can be adjusted to a suitable value by substituting non-magnetic ions such as Ga, Ge, Al and so on, into tetrahedral sites.

In garnet films epitaxially grown by the LPE method, a large uniaxial magnetic anisotropy K_u (10^4-10^5 erg/cm^3) is induced, whose easy axis is perpendicular to the film plane. Pairs of rare earth ions such as Sm–Lu, Sm–Tm, Eu–Lu are very effective for inducing K_u. Among non-magnetic ions, the pair Bi–Y induces a large K_u. This kind of anisotropy depends on the direction of the growth of the garnet and called growth-induced anisotropy. As mentioned above, there are 24 dodecahedral sites which rare-earth ions should occupy in one unit cell. These 24 sites are not spatially equivalent with respect to the film growth direction. Therefore two kinds of ions with different ion sizes such as Sm and Lu are apt to occupy different sites during film growth. Thus each rare earth ion, including magnetic ions such as Sm, occupies some dodecahedral site preferentially. This preferential site occupation of magnetic rare-earth ions will induce the uniaxial anisotropy through the super-exchange interaction with Fe ions [11.14].

The growth induced anisotropy disappears after garnet films are heated up to about 1200 °C. This is explained in terms of a random distribution of rare earth ions caused by annealing. This annealing effect in growth-induced anisotropy is very different from the annealing of crystalline anisotropy due to crystal structure. Although the above model is very plausible, the preferential site occupation has not yet been demonstrated experimentally.

In actual devices $(YSmLuCa)_3(FeGe)_5O_{12}$ garnet films supporting 1–3 µm magnetic bubbles are used. The Sm and Lu ions are used for inducing K_u, the Ge ion is used for adjusting $4\pi M_s$, the divalent Ca ion is used for charge compensation, since the Ge ion is tetravalent, and the Y ion is used mostly for adjusting the lattice constant to that of the GGG substrate, 12.383 Å.

The characterization of the magnetic properties of the garnet films is carried out as follows. The width of the stripe domains in the demagnetized state, W, and the collapse field, H_0, are obtained by observing magnetic domains through a polarized microscope. Magnetic bubbles observed in this way are simple $S = 0$ or $S = 1$ and H_0 can be obtained accurately. From W and H_0, the wall energy density, σ_w, the characteristic length $l(\equiv \sigma_w/4\pi M_s^2)$, $4\pi M_s$ and other physical parameters [11.15, 16] can be obtained using the equations

$$\frac{l}{h} = \frac{4}{\pi^3}\left(\frac{W}{h}\right)^2 \sum_{n=\text{odd}} \frac{1}{n^3}[1 - e^{-n\pi h/W}(1 + n\pi h/W)] \qquad (11.11)$$

$$H_0 = 4\pi M_s\left[1 + \frac{3}{4}\frac{l}{h} - \left(\frac{3l}{h}\right)^{1/2}\right]. \qquad (11.12)$$

Here h is the thickness of the film obtained using an interferometer. The uniaxial anisotropy field $H_k(\equiv 2K_u/M_s)$ is obtained by ferromagnetic resonance. The physical parameters obtained by the methods mentioned above for typical garnet films with 1–5 µm diameter magnetic bubbles are shown in Table 11.1. The exchange stiffness constant A (Appendix A.8) is obtained from the relation $\sigma_w = 4(AK_u)^{1/2}$. The Curie temperature T_C is obtained using a magnetic balance.

11.2.3 Suppression of Hard Magnetic Bubbles

As mentioned in Sect. 11.2.2 hard magnetic bubbles having many Bloch lines in the domain wall exist among the magnetic bubbles in a garnet film. These hard bubbles harm the device operation, since they move very slowly and are deflected from the bias field gradient. Thus in actual devices hard magnetic bubbles have to be removed.

Three methods for suppressing hard magnetic bubbles in garnet films have been developed: 1. the ion-implantation method [11.17], 2. the permalloy

Table 11.1. Characteristic parameters of typical garnet films

Composition	Bubble diameter d [μm]	Characteristic length l [μm]	Magnetic wall energy density σ_w [erg/cm^2]	Exchange stiffness constant A [erg/cm]	Saturation induction $4\pi M_s$ [G]	Anisotropy field H_k [Oe]	Curie temperature T_c [°C]
$(Y_{1.4}Sm_{0.3}Lu_{0.4}Ca_{0.9})$ $(Fe_{4.1}Ge_{0.9})O_{12}$	5	0.53	0.22	2.0×10^{-7}	240	1600	195
$(Y_{1.1}Sm_{0.3}Lu_{0.7}Ca_{0.9})$ $(Fe_{4.1}Ge_{0.9})O_{12}$	3	0.32	0.26	2.1×10^{-7}	330	1600	200
$(Y_{1.0}Sm_{0.5}Lu_{0.7}Ca_{0.8})$ $(Fe_{4.2}Ge_{0.8})O_{12}$	2	0.21	0.33	2.2×10^{-7}	430	1800	210
$(Y_{0.4}Sm_{0.7}Lu_{1.3}Ca_{0.6})$ $(Fe_{4.4}Ge_{0.6})O_{12}$	1	0.11	0.50	2.4×10^{-7}	840	2000	220

deposition method [11.18] and 3. the double garnet epitaxy method [11.19]. In all cases, a second magnetic layer in which the magnetization is in the plane is laid on the top surface of the magnetic bubble layer as shown in Fig. 11.12. The magnetization of the second layer interacts strongly with the magnetization in the wall of a magnetic bubble through the exchange interaction. This interaction should make a magnetic bubble having a simple wall structure ($S = 0$) energetically favorable as shown in Fig. 11.13a. Hard magnetic bubbles as shown in Fig. 11.13b should hardly exist because of the high exchange energy. Actually planar domain structures appear in the second layer as shown in Fig. 11.14a, which makes $S = 1$ magnetic bubbles without Bloch lines the most stable [11.20] as shown in Fig. 11.14b. When the magnetic field is applied parallel to the film plane, the planar domain structure disappears and an $S = 0$ magnetic bubble appears [11.20].

In actual devices, $S = 1$ and $S = 0$ magnetic bubbles will coexist. $S = 1$ magnetic bubbles are deflected from the bias field gradient by some definite angle, but this has not been observed to harm the device operation. As a method for suppressing hard magnetic bubbles, ion implantation is now most used in practice because of its simplicity.

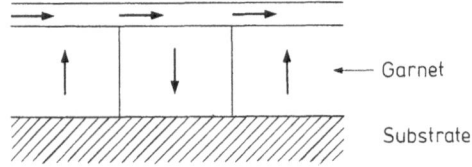

Fig. 11.12. Garnet film with a capping magnetic layer on the top surface

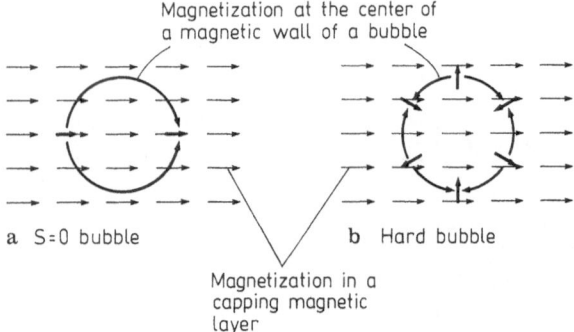

Fig. 11.13a, b. Domain wall structure of a bubble (thicker line) and magnetization in a capping magnetic layer (thinner line)

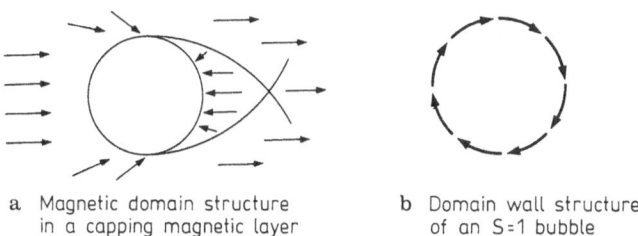

a Magnetic domain structure
  in a capping magnetic layer

b Domain wall structure
  of an S=1 bubble

Fig. 11.14a, b. Planar domain wall structure in a capping magnetic layer and domain wall structure of a bubble

11.3 Magnetic Bubble Devices

11.3.1 Outline of Devices

Magnetic bubble devices are shift register type or sequential memory devices in which the binary codes '1' and '0' are represented by the presence and absence of a magnetic bubble respectively, and consequently information is stored as a train of magnetic bubbles propagating in a garnet film.

Methods for propagating magnetic bubbles in a garnet film are classified into field access [11.13] and current access [11.21, 22] drives. In the field access drive, a magnetic field for propagating magnetic bubbles is applied to a device by supplying a current to a coil surrounding the device. On the other hand, in the current access drive, current is supplied to fine conductor patterns formed on a garnet film and the magnetic driving field is generated. The current access drive has the big disadvantage of high joule heating. Consequently field access drive is now used in practice.

In the field access drive, a propagating track is needed for stabilizing the bit positions and driving the magnetic bubbles. As a propagating track, permalloy patterns deposited on a garnet film (Fig. 11.15) or ion implanted patterns (Fig. 11.17) are now used. A permalloy track is composed of many small permalloy (80Ni–20Fe alloy, ~ 4000 Å) patterns periodically formed on a garnet film (Fig. 11.15). When a magnetic field is applied parallel to the film plane, small permalloy patterns are magnetized and magnetic N poles appear at specified positions on the permalloy patterns. As the direction of the applied magnetic field is rotated, the N poles shift along the permalloy track. The magnetic bubbles are attracted by the N poles and move along the permalloy track (Fig. 11.16), since magnetic bubbles have S poles on their top surfaces. With one revolution of the rotating magnetic field, magnetic bubbles transfer along the permalloy track by one period. This is the basic function of the device.

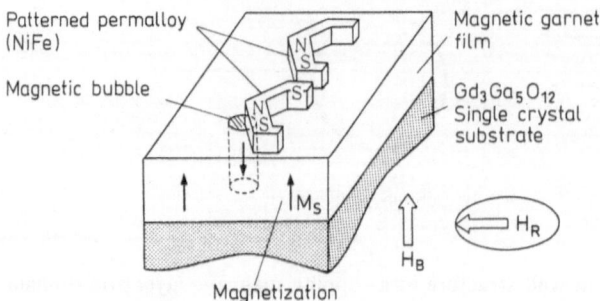

Fig. 11.15. Permalloy track

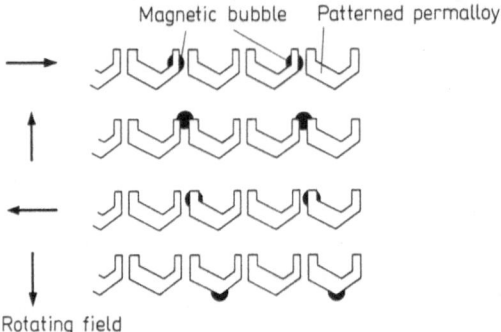

Fig. 11.16. Direction of inplane field vs position of magnetic bubbles

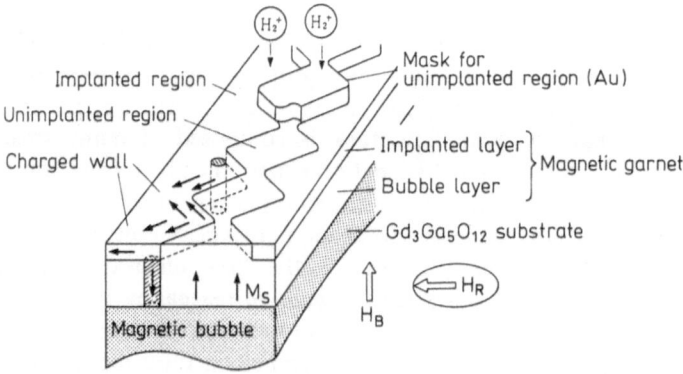

Fig. 11.17. Ion-implanted track

Ion implanted tracks are formed in the top layer of a garnet film by implanting Ne or H_2 ions through masks with contiguous patterns (Fig. 11.17). In the ion implanted region, strain is induced in the top layer of the garnet film and the magnetization switches its direction from perpendicular to parallel to the film plane through magnetostriction. When the magnetic field is applied parallel to the film plane, N poles appear locally at the positions where neighboring magnetizations in the ion implanted region collide since they tend to be directed parallel to the edge of the contiguous patterns. This local place where N poles appear is called a charged wall. Magnetic bubbles are attracted by charged walls and are transfered with a shift of the charged wall along the contiguous patterns caused by the rotating magnetic field. Since these contiguous patterns have no gaps between patterns, defining patterns with shorter periods is easy. Therefore ion implanted tracks are now being developed for high storage density devices.

The organization of memory devices is shown in Fig. 11.18. The write operation is carried out by generating a train of magnetic bubbles at a bubble generator located at one end of write major line, transfering them along the major line to swap gates (Appendix A.36) and then gating them out to many minor loops with one magnetic bubble stored in one minor loop in one operation. A train of magnetic bubbles stored in many minor loops corresponds to information for one address. By carrying out write operations sequentially, information for many addresses is stored by many trains of magnetic bubbles stored in minor loops. The read operation is carried out by replicating magnetic bubbles at block replicators located at each junction between minor loops and read major lines, transfering replicated magnetic bubbles along the read major line and detecting the presence or absence of magnetic bubbles with a bubble detector. Even after replicating, magnetic bubbles are still stored in minor loops, which leads to non-volatility of stored information.

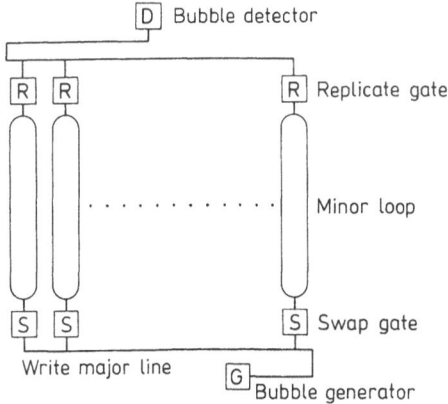

Fig. 11.18. Organization of a magnetic bubble device

This major line–minor loop organization has several advantages. 1. Access time is determined mainly by the length of the minor loops and can be much shorter than single loop organization. 2. Redundant minor loops can be used instead of bad minor loops.

Replicate and swap gates, bubble generators and detectors perform complicated functions. They are called functional parts. Functional parts based on permalloy tracks work well and have found practical use. On the other hand, functional parts based on ion implanted tracks are still at the laboratory level.

11.3.2 Permalloy Devices

Several patterns have been devised as permalloy tracks. In Fig. 11.19, some typical patterns are shown. In each case, the period of the track λ is about four times as long as the diameter of a magnetic bubble d, since the interactions between magnetic bubbles should be sufficiently weak. At an early stage of research, T bar patterns [11.3] (Fig. 11.19a) were used, but nowadays, half discs [11.24] or asymmetric chevron patterns [11.25] with wide gaps between patterns are in practical use (Fig. 11.19b and c). At present 1–4 Mbit devices with $d = 1.5-2\,\mu m$, $\lambda = 6-8\,\mu m$ and a gap of $1\,\mu m$ are commercially available. Recently, wide gap patterns [11.26] (Fig. 11.19d) have been devised and devices with a λ of $4\,\mu m$ are now being developed. For λ shorter than $4\,\mu m$, permalloy patterns are too small and magnetic poles are too weak to propagate magnetic bubbles.

Functional parts such as bubble generators are shown in Fig. 11.20. The bubble generator is composed of a hair-pin conductor (Al–Cu, Au/Mo) formed under a permalloy pattern with an intermediate dielectric layer between them. A strong magnetic field locally generated at the center of a hair-pin by applying a

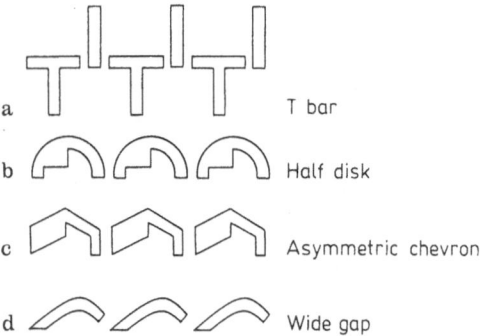

a   T bar

b   Half disk

c   Asymmetric chevron

d   Wide gap

Fig. 11.19a–d. Various types of permalloy tracks

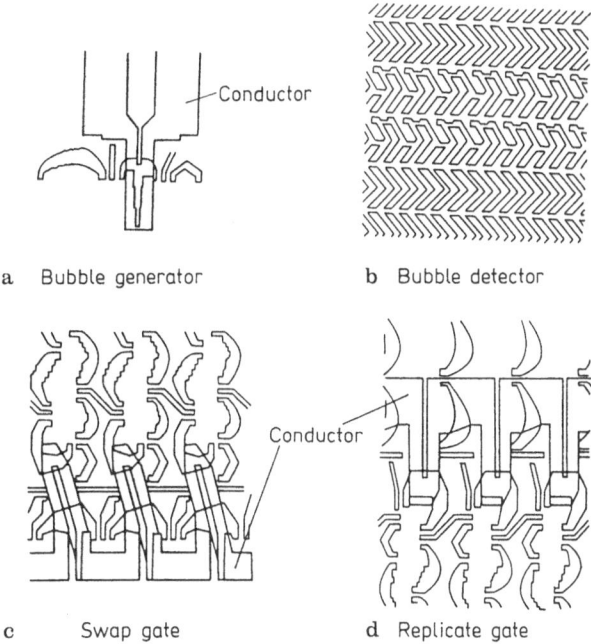

a   Bubble generator   b   Bubble detector

c   Swap gate   d   Replicate gate

Fig. 11.20a–d. Functional parts for a permalloy device

pulsed current to the conductor nucleates a magnetic bubble. Swap and replicate gates are also composed of permalloy and conductor patterns. In these
gates, local magnetic fields are generated by applying a pulsed current to
conductors and these stop the propagation of magnetic bubbles. A bubble
detector is composed of many trains of chevron permalloy patterns. These
patterns elongate a magnetic bubble to a long horizontal strip domain of about
1 mm in length. A magnetoresistive permalloy sensor connecting the chevron
patterns is laid along the track to detect fringing magnetic fields from the
propagated strip domain. Thus the presence or absence of magnetic bubbles can
be detected.

The cross section of a permalloy device is shown schematically in Fig. 11.21.
Conductors such as Al–Cu or Au/Mo and permalloy patterns are deposited
sequentially onto garnet with intermediate layers of SiO_2 or polyimide. It
should be noticed that almost all of the device area is occupied by minor loops
composed of permalloy patterns only. Therefore device fabrication is easy and
1–4 Mbit devices with minimum features of 1 μm are now produced using
conventional UV lithography.

Devices are usually operated by a rotating magnetic field H_R whose frequency is 100–200 kHz. Device characteristics are represented by a bias magnetic field $H_B - H_R$ operation region (Fig. 11.22). When H_B and H_R are in the

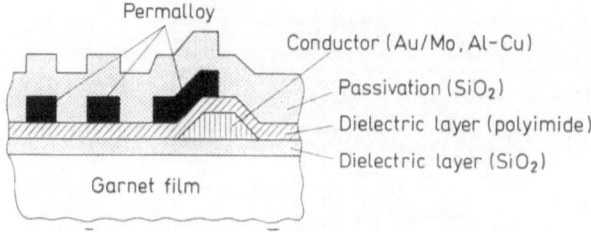

Fig. 11.21. Cross-sectional structure of a permalloy device

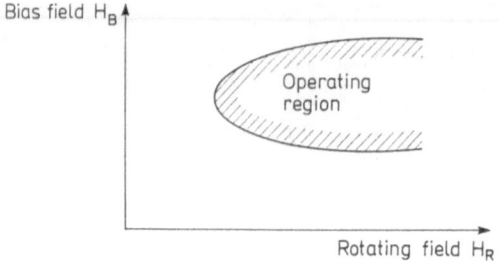

Fig. 11.22. Operating characteristics of a device

appropriate hatched region, the device operation is correct. When H_R is smaller, the magnetic pole on the permalloy track is so weak as to lead to mis-operation in the propagation of magnetic bubbles. When H_B is larger, magnetic bubbles are apt to collapse. On the other hand, when H_B is smaller, magnetic bubbles tend to run out to labyrinth domains. Both of these effects lead to mis-operation. The larger the operation region of $H_B - H_R$, the more stable the device operation.

The speed of device operation is limited by the drive circuit for generating the rotating magnetic field or by joule heating at the coils rather than motion of magnetic bubbles. This will be described in Sect. 11.5. At present the upper limit of frequency for practical operation is 200–300 kHz.

11.3.3 Ion Implanted Devices

As mentioned above, permalloy devices are commercially available. However, devices with bit periods less than 4 μm cannot really be achieved using the permalloy track approach. On the other hand, ion implanted tracks can transport submicron diameter magnetic bubbles which opens up the possibility of 4–16 Mbit devices with a bit period of 2.0–4.0 μm [4.27]. Figure 11.23 shows

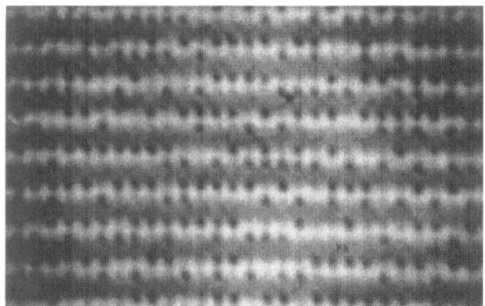

Fig. 11.23. 0.5 μm bubble propagating on an ion-implanted track of 2 μm period observed by a polarized microscope

the propagation of 0.5 μm diameter magnetic bubbles along an ion implanted track with a 2.0 μm period formed in $(SmLuGd)_3(FeAl)_5O_{12}$ garnet films. This is a photograph of magnetic bubbles and an ion implanted track observed by a polarized microscope. Here the black circles are magnetic bubbles and the dark region is the H_2 implanted region.

The important factor for improving the characteristics of propagation tracks is to find out the right conditions for ion implantation. When ions such as Ne, He or H_2 are implanted into a garnet film, strain is induced in the implanted region. The easy axis of uniaxial anisotropy is switched from the perpendicular to the direction parallel to the film plane through magnetostriction. The change in the anisotropy field ΔH_k in this process is given by

$$\Delta H_k = \frac{3\lambda}{M_s} \cdot \frac{E\,\Delta a}{(1+v)a} \ . \tag{11.13}$$

Here λ is the magnetostriction constant, E, Young's modulus, v, Poisson's ratio, and a and Δa, the lattice spacing and its change respectively. From (11.13), ΔH_k should be proportional to the strain $\Delta a/a$. Experimental data on ΔH_k are shown in Fig. 11.24 [11.28]. For Ne and He, ΔH_k is saturated at larger strains. On the other hand, for H_2, ΔH_k is proportional to strain according to (11.13). The reason for this different behavior of ΔH_k for H_2 and for other ions is not yet clear.

The thickness of an ion implanted layer is roughly one-third of that of a garnet film, and the strain profile along the depth of a film due to single ion implantation is of almost Gaussian type. Therefore multiple ion implantations with different ion doses and energies are performed to obtain a flat strain profile (Fig. 11.25), which leads to good characteristics of bubble propagation.

Bubble generators, bubble detectors using thin permalloy films and transfer gates with good performance have been extensively developed as functional parts for ion implanted devices [11.29, 30] (Fig. 11.26). However, replicate and

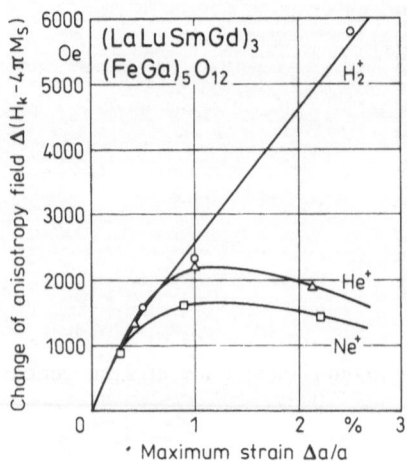

Fig. 11.24. Change in effective anisotropy field vs strain induced by ion implantation

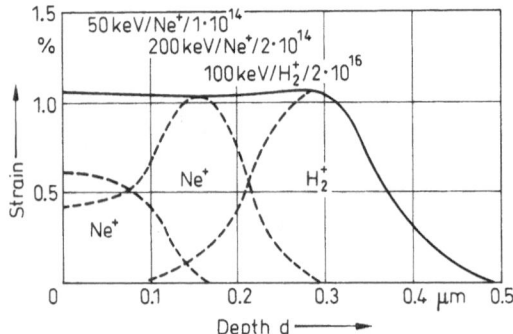

Fig. 11.25. Profile of strain induced by multiple ion implantation

swap gates are not good enough yet for practical use. These two gates are now under development, since they are necessary for non-volatility of information.

Recently, hybrid devices have been proposed and developed [11.31] in which minor loops are composed of ion-implanted tracks, and major lines and functional parts including replicate and swap gates are composed of permalloy tracks. These devices have the advantages of high storage density due to the ion-implanted tracks used for the minor loops, and of non-volatility of stored information due to block replicate and swap gates in the major line based on permalloy tracks.

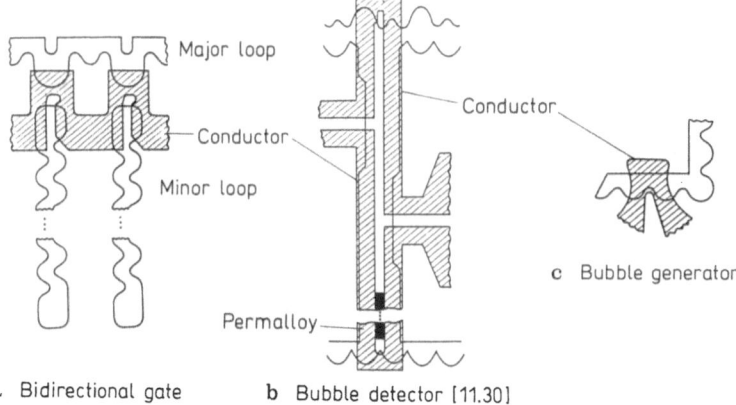

Major loop

Conductor

Minor loop

Conductor

Permalloy

a   Bidirectional gate

b   Bubble detector [11.30]

c   Bubble generator

Fig. 11.26a, b. Functional parts for an ion-implanted device

11.4   Magnetic Bubble Memories and Applications

11.4.1   Memory Modules and Drive Circuits

In order to drive permalloy or ion implanted magnetic bubble devices, it is necessary to apply a dc bias magnetic field and rotating field perpendicular and parallel to the film plane, respectively. For this purpose a magnetic bubble device is placed in a structure or module (Fig. 11.27). The bias field is applied by a pair of permanent magnetic plates set on and under the device. Since the magnets are permanent, a definite value of the bias field is always applied

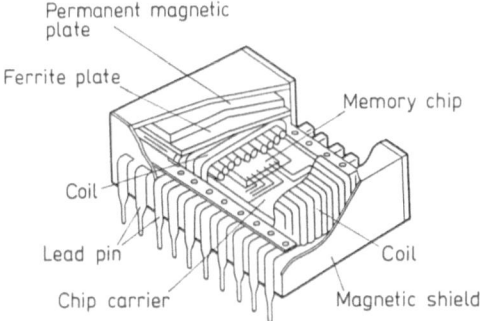

Permanent magnetic plate

Ferrite plate

Memory chip

Coil

Lead pin

Coil

Chip carrier

Magnetic shield

Fig. 11.27. Structure of a magnetic bubble module

perpendicular to the plane of the garnet film to stabilize the magnetic bubbles. Non-volatility of stored information is based on this stability of magnetic bubbles. When the surrounding temperature is changed, the bias field stabilizing the magnetic bubbles is changed according to the temperature dependence of the $4\pi M_s$ of a garnet film. The stability is achieved automatically by making the temperature dependences of $4\pi M_s$ for garnet and permanent magnets as similar as possible. In most cases, barium ferrite is used as a bias permanent magnet. Consequently the temperature dependences of the $4\pi M_s$ of garnet is matched to that of barium ferrite by adjusting the composition of the garnet.

The rotating magnetic field parallel to the film plane is applied by supplying sinusoidal or triangular currents with phases differing by 90° to two coils set orthogonal to each other surrounding the device (Fig. 11.28). The advantage of this method is that the start/stop operation is possible at any time. So a rotating field is applied only when a write or read operation is carried out, which leads to low power consumption.

With the start/stop operation, millions of magnetic bubbles simultaneously start/stop propagating. At this time not even a single magnetic bubble can be allowed to behave in the wrong manner. The start/stop operation is stable if a dc magnetic field of several Oersteds is applied in a specified direction parallel to the garnet film plane. This dc field is applied by making the device tilt from the permanent magnet plane by several degrees.

A memory module is magnetically shielded by permalloy plates so that malfunction due to spurious magnetic field is prevented. The pulsed currents for bubble generators and gates and output signals from bubble detectors are supplied or taken out through lead-pins. The electronic circuits for pulsed current suppliers and signal amplifiers are conventional ones.

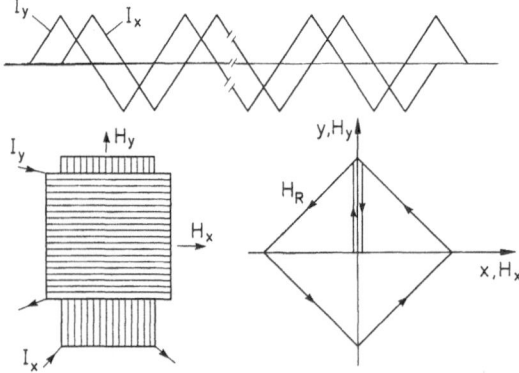

Fig. 11.28. Coils and drive currents for generating a rotating field

11.4.2   Magnetic Bubble Memories

Magnetic bubble memories are composed of modules, drive circuits, controllers and power suppliers. A typical example [11.32] is shown in Fig. 11.29. On the one board, there are four 1 Mbit modules and drive circuits along with controllers and a power supply. The controllers and some of the drive circuits are LSI, which helps to make the volume of the memory much smaller.

The above mentioned modules on the board cannot be taken on and off once they are mounted, but other cassette type modules [11.33] capable of being taken on and off have been developed.

The capacity of the memories most in use is rather small, ranging from 1 Mbit to 16 Mbit, but memory capabilities of 64 Mbit or more are also used [11.34].

The operating frequency of magnetic bubble memories is 100–300 kHz. This is because at frequencies higher than 300 kHz the voltage applied to the transistors used for the rotating field drive circuits exceeds the allowed upper limit for the transistors.

Performances and specifications of magnetic bubble memories vary considerably and an example is shown in Table 11.2. The average access time is 4–13 ms and the transfer rate is 100–200 kbit/s. Transfer rates can be increased by parallel device operation. Power consumption depends on memory capacity, but roughly it is 3–20 W. The ambient temperature is 0–50°C.

Fig. 11.29. 512 kB memory using four 1 Mbit module (150 mm × 240 mm)

Table 11.2. Performance of bubble memory boards. Performances of memory boards of 1 Mbit bubble devices using a +5V mono-power supply

	BEL 0830	BEM 0830	BEN 0830
Memory capacity	128 kB	256 kB	512 kB
Number of pages	4096	8192	16384
Data bits/page	256 bit/page		
Power supply voltage	+5V + 5%		
Power consumption			
in operation	11 W		
non-operation	3.5 W		
Data transfer rate	100 kbit/s max		
Average access time	13 ms		
Interface	8 bit parallel, TTL compatible		
Temperature range			
in operation	0–50 °C		
non-operation	− 40–80 °C		
Board size	150 mm × 240 mm		
Transfer mode	PIO mode, DMA mode		
Access page number	Single page, multiple page		

Note: PIO (Program input output), DMA (Direct memory access), TTL (Transistor transistor logic)

11.4.3 Applications of Magnetic Bubble Memories

The advantages (circled) and disadvantages of the magnetic bubble memories described here are compared below with semiconductor memories and with magnetic recording.

Semiconductor
1 Volatile
2 Device fabrication is complicated (more than 6 photomasks needed)
③ Access time is very short (~ 100 ns)

Magnetic bubble
① Non-volatile
② Device fabrication is simple and easy only 2–3 photomasks needed
3 Access time is long (> 1 ms)

Magnetic recording
1 Mechanical moving parts leading to low reliability
2 Not resistant to vibration and not portable
3 Maintenance necessary
④ Cost per bit very low
5 Noisy

Magnetic bubble
① All solid state components leading to high reliability
② Small and portable
③ Maintenance-free
4 Cost per bit high
⑤ Noise-free

Some applications of magnetic bubble memories are the following:

1. Electronic switching systems [11.34] (high reliability desirable)
2. Numerical control machines, robots (high reliability and high resistivity to surroundings required)
3. POS, work stations (require small volume, low power consumption, no noise)
4. Voice recorder [11.35] (should be maintenance-free)

11.5 Future Trends in Magnetic Bubble Devices and Memories

As mentioned in Sect. 11.4, 1–4 Mbit magnetic bubble devices using 1.5–2.0 μm magnetic bubbles are at present commercially available. In these devices permalloy tracks of 6–8 μm in bit period and with 1 μm minimum features are used. The future of magnetic bubble memories depends on how large a storage density or capacity can be achieved. That is, how short a bit period can be achieved is a key for the development of magnetic bubble devices. The evolution and expectation of the bit period λ, the diameter of the magnetic bubbles d, and the minimum feature size W necessary to fabricate devices are shown in Fig. 11.30. In order to shorten the bit period, it is necessary to develop submicron diameter magnetic bubble garnets and submicron lithography as well as device technology for manipulating submicron magnetic bubbles such as transfer, detection and replication. Research on these subjects is being intensively carried out in Japan and France, and may lead to the realization of 16–64 Mbit devices in the future.

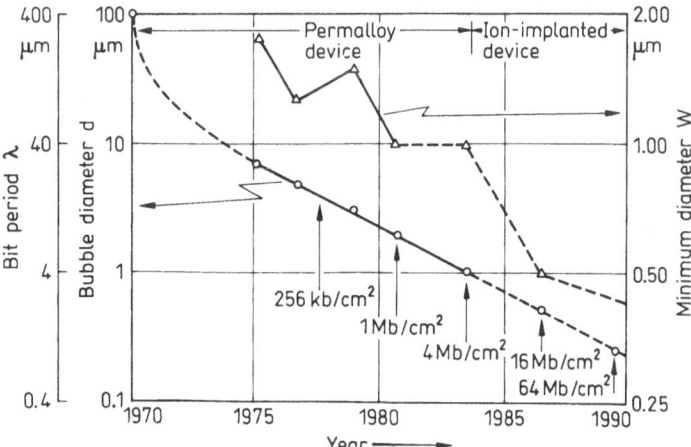

Fig. 11.30. Evolution of bit period, bubble diameter and minimum feature size

For submicron magnetic bubble garnets, $(SmLuBi)_3(FeSc)_5O_{12}$ garnet films supporting 0.35 µm magnetic bubbles have been developed [11.36]. For submicron lithography, new techniques such as far UV, X-ray, and excimer lasers are now in progress. In the light of this, 0.2–0.5 µm minimum features will be in practical use in the near future.

For device technology, ion implanted device technology, including replicate and swap gates, is now under development, which will lead to 16–64 Mbit devices with 1 µm bit periods.

Very recently, a new device for high storage density has been proposed [11.37] based on magnetic bubble technology. In this device, binary digits '1' and '0' are represented by the presence and absence of Bloch lines in the domain wall. At present, basic research on this Bloch line memory device is in progress, and may contribute to the appearance of 100 Mbit or more devices in the future.

11.6 Summary

Magnetic bubbles have been reviewed from the materials and physics viewpoint through to devices and their applications. As described above, the development of magnetic bubble devices has been based on the results of research on the physics of magnetism such as the magnetism of garnets, magnetic domain structures and so on. Therefore magnetic bubble devices can be said to be typical examples of the application of the physics of magnetism. Research on devices, in turn, has led to attractive new results in such basic areas as growth-induced uniaxial anisotropy in garnets, domain wall statics and dynamics, ion implantation effects in garnets and so on, which have contributed much to the physics of magnetism.

References

11.1.   A.H. Bobeck: Bell Syst. Tech. J. **46**, 1901 (1967)
11.2.   A.H. Bobeck, R.F. Fischer, A.J. Perneski, J.P. Remeika, L.G. Van Uitert: IEEE Trans. Magn. MAG-**5**, 544 (1969)
11.3.   A.J. Perneski: IEEE Trans. Magn. MAG-**5**, 554 (1969)
11.4.   A.A. Thiele: Bell Syst. Tech. J. **48**, 3287 (1969)
11.5.   A.A. Thiele: J. Appl. Phys. **41**, 1139 (1970)
11.6.   A.P. Malozemoff: Appl. Phys. Lett. **21**, 149 (1972)
11.7.   W.J. Tabor, A.H. Bobeck, G.P. Vella-Coleiro, A. Rosencwaig: AIP Conf. Proc. No. 10, 442 (1973)
11.8.   H. Nishida, T. Kobayashi, Y. Sugita: AIP Conf. Proc. No. 10, 493 (1973)

11.9.   J.C. Slonczewski, A.P. Malozemoff, O. Voegeli: AIP Conf. Proc. No. 10, 458 (1973)
11.10. A.A. Thiele: Bell Syst. Tech. J. **50**, 725 (1971)
11.11. J.C. Slonczewski: J. Appl. Phys. **45**, 2705 (1974)
11.12. J.C. Slonczewski: J. Appl. Phys. **44**, 1758 (1973)
11.13. H.J. Levinstein, S. Licht, R.W. Landorf, S.L. Blank: Appl. Phys. Lett. **19**, 486 (1971)
11.14. H. Callen: Appl. Phys. Lett. **18**, 311 (1971)
11.15. D.C. Fowlis, J.A. Copeland: AIP Conf. Proc. No. 5, 240 (1972)
11.16. H. Callen, R.M. Josephs: J. Appl. Phys. **42**, 1977 (1971)
11.17. R. Wolfe, J.C. North: Bell Syst. Tech. J. **51**, 1436 (1972)
11.18. M. Takahashi, H. Nishida, T. Kobayashi, Y. Sugita: J. Phys. Soc. Jpn. **34**, 1416 (1973)
11.19. A.H. Bobeck, S.L. Blank, H.J. Levinstein: Bell Syst. Tech. J. **51**, 1431 (1972)
11.20. R. Suzuki, M. Takahashi, T. Kobayashi, Y. Sugita: Appl. Phys. Lett. **26**, 342 (1975)
11.21. J.A. Copeland, J.P. Elward, W.A. Johnson, J.G. Ruch: J. Appl. Phys. **42**, 1266 (1971)
11.22. A.H. Bobeck, S.L. Blank, A.D. Butherus, F.J. Ciak, W. Strauss: Bell Syst. Tech. J. **58**, 1453 (1979)
11.23. R. Wolfe, J.C. North, W.A. Johnson, R.R. Spiwak, L.J. Varnerin, R.F. Fisher: AIP Conf. Proc. No. 10, 339 (1973)
11.24. P.I. Bonyhard, J.L. Smith: IEEE Trans. Magn. MAG-12, 614 (1976)
11.25. A.H. Bobeck, I. Danylchuk: IEEE Trans. Magn. MAG-13, 1370 (1979)
11.26. A.H. Bobeck: 27th Conf. on MMM EA-1 (1981)
11.27. Y. Sugita, R. Imura, T. Takeuchi, T. Ikeda, R. Suzuki, N. Ohta, H. Umezaki: 27th Conf. on MMM BA-1 (1982)
11.28. T. Takeuchi, N. Ohta, Y. Sugita: IEEE Trans. Magn. MAG-20, 1108 (1984)
11.29. T.J. Nelson: 21st Intermag. Conf. HA-1 (1983)
11.30. D.T. Ekholm, P.I. Bonyhard, D.J. Muehlner, T.J. Nelson: J. Appl. Phys. **53**, 2525 (1982)
11.31. Y. Sugita, R. Suzuki, T. Ikeda, T. Takeuchi, N. Kodama, M. Takeshita, R. Imura, T. Satoh, H. Umezaki, N. Koyama: IEEE Trans. Magn. MAG-22, 239 (1986)
11.32. Hitachi Magnetic Bubble Memory Card Catalog.
11.33. H. Maekawa et al.: IECE Report
11.34. K. Iida, M. Saito, K. Furukawa: IEEE Trans. Magn. MAG-15, 1892 (1979).
11.35. J.E. Rowley, J. Bernardini: Bell Syst. Tech. J. **61**, 1841 (1982)
11.36. Y. Hosoe, K. Andoh, N. Ohta, Y. Sugita: J. Appl. Phys. **55**, 2542 (1984)
11.37. S. Konishi: IEEE Trans. Magn. MAG-19, 1838 (1983)

12. High Density Magnetic Recording. Recent Developments in Magnetic Tapes, Discs and Heads

Eiichi Hirota

Magnetic recording technology, i.e. electrical signals recorded onto a magnetic medium in the form of magnetic remanence signals produced by a magnetic head, was primarily invented by Poulsen at the end of the 19th century. He recorded audio signals on a steel wire by a pole-type electromagnet. Present day audio-tape recorders (ATR) were completed and put into practical use just after the invention of magnetic tapes and ring-type magnetic heads in the 1930s. In the 1950s, video-tape recorders (VTR) and magnetic disc drives (MDD) used as computer peripherals were invented and used widely in industry. Today VTR and MDD have become popular and are widely used consumer products. In many fields of social activities, one has to appreciate the convenient tools arising from magnetic recording technology such as magnetic cards for various vending machines, tickets, certificates, magnetic scales for NC-machines, as well as the ATR, VTR and MDD mentioned above.

Magnetic recording technologies are superior to other recording or memory technologies with respect to (1) the simplicity of recording, reading and erasing operations and to (2) the non-volatility of stored memories without any energy consumption. In Fig. 12.1, the basic principles of magnetic recording are shown.

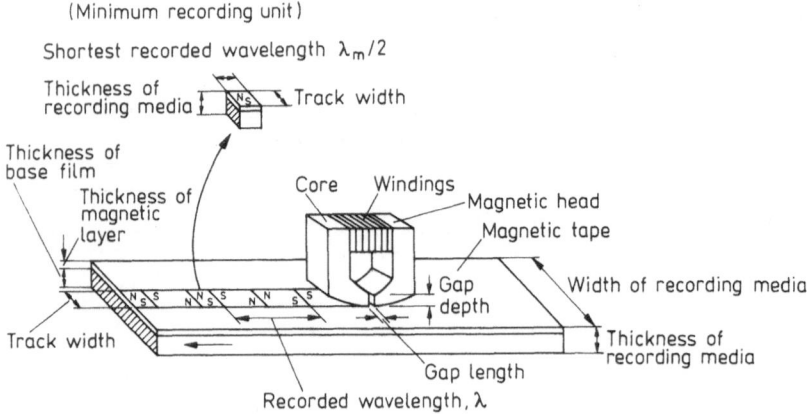

Fig. 12.1. Principle of magnetic recording: Recorded signals on a magnetic recording media written by a magnetic head

High density magnetic recording depends entirely on how to shorten the shortest recording wavelength λ_m and also how to narrow the recording track width w. The highest density in practice is realized by VTR's of which the λ_m and the w are about 1 µm and 20 µm, respectively. Using these, the maximum video-recording time can be extended up to 8 hours using a standard magnetic tape 1/2″ wide and about 250 m long. This VTR has been implemented after numerous improvements in materials and fabrication processes for magnetic tapes and heads, together with practical solutions for the tribological problem of tape-head contact.

The progress in recording density is shown in Fig. 12.2. In this century the density has become 1000 times higher than that of Poulsen's recorder. This progress has in part been realized by advances in magnetic media, i.e. the invention of magnetic tape and the improvement of magnetic fine powders with large coercivity H_c and remanence B_r. The isotropic iron oxide powders (Fe_3O_4 and γ-Fe_2O_3) of the early days were replaced by the needle-like iron oxide powders invented by Camras in the 1950s when the first generation of magnetic recording technology was assumed to be completed as far as ATRs were concerned. In the 1960s, DuPont reported that CrO_2-tape shows excellent high

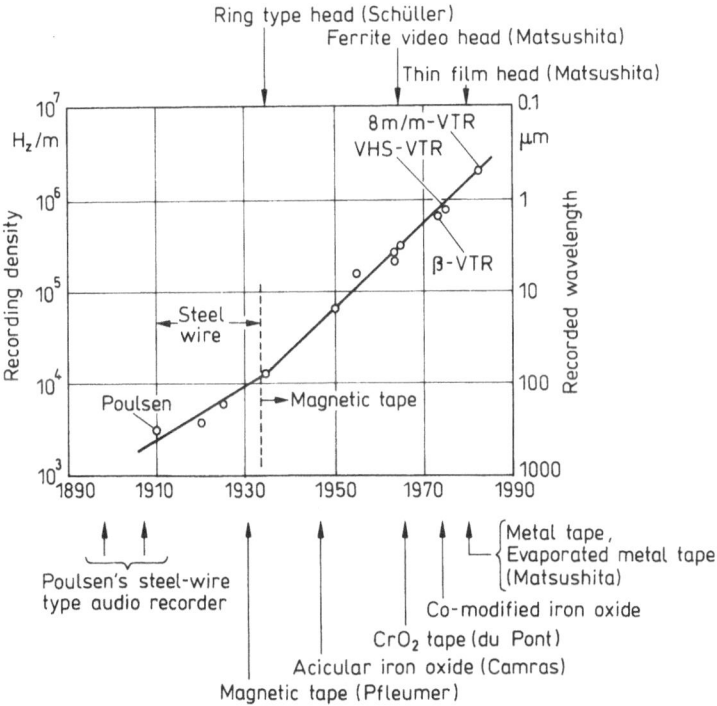

Fig. 12.2. Progress of magnetic recording technology, magnetic tapes, heads and recording density

density recording characteristics due to the well-aligned crystalline CrO_2 powders with higher H_c and higher B_r than Camras's iron oxide powder. With the stimulus of these high energy tapes (i.e. high H_c and high B_r tapes), much work was carried out on various types of Co-modified iron oxide powders in the 1970s. The great success of VTRs, β- or VHS-type (Appendix A.3) on the consumer market today, is partially due to the development of the high energy Co-modified iron oxide tapes. In the 1980's Fe or Fe–Co alloy powders with even higher H_c and B_r have been introduced and are generally known as "metal tapes". Other types of metal tapes, such as continuously evaporated Co–Ni alloy thin film tapes are assumed to be the super high density recording tapes of the next generation and have been developed for application to the compact VTR such as the 8 m/m-VTR. Research on these thin film magnetic media is being carried out quite intensively all over the world with the clear indication that λ_m is now heading into the submicron region.

Magnetic heads have also been continuously improved. For the ATR, magnetic head cores are fabricated by stacking thin Fe–Ni alloy sheets of high permeability. The alloy is modified with Nb, Ti or other metal additives which harden the alloy and improve the wear resistance of the metal heads. The wear of VTR heads is even worse since the relative speed of tape and head is several meters per second and faster by 100 times than that of ATRs. In order to improve this wear problem and also the high frequency characteristics, the VTR heads are made of single crystal Mn–Zn ferrites which have sufficiently high mechanical hardness H_v and electrical resistivity ρ. However, the saturation magnetization B_s of the ferrites is not large enough to fully magnetize the high H_c metal tapes described above. There have already been heads developed for this purpose, fabricated by stacking thin films of an Fe–Al–Si alloy or an amorphous Co–Nb–Zr alloy with satisfactory hardness and B_s.

The head signals reproduced from magnetic tapes decrease if there is a space between tapes and heads, as described below (the spacing loss). In practice the space d between tape and head must be less than $1/20$ of the λ_m. Also the magnetic gap length g of the heads must be about $1/3$ of the λ_m. Precise machining technologies with a tolerance of less than 0.1 µm are used for the farbrication of these heads. Furthermore, in future sub-micron recording, a machining tolerance of the order of 100 Å is required for the gap length g and also for the space d. Thus a better understanding of the tribology of the head-tape interface is very important to improve tape and head materials.

Figure 12.3 shows the progress of MDDs, mainly IBM's disc systems for computers. Magnetic discs consist generally of magnetic particulate layers of about 1 µm thick on an aluminum disc and are called hard discs. Magnetic layers are applied to an Al-disc by spin-coating of acicular γ-Fe_2O_3 powders dispersed in a dilute solution of resins and drying. For computer use, a very high reliability for recorded data is required and the bit error rate (BER; a probability to make errors in one cycle of the read and write of one bit signal) is assumed to be less than 10^{-12}. The read-write operation is made by the flying magnetic heads which glide over a disc with a constant space between head and disc,

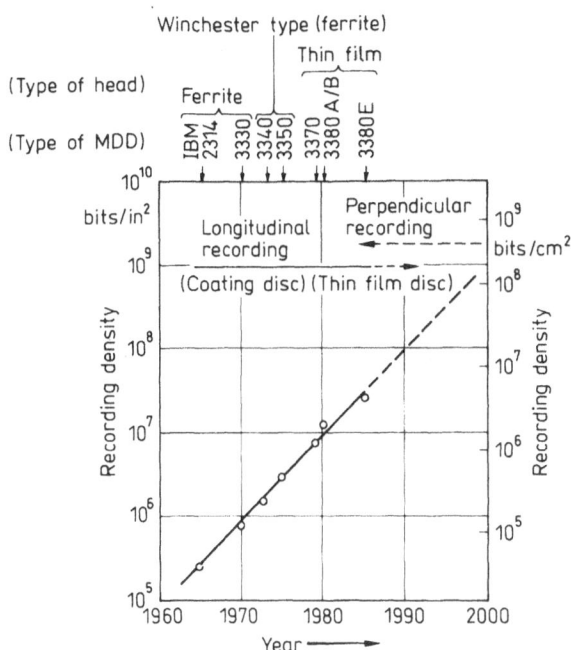

Fig. 12.3. Progress of recording density in magnetic discs

namely, the flying height h_d. The recording densities are about 1/10 of those of the ATRs and VTRs described above and yet they are constantly improved by about 10 times every 10 year period. To realize the higher density recording in the MDD various thin film discs have been intensively studied. Reported so far are a plated film of Co–P alloys, an evaporated film of Co–Ni alloys and a reactive-sputtered film of iron oxides. Although today's flying heads consist of ferrites, thin film heads have been introduced recently for increasing the recording density.

Various types of floppy discs are currently used as memory devices for personal computers, word processors and so on. These discs are made of particulate media consisting of flexible plastic films and coated magnetic layers. Contrary to hard discs, the floppy disc makes intimate contact with the floppy heads during recording and reading data signals. Also with this application, intensive research efforts are being made to miniaturize the memory devices, namely, to realize higher recording density. Recently increased interest and work have been reported on the development of so-called perpendicular recording technology where the recorded signals are the magnetic remanences perpendicular to the magnetic media. Since sputtered or evaporated Co–Cr alloy thin films have a large magnetic anisotropy perpendicular to the film, these alloy films have been studied as typical media for perpendicular recording. In normal magnetic recording, i.e., the longitudinal recording described above, the self-demagnetizing effect decreases the recorded remanence signals as λ_m becomes shorter [see (12.3)]. Therefore, in the high density longitudinal magnetic recording magnetic media, very high H_c and very thin magnetic layers ($\lambda_m/\delta > 1$, where δ is the thickness of a magnetic layer) are required. However, perpendicular recording in principle is not restricted to these requirements and shows extremely sharp magnetization reversal. Thus it is the most suitable technology for very high density digital recording. Actually it has been reported that read/write operations with short wavelengths of about 0.1 µm are experimentally plausible and that the recording density is assumed to be increased by several to 10 times compared with the usual longitudinal recording. Perpendicular recording technology is the most important one for the next generation of recording technology, since not only computer signals but also audio and video signals will be recorded as digital signals on the recording media.

The following sections are concerned with, first, the basic physics of magnetic recording and second, the progress of magnetic media and heads, selected from the various improvements made, especially in the past 10 years of magnetic recording technology described above. Finally magnetic discs and optical discs are described not as competing technologies, but coexisting in the high density recording technology of the future.

12.1 Physics of Magnetic Recording

12.1.1 Recording Process

Magnetic recording technology is based on the process of storing signals in the recording medium in the form of remanent magnetization which is created by the leakage magnetic field in the air gap of the high permeability magnetic head cores, and of reproducing the signals thus recorded in the recording medium. As shown in Fig. 12.4a, the magnetomotive force, NI, in the windings, produces the magnetic potential difference $V_0 = gH_0$ across the gap of the head and, furthermore, generates the head field proportional to V_0 at the gap. V_0/NI gives the recording efficiency, α, of the head. On the other hand, the reproducing efficiency is defined by ϕ_h/ϕ_s, where ϕ_s is a surface flux on the medium which flows into the head gap and ϕ_h is the effective part of the ϕ_s which interacts with the head windings. The recording and reproduction efficiencies are shown to be the same and can be described as the sum of magnetic reluctances of head core portions:

$$\alpha = 1/(1 + R_c/R_g + R_c/R_l) = 1/\{1 + (l_c/g)(A_g/A_c)(1/\mu_e)\} \qquad (12.1)$$

assuming that $R_c/R_l \ll 1$, where l_c is the effective length, A_c is the effective cross-sectional area of the magnetic path, A_g is the cross-sectional area of the gap and

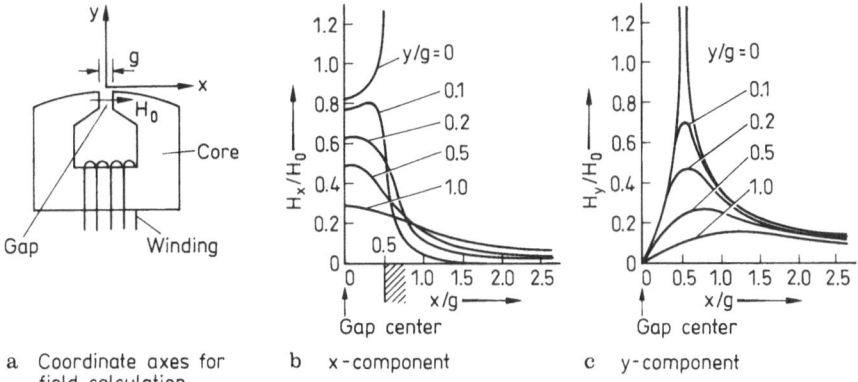

a Coordinate axes for   b x-component   c y-component
  field calculation

Fig. 12.4a–c. Magnetic field distribution around a head gap. **a** Coordinate axes for field calculation; **b** x-component; **c** y-component

μ_e is the effective permeability of the head core. The magnetic field at the gap of the recording head is given by $H_0 = (4\pi/10)\,\alpha NI/g$. In order to make the core efficiency as close as possible to 1, it is clear that l_c/g and A_g/A_c must be small and μ_e must be large.

When the coordinate system shown in Fig. 12.4a is used, the head field distribution can be described by Fan's formula (12.2) and is shown in Fig. 12.4b and c.

$$H_x(x, y) = \partial V_c/\partial x \ ,$$
$$H_y(x, y) = \partial V_c/\partial y \ ,$$

(12.2a)

and

$$V_c(x, y) = (V_0/\pi) \int_0^\infty \{\sin k \, \sin(kx/k^2)\} \exp(-ky)dk$$

$$+ \sum_{n=1}^\infty A_n 2n(-1)^n \int_0^\infty [\sin k \sin\{kx/[k^2 - (n\pi)^2]\}]$$

$$\times \exp(-ky)dk \ ,$$

$$A_1 = -0.082(V_0/2), \quad A_2 = +0.027(V_0/2), \quad A_3 = -0.014(V_0/2), \ldots$$

(12.2b)

In this description, it is assumed that $Z = \pm\infty$ because the head core thickness is far larger than the gap length g, that $\mu_e = \infty$, and the corner of the gap is a right angle. The figures show that recording media will be magnetized by the head field sharply distributed within the narrow area of $y/g \sim 1$ from the head surface and $x/g \sim 1$ from the gap center.

For a sufficiently thin recording medium (i.e. $\delta/g \sim 1$) with a strong magnetic anisotropy along the x-axis, one can assume that the magnetization along the

y-axis is uniform and only H_x is variable. As the medium runs over the head in intimate contact, the medium experiences the maximum head field at the gap center, and then finally retains its remanent magnetization as it moves away from the gap region. When a sinusoidal signal of frequency f is applied to the head, the remanent magnetization of the medium will be (by using the coordinate system (x', y', z') fixed to the medium running in x-axis) $I'_x = I_m \sin(2\pi x'/\lambda)$, $I'_y = 0$ and $I'_z = 0$, where $\lambda = v/f$, and v is the relative velocity of medium to head. This remanence produces the magnetic charge distribution $-\mathrm{div}(I'_x)$, which act as the demagnetization field resulting in the reduction of the remanent magnetization (i.e., self-demagnetization effect of the recording process). The actual magnetization recorded on the medium can be determined by the demagnetization curve of the medium and the average demagnetization field in (12.3). Thus it becomes I'_m as shown in Fig. 12.5. In this figure

$$H_d = -N'_x I'_x \, , \tag{12.3}$$

where $N'_x = 4\pi[1 - \lambda/2\pi\delta\{1 - \exp(-2\pi\delta/\lambda)\}]$. Therefore, it is clearly necessary that the medium has a large $B_r (= 4\pi I_m)$, H_c and squareness ratio, and it further has a smaller δ so that λ/δ becomes large enough.

Since in recent VTRs the recording wavelength $\lambda \approx 1$ μm, the media thickness $\delta \approx 5$ μm, and the gap length $g \approx 0.3$ μm, the perpendicular component of the head field, H_y, and the magnetization of the medium recorded by the field cannot be ignored. However, the evaporated or plated media generally have a thickness $\delta \approx 0.1$ μm. Since the condition $\delta \ll \lambda$ holds, the simplified magnetization process described above generally applies.

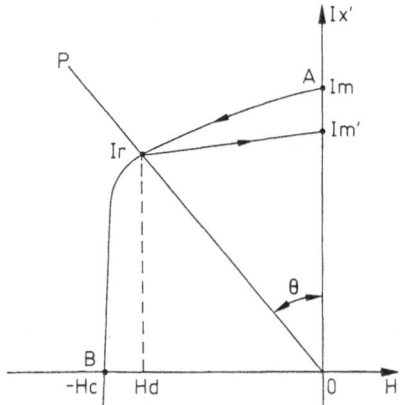

Fig. 12.5. Self-demagnetizing effect: Demagnetization curve (AB) of a magnetic tape and the line of permeance (OP) give an intersection I_r, where the reversible magnetization curve rises to the remanence I'_m at $H = 0$, where $\theta = \tan^{-1} N'_x$ and is a function of δ/λ

12.1.2 The Reproduction Process

As described above, the medium is assumed to be magnetized as $I'_x = I'_m \sin(2\pi x'/\lambda)$, $I'_y = I'_z = 0$ and uniformly in thickness, and further, the medium is assumed to be transported in the x-axis floating on the head surface with the spacing d. According to the reciprocal theorem, the element of flux $d\phi$ induced in the head winding by the element magnetization $I(x, y)dV$ in the medium near the head gap is the scalar product of $I(x, y)dV$ and the head field $H(x, y)$ which is excited at the point of interest (x, y) by the unit recording current. Thus, the total flux ϕ_h is given by

$$\phi_h = \alpha w \{2\pi/(V_0/2)\} \int\limits_{d}^{d+\delta} dy \int\limits_{-\infty}^{+\infty} I(x, y)H(x, y)dx , \qquad (12.4)$$

where α is the reproducing efficiency and w the track width. Applying Fan's formula (12.2), the flux ϕ_h at $x = x_0$ can be described as:

$$\phi_h(x_0) = (\alpha\delta w)4\pi I'_m \sin(2\pi x_0/\lambda)$$

$$\times [\{1 - \exp(-2\pi\delta/\lambda)\}/(2\pi\delta/\lambda)] \times \exp(-2\pi d/\lambda)$$

$$\times [\sin(\pi g/\lambda)/(\pi g/\lambda)]\left[1 + \sum_{n=1}^{\infty} \frac{A_n}{V_0/2}\frac{4\pi n(-1)^n}{4 - (2n\lambda/g)^2}\right]. \qquad (12.5)$$

The time derivative at $x = x_0$, gives the reproduction output E_p,

$$E_p = -N(d\phi_h/dt) \times 10^{-8} = -Nv(d\phi_h/dx_0) \times 10^{-8} \quad \text{(Volt)} , \qquad (12.6)$$

where N is the number of head windings. The ratio of $E_0 = 4\pi I'_m(Nv)(\alpha\delta w) \times 10^{-8}$ without any losses to the magnitude $|E_p|$ can be described as the sum of losses as follows:

$$-20\log|E_p/E_0| = L_g + L_d + L_t \quad \text{(dB)} , \qquad (12.7)$$

where

$$L_g = 20\log[\{(\pi g/\lambda)/\sin(\pi g/\lambda)\}\{1 + \sum{}''\}^{-1}] \quad \text{(gap loss)} \qquad (12.7a)$$

$$L_d = 20\log\exp(2\pi d/\lambda) = 54.6(d/\lambda) \quad \text{(spacing loss)} \qquad (12.7b)$$

$$L_t = 20\log[(2\pi\delta/\lambda)/\{1 - \exp(-2\pi\delta/\lambda)\}] \quad \text{(thickness loss)} . \qquad (12.7c)$$

Since these losses decrease with decreasing g/λ, d/λ and δ/λ, a smaller gap length g, a smaller medium to head spacing d and a smaller medium thickness δ are clearly necessary for realizing higher output in short wavelength recording.

12.2 Magnetic Tapes and Discs

Figure 12.6 shows the characteristics of magnetic recording media commercially available or under development. The evolution of particulate recording media in the past ten years or so is characterized by the research and development of fine particle magnetic materials such as CrO_2, Co-modified iron oxide and metal pigments. Thin film media which have been extensively investigated in recent years are aggregates of fine crystalline particles. For example, a metal tape evaporated by an oblique incident method consists of fine columnar micro-crystals of Co–Ni alloy and plated or sputtered Co films composed of fine particles with a single domain size. Coupled with the single domain particle size effect, the shape anisotropy of acicular magnetic particles or the magnetocrystal-line anisotropy in the case of Co-based alloys results in a very high H_c.

Magnetic recording tapes and discs are divided into two main categories: particulate media and continuous thin film media. The former are composed of plastic base films or Al substrates coated with magnetic inks which are a mixture of fine magnetic powders and resin binders. The latter are produced by vacuum depositing or plating magnetic thin films on the base. Particulate type tapes in general use today have a magnetic layer thickness of several micrometers. On the other hand, the spin coating method provides thinner layers of 0.5–1.0 μm for magnetic discs. Continuous thin film tapes with only a 0.1 μm thick Co–Ni alloy layer are made by a continuous evaporation method on PET films and are

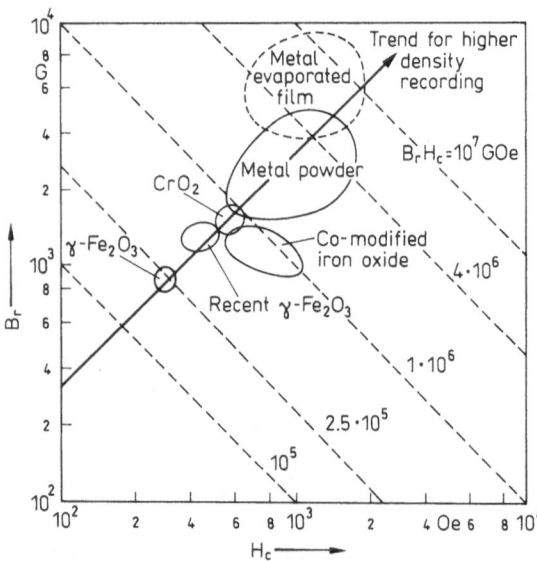

Fig. 12.6. Advances in magnetic tapes

now being intensively studied for high density audio and video tapes. Thin film discs with vacuum deposited or plated Co–Ni alloy layers of about 0.1 μm thick are also being developed.

In the following sections, recent magnetic materials commercially available and under development are briefly discussed (Table 12.1).

12.2.1   Magnetic Powders for Particulate Media [12.1]

(1) Iron Oxide Particles

Iron oxide fine particles are the most widely used materials for magnetic media. They are spinel ferrites such as Fe_3O_4, γ-Fe_2O_3, $CoFe_2O_4$, and composites of these ferrites in either granular or acicular (needle-like) shape. Although granular Fe_3O_4 particles were employed in the early days of recording history, they have been replaced by needle-like Fe_3O_4 and γ-Fe_2O_3 particles because the granular particles have instabilities such as magnetic after-effects and stress-induced demagnetization. Particles of a $CoFe_2O_4$–Fe_3O_4 solid solution, a permanent magnetic material, have not been used for recording media although the particles have a high H_c. This was due to the large temperature coefficient of H_c, the magnetic after-effects and the stress-induced demagnetization. Recently, Co-modified iron oxide composites of acicular γ-Fe_2O_3 or Fe_3O_4 particles coated with Co-iron oxides have been developed to cope with the above difficulties and have been put into practical use.

Iron oxide particles are chemically stable, pollution free, inexpensive and industrially excellent materials. More detailed explanations of (a) acicular γ-Fe_2O_3 and Fe_3O_4 and (b) Co-modified iron oxide composites are given below.

Table 12.1. Characteristics of Magnetic Recording Materials

Phase	Form	Recording material	Saturation magnetization	Coercivity	Curie temp.
			B_s[G]	H_c[Oe]	T_c[°C]
Oxide	Powder	γ-Fe_2O_3 (acicular)	5000	200–350	675
		Co-modified γ-Fe_2O_3 (acicular)	5000	400–1000	—
		CrO_2 (acicular)	6000	300–600	120
Metal	Powder	Fe (acicular)	20000	1000–1500	770
	Thin film	Co–Ni	16000	500–1000	~ 1000
		Co–Cr	5000	—	500

(a) Acicular (needle-like) γ-Fe_2O_3 and Fe_3O_4: Dehydration, oxidation or reduction of acicular α-FeOOH particles provides γ-Fe_2O_3 or Fe_3O_4 particles in the acicular shape with a length of 0.3–0.7 µm and a diameter of about 0.05 µm. α-FeOOH is an acicular single crystal particle and shows topotactic reactions on dehydration and oxidation. As shown in Fig. 12.7a a resultant acicular iron oxide is composed of smaller grains, with their $\langle 110 \rangle$ crystal axes aligned along the longitudinal direction of the resultant acicular particle. The overall shape of the acicular particle provides an easy direction of magnetization along the long axis of the particle (shape anisotropy) and thus these acicular particles align along the direction of the applied magnetic field in the tape coating process, which becomes the easy direction of magnetization of the tape. Figure 12.7a shows that the acicular particles contain voids. Since magnetic charge arises at voids, the homogeneous dispersion of magnetic particles in the magnetic ink is disrupted, resulting in lower orientation and a lower H_c. Improved processes have been developed recently to reduce the number of voids (Fig. 12.7b).

Fine particles of about 0.3 µm in length have been introduced in order to suppress the tape noise. Finer particles are desirable for minimal wear of magnetic heads, but too fine particles degrade the tape H_c because of superparamagnetism. Thus, precise control of the particle size distribution is required.

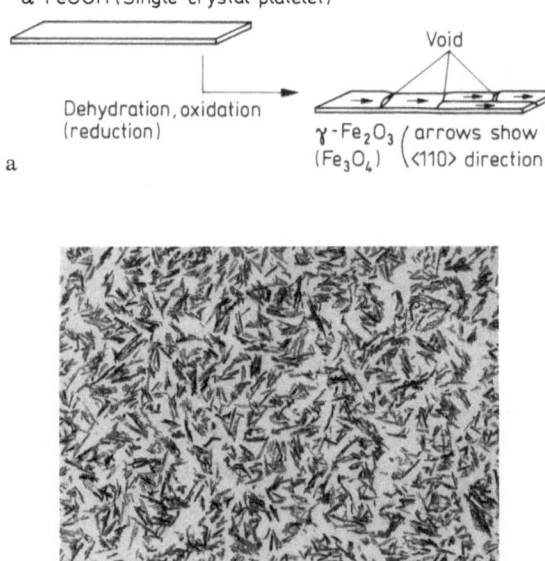

Fig. 12.7a, b. Iron oxide powders for magnetic tapes. ($\times$ 20000) a A model of iron oxide made by dehydration, oxidation (and reduction) of α-FeOOH; b Particle shape of iron oxide powders (by courtesy of Titan Kogyo Co., Ltd.)

As compared with γ-Fe_2O_3, Fe_3O_4 has a higher H_c and a lower resistivity which prevents harmful effects caused by frictional charging due to tape-head contact. However, Fe_3O_4 has instabilities, such as a larger temperature coefficient of H_c, magnetic after effects and bad print-through characteristics. Thus, γ-Fe_2O_3, which is free of these shortcomings, has become the most popular tape material. More recently, Fe_3O_4 has been revived as a core material for Co-modified iron oxide particles in the form of either Fe_3O_4 or a solid solution of Fe_3O_4 and γ-Fe_2O_3.

(b) Co-modified iron oxide: Many studies have been carried out on solid solutions of γ-Fe_2O_3, Fe_3O_4 and $CoFe_2O_4$, because Co-ferrite, $CoFe_2O_4$, has a large H_c. But these solid solutions have large instabilities similar to Fe_3O_4 and are difficult to apply as tape materials. To meet recent needs for high energy magnetic tapes of higher H_c and B_r and also without the above instabilities, a new Co-modified iron oxide has been developed. It is not a homogeneous solid solution but an iron oxide core particle covered with Co-ferrite which is grown epitaxially on the core surface by dipping core oxide particles into an aqueous solution of Co^{2+} and then heating these dipped particles. This Co-modified iron oxide is free from the above instabilities peculiar to Co-iron oxide and is widely used today. Various processes to produce this material have been reported [12.2].

(2) CrO_2 Particles

Generally, CrO_2 is produced by reducing CrO_3 in a hydrothermal reaction at a temperature between 400 and 500 °C under a pressure of 300 atm or more [12.3]. The pure CrO_2 product is in the form of plate-shaped crystals of a few μm in length and 0.5–1 μm in width. Since H_c is several tens of Oersted, it is not applicable as a recording medium. For application as a recording medium, a few percent of Te, Sn, or Sb oxide is added to the CrO_3 in a hydrothermal reaction in order to produce fine acicular single crystal particles of 0.3–0.7 μm in length and about 0.05 μm in width and with an axial ratio (length to width ratio) of over 10. These particles have an H_c larger than 400 Oe. Because CrO_2 is a ferromagnetic material, unlike the ferrimagnetic iron oxides described above, the temperature coefficient of $4\pi I_s$ or B_r is comparatively small in spite of its low T_c. Since the CrO_2 is composed of single crystals, and is therefore well oriented, magnetic tapes of high H_c and a large squareness ratio are obtained. As a result they also have excellent high frequency characteristics. This material, however, has a problem of heavy head wear, because the CrO_2 crystal particle is a mechanically hard material. Surface stabilization of the particles is introduced to prevent the problem of Cr^{6+} pollution.

(3) Fine Metallic Particles

Fine particles of Fe, Co and their alloys have been studied for the most plausible high H_c and high B_r media [12.4]. Very recently, fine Fe particles have been developed for recording media. Three main production methods have been reported: (1) hydrogen reduction of metal compounds, such as α-FeOOH, oxides or oxalates, (2) reduction of the above compounds by strong reducing agents such as sodium borohydride ($NaBH_4$), (3) evaporation of metals or alloys in a low pressure Ar atmosphere. Because of the processing costs, method (1) is the most favored for practical use. Starting materials such as α-FeOOH are reduced into Fe particles (Fig. 12.8) for 4–5 hours at a temperature of 340–360 °C. In order to prevent the sintering deformation of particle shape or the degradation of the particle by oxidation, thin oxide films of Al or Si are coated onto the particle surface.

12.2.2 Thin Film Media

(1) Metal Thin Films

Metal thin films are expected to be ideal magnetic recording media, since the highest magnetic flux density can be realized. However, the thin film media actually developed today are not ideally continuous films but the aggregates of fine metal particles which are separated magnetically from each other. In order to obtain high H_c media, the volume fraction of metal particles in the films is only about 70%. Metal thin films already developed are made (1) by electroplating of Co-P or Co-Ni-P, and (2) by vacuum deposition of Co or Co-Ni alloys.

Plated films are used for rigid discs. The magnetic characteristics of these films depend on the conditions of the plating process, such as electrolytes, ion concentration, plating current, and the temperature or pH of the plating bath. The media are plated on Al substrates and have protective overcoats. An H_c of 500 Oe and a total film thickness of 0.1–1 μm are generally adopted. The films

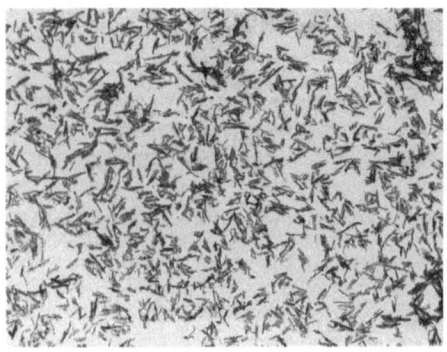

Fig. 12.8. Fe-powder made by hydrogen reduction of α-FeOOH. (× 20000) (by courtesy of Titan Kogyo Co., Ltd.)

show good magnetic characteristics at a composition of about 70% Co–30% Ni in the Co–Ni alloy system, which lies at the boundary of the hexagonal and cubic structures of the alloy. They consist of acicular fine particles of which each hexagonal *c*-axis (axis of easy magnetization) aligns in the film plane [12.5].

Recently, sputtered Co–Ni alloy thin films have been developed and used for small rigid discs of diameter 5.25″ or less. After Cr underlayers are deposited on Al substrates and treated for magnetic orientation, sputtered films of the above alloy form magnetic thin films whose easy axis is in the film plane. The discs are further coated with protective and lubricating layers on the magnetic film surfaces.

Evaporated films of type (2) have been developed for high density magnetic recording tapes. The films consist of oblique deposited Co–Ni fine particles which have their easy axis of magnetization in the film plane and are stabilized by surface treatment.

An outline of the continuous oblique evaporation method is shown in Fig. 12.9. A Co–Ni magnetic layer of 0.1 µm thickness is formed by oblique incidence evaporation on a PET film of about 6 µm thickness. The magnetic layer is an aggregate of fine columnar crystals of about 200 Å in diameter which elongate in the direction of incidence. Also being studied are evaporated tapes for perpendicular recording in which the Co–Cr alloy film is deposited on a heated polymer film substrate. In the near future, digital recording of audio and video signals will come into general use, and evaporated perpendicular recording tapes are expected to play an important role [12.6].

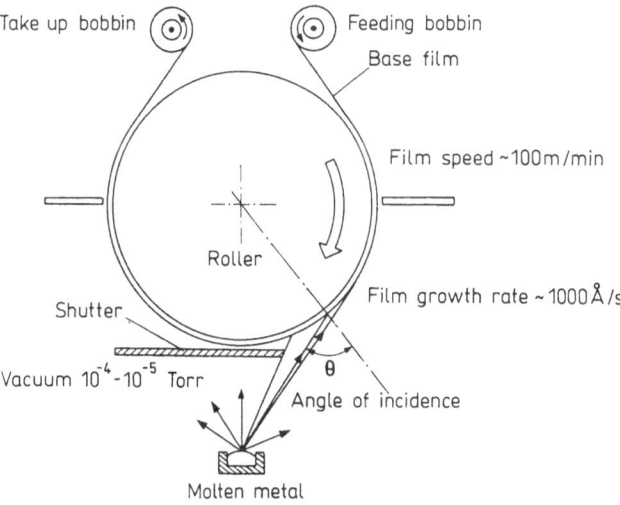

Fig. 12.9. Schematic of continuous vacuum evaporator

(2) Iron Oxide Thin Films

Iron oxide thin films of 1 μm thick or less have been studied for use as disc media. Thin films of $CoFe_2O_4$–Fe_3O_4 solid solution or γ-Fe_2O_3 are formed on substrates by reactive evaporation or sputtering methods. The films have an H_c of 300–700 Oe, a B_r of 1500–4000 G, and have superior characteristics against corrosion and abrasion by head contacts. However, they suffer from stress-induced demagnetization [12.7].

12.2.3 Perpendicular Recording Media

Recently, perpendicular recording films have been extensively studied for the high density recording technology of the next generation [12.8]. As shown in Fig. 12.10, in the case of conventional longitudinal media, the shorter the recording wavelength λ, the larger the demagnetization of the flux by its demagnetizing field H_d. Thus it is necessary to have a smaller film thickness δ, a larger H_c of the magnetic material, or smaller magnetization I_s for higher density recording. In perpendicular recording, on the other hand, there is no such restriction, because H_d becomes zero when λ approaches zero. Therefore, materials with an appropriate H_c and a large I_s can be applied. Perpendicular recording films require a magnetic anisotropy which stabilizes the magnetization perpendicular to the film surface. Sputtered or evaporated Co films have the required structure consisting of fine columnar crystals whose c-axes elongate perpendicular to the film surface. They therefore have a relatively large perpendicular magnetic anisotropy K from both magnetocrystalline anisotropy and shape anisotropy. To fulfil the condition of perpendicular magnetization $K > 2\pi I_s^2$, I_s is reduced by alloying Co with Cr. In practice, 10–20% Cr–Co alloys have been the most studied. Perpendicular recording media for floppy

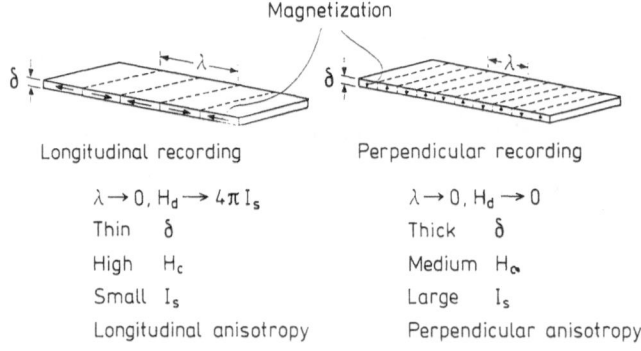

Fig. 12.10. A comparison of longitudinal and perpendicular recording

discs are also under development. For this case, a soft magnetic Fe–Ni alloy thin film and then a Co–Cr thin film are deposited on a substrate such as a polyimide film.

Ba-ferrite fine particle coatings have also been developed as perpendicular recording media for floppy discs. Fine particles of Ba-ferrite are now synthesized in platelet form with an extended c-plane and large magnetic anisotropy perpendicular to the c-plane.

12.3 Magnetic Heads

A magnetic head is composed of a magnetic core, a winding coil and a magnetic gap whose shape is extremely important. The gap length g should meet the condition $g/\lambda_m \sim 0.3$ (λ_m is the minimum recorded wavelength). For a high recording density of $\lambda_m \sim 1\ \mu m$, g is then about 0.3 μm and a precise machining technology is therefore needed to furnish a precise gap length with a tolerance of 0.03 μm. Furthermore, the glass bonding technology needed to form a mechanically strong gap has been developed and put into use, as the shape of the gap must not deteriorate by wear against the tape. Although it is necessary to have a narrower track width w for higher recording density, the narrowest track width is presently about 10 μm in practice, in order to optimize the signal to noise ratio and the output of the head. As described above, a smaller l_c/g, A_g/A_c and $1/\mu_e$ result in a better head efficiency. It is therefore necessary to shorten the magnetic core path l_c, to make the cross section of the gap plane smaller than that of the magnetic core (in another words, to make the gap depth shallower within the limit of the head wear) and to make the permeability of the magnetic core material higher in the relevant frequency range. As magnetic heads are constantly rubbed by magnetic tapes at a relative speed of, for instance, several meters per second, core materials with high wear resistance, such as ferrites, Fe–Al–Si alloys and etc. are the most practicable. Metallic core materials are used in the form of laminated cores where the thickness of each laminated sheet is equal to, or less than, its skin depth, in order to avoid degradation of the core permeability in the high frequency range. The core materials presently used for magnetic heads are listed in Table 12.2.

In general, $\mu_e \propto I_s^2/(K + a\lambda_s\sigma) \propto 1/H_c$, where K and λ_s are the magnetic anisotropy and the saturation magnetostriction constant, respectively. Therefore magnetic materials for head application are made at the optimum composition with K and λ_s nearly zero and with I_s as large as possible.

(1) Video Heads

The minimum recorded wavelength λ_m actually utilized in commercial 1/2" VTRs is about 1 μm. Magnetic heads consist of Mn–Zn ferrite single crystals

Table 12.2. Core Materials for Magnetic Heads

Material	Metal				Ferrite		Thin Film (Amorphous metal)
	Permalloy	Hard permalloy	Sendust	Amorphous	Ni–Zn	Mn–Zn	Co–Nb–Zr
Main composition [at %]	80 Ni 20 Fe	80 Ni 20 Fe + 3% Ti or Nb	6 Al 9 Si 85 Fe	$(CoFeMn)_{80}(SiB)_{20}$			
μ_0	20,000	30,000	20,000	20,000	2,000	20,000	10,000
B_m [G]	7,000	6,000	9,000	9,500	3,500	5,000	10,000
H_c [Oe]	0.02	0.01	0.01	0.01	0.1	0.02	0.01
ρ [Ω-cm]	5×10^{-5}	10^{-4}	8×10^{-5}	2×10^{-4}	10^4	10	12×10^{-5}
T_C [°C]	460	280	500	400	150	250	$T_C > T_x^* = 500$
Hardness [H_v]	130	230	500	900	700	550	700
Density [g/cm³]	8.7	8.6	8.8	8.3	5.3	5.1	8.6

* T_x: Crystallization Temp.

with a gap length g and track width w of $\sim(0.3 \pm 0.03)\,\mu\mathrm{m}$ and $\sim 20\,\mu\mathrm{m}$, respectively. An example of a video head is shown in Fig. 12.11. The processing technologies of slicing, grinding and lapping as well as glass bonding technology are all important to the manufacture of magnetic heads [12.9]. In order to manufacture high performance heads with the above-mentioned precision, it is important that the processing technologies are optimized both mechanically and physically and do not leave a Beilby (magnetically dead) layer at the gap plane.

Figure 12.12 shows the anisotropic wear characteristics of a Mn–Zn ferrite single crystal. Core materials are selected with these anisotropies in mind in order to produce a magnetic head with high core efficiency. Super compact VTRs (so-called 8 m/m VTRs) use the lower recorded wavelength λ_m of $\sim 0.7\,\mu\mathrm{m}$. The magnetic tapes for the 8 m/m VTRs are not iron-oxide particulate tapes but metal tapes with coercivities higher than 1000 Oe. The saturation magnetization of single crystal ferrites is not large enough to magnetize such high H_c tapes. Magnetic heads with exotic structures whose magnetic cores consist of thin films formed by sputtering high B_s materials such as Fe–Si–Al alloys or Co–Nb–Zr amorphous alloys have been developed recently for use with metal tapes. This is just the beginning of a trend of shifting from bulk materials to thin films for magnetic cores.

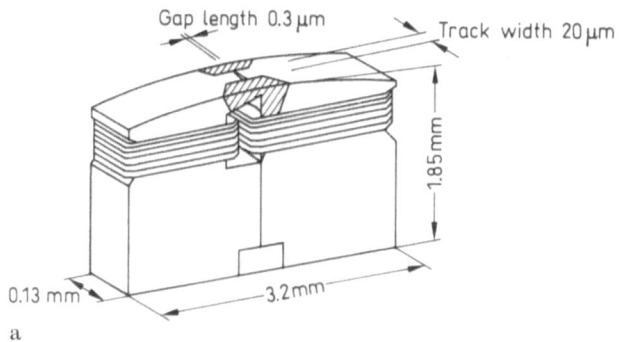

a

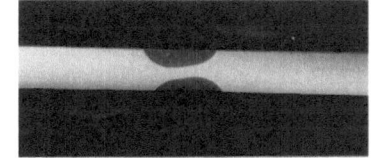

b

Fig. 12.11a, b. The VHS-type video head. **a** An overview of the head; **b** a tape-touching surface

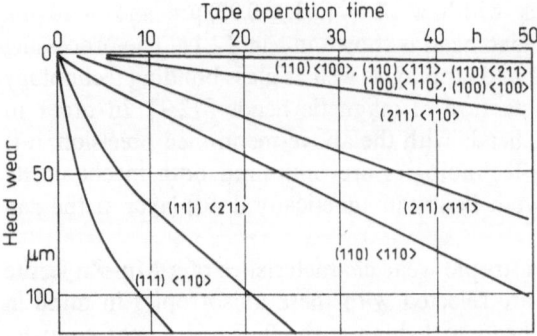

Fig. 12.12. Wear of Mn–Zn ferrite single crystal video heads having various crystal planes and axes as tape-touching surface ( ), tape-touching surface; ⟨ ⟩, tape running direction. Tape: CrO_2. Head: core width, 110 μm; penetration depth, 180 μm. Relative tape speed, 11 m/s

(2) Flying Heads

The heads for the disc drives of computer systems are called flying heads as they fly above the magnetic discs at a certain height (the flying height: h_f) during the read/write operation. The recording density realized by flying heads is about one-tenth of that achieved by contact type heads like video heads. Figure 12.13 shows a winchester type disc head. The heads made of Mn–Zn or Ni–Zn ferrites consist of a magnetic gap and a slider part which keeps the flying height stable. The shape of the slider part is designed and fabricated on the basis of aerodynamic calculations, so that it flies above the disc at h_f during operation. The flying heads already in use achieve a linear recording density of ∼ 15 Kfrpi ($\lambda_m \sim 3.4$ μm) with a track width w of 20 μm and a flying height h_f of ∼0.2 μm. A higher recording density depends critically on making h_f smaller, and thus improvements are being made on the flatness of the head and disc surfaces as

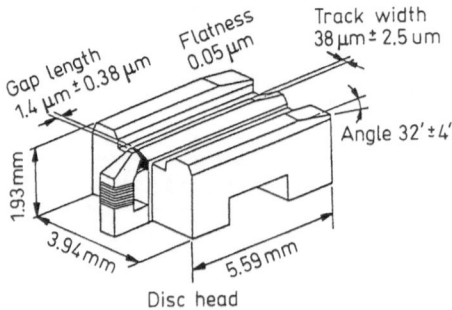

Fig. 12.13. A view of a magnetic disc head

well as the wear characteristics. The possibility that the flying height can be reduced to less than 0.1 μm has been reported.

(3) Thin Film Heads

Thin film heads for computer disc drives have recently been developed and applied for commercial use [12.10]. The heads show better frequency response in the high frequency range because of the shape effect of the thin film magnetic cores. Thin film heads have been extensively studied for perpendicular recording which is widely recognized as the high recording density technology of the next generation. These thin film heads are fabricated by photolithographic techniques, i.e., etching high permeability thin films made of Fe–Ni alloys or amorphous Co–Nb–Zr alloys into desired core patterns and conductive thin film materials into winding coils. Thin film heads are expected to be used for various types of heads since the cross-talk characteristics are far better and precision processing is much easier. Thin film integrated heads, such as multi-channel tape heads for computer tape drives and video floppy heads have been reported. Furthermore, due to the superior precision in their fabrication and their superior high frequency characteristics, thin film magnetic heads are expected to be used for high density recording film heads.

12.4 Future Trends

In magnetic recording technology, digital recording has made progress not only in the area of computer memories but also in audio and video signal processing which is presently analogue recording. Along with this trend, both magnetic

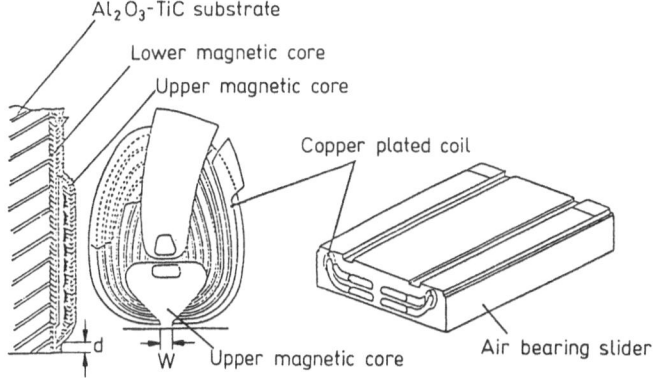

Fig. 12.14. A view of a thin film disc head

heads and media are expected to be of the thin film type. In Fig. 12.3, the progress of magnetic disc technology and its future trends are shown and clearly indicate that longitudinal recording technology with its combination of particulate media and ferrite heads is about to be replaced by high density recording technology with thin film media and thin film magnetic heads, and by perpendicular recording technology. In perpendicular recording, it has been reported that a read/write operation of 200 Kbpi ($\lambda_m \sim 0.13$ µm) has been carried out and that recording with a track width of 2 µm has been achieved. As research and development efforts realize high density magnetic recording of more than 10^8 bits/cm^2 and shorter recording wavelengths of sub-microns, the most important and critical problems are the development of precise and fine mechanical control, such as high speed head/media tracking and reliable air bearing, and durable head materials.

A recording density of 10^8 bits/cm^2 is equivalent to the optical memory achieved with semiconductor lasers. In general, an optical memory has an advantage over a magnetic memory since optical heads do not come into contact with the media in the read/write operation. However, the wavelengths of semi-conductor lasers are not short enough, resulting in an upper limit of recording density and it is difficult to realize facilities like over-write and high speed signal transfer. Thus optical memories will be applied for very large memories which are used differently from magnetic disc memories.

It is well known that the human brain has a memory capacity of 10^{15} bits. High density magnetic recording is one of the most important among many technologies that are being extensively studied with the aim of realizing such a super high recording density.

References

12.1. E. Hirota: J. Japan Inst. Metals, Vol. 21, No. 11 (1982) 837 (in Japanese)
        G. Bate: Ferrites, Proc. Int. Conf., Japan (1980) p. 509
        Y. Imaoka, K. Takada, T. Hamabata, F. Maruta: ibid., p. 516
12.2. Y. Imaoka, S. Umeki, Y. Kubota, Y. Tokuoka: IEEE Trans. Magn. MAG-14, 649 (1978)
        M. Amemiya, M. Kishimoto, F. Hayama: IEEE Trans. Magn. MAG-16, 17 (1980)
12.3. E. Hirota, T. Mihara, T. Kawamata, Y. Terada: National Tech. Rep., 20, 639 (1974)
12.4. G. Akashi: Ferrites, Proc. Int. Conf., Sept-Oct. Japan p. 548 (1980)
12.5. M. Nagao, Y. Suganuma, H. Tanaka, F. Goto, M. Yanagisawa: 8th Colloquium of Mag. Soc. of Japan, 51 (1978)
12.6. Y. Maezawa, M. Takao, H. Hibino, M. Odagiri, K. Shinohara: 4th Int. Conf. Video & Data Rec., 20 (1982)
        T. Fujita, M. Odagiri, K. Shinohara: Nat. Tech. Rep. 28 [3] 502 (1982)
12.7. S. Hattori, A. Tago, S. Yoshii: 18th Colloquium of Mag. Soc. of Japan, 25 (1977)

12.8. S. Iwasaki, Y. Nakamura, K. Ouchi: IEEE Trans. Mag. MAG-15, 1456 (1979)
       S. Iwasaki, K. Ouchi, N. Honda: IEEE Trans. Mag. MAG-16, 1111 (1980)
       Collected Papers on "Perpendicular Recording"; 18th Symp. of Res. Inst. Elect.
       Comm. Tohoku Univ. (1982)
12.9. E. Hirota, K. Kugimiya, K. Hirota: Ferrite, Proc. Int. Conf., Sept.-Oct. Japan
       (1980) p. 667
12.10. R.E. Jones, Jr.: IBM disk storage technology, p. 6–9; Feb. (1980)
       K. Kanni, N. Kaminaka, N. Nomura: IEEE Trans. Mag., MAG-15, 1130

13. Magnetic Domains Observed by Electron Holography

Akira Tonomura

This chapter describes a newly developed method for observing magnetic domain structures in thin films. In this method, the wavefront of an electron beam transmitted through the films is displayed as a contour map using electron holography. Adjacent contour lines enclose a constant magnetic flux of h/e ($= 4.1 \times 10^{-15}$ Wb) and are along the in-plane magnetic lines of force.

Optical methods have often been used to observe the magnetic domain structures of ferromagnets. However, the use of electron beams permits higher resolution because their wavelengths are much smaller than that of light. Furthermore, the distinctive features of electron beams are not only limited to the resolution, but extend to the observation principle itself. Specifically, these beams are deflected by a Lorentz force during passage through magnetic fields. These fields can be visualized, just as objects can be recognized from the scattered light. For an electron beam, magnetic fields are real observable objects.

The question is how an electron beam stores information collected from magnetic fields. For weak fields, information is contained in the phase distribution of the electron beam, which experiences them as pure phase objects. When drawn on an electron micrograph, the magnetic field distribution can thus be observed with the high resolution of electron microscopy.

Another development, interference microscopy, displays the contour lines of the phase distribution as interference fringes. However, it had not been practically applied to electron microscopy until the recent development of electron holography [13.1]. Using this holography technique, the electron wavefronts can be systematically reproduced as light wavefronts. Consequently, interference electron microscopy becomes possible by using an interferometer during the optical reconstruction stage.

The interpretation of an interference electron micrograph is extremely simple [13.2] since the contour fringes are along the magnetic lines of force. In addition, quantitative measurements are possible, since a constant magnetic flux of h/e (h: Planck's constant, e: electron charge) flows between two adjacent contour fringes. The present chapter deals with this new method for observing microscopic magnetic fields using electron holography. To begin with, the principles of electron holography will be discussed.

13.1   Principles of Electron Holography

Holography is well known as a type of stereoscopic photography. However, it was originally devised by Gabor as a means to improve electron microscopic resolution by compensating for electron lens aberrations. In simple terms, this ingenious idea can be expressed as "a two-step imaging method". First, an electron beam illuminates an object, and the scattered wave is recorded on film as an interference pattern by superposing a reference wave. This "hologram" contains information about both the intensity and phase of the scattered electron wave. When the hologram is then illuminated with light, the image of the object is three-dimensionally reconstructed because electron wavefronts are reproduced by light wavefronts. An additional image called a "conjugate image" is produced beside the image, whose amplitude is the complex conjugate of that for the reconstructed image.

Holography requires waves which have clearly defined wavefronts. With the advent of lasers in 1960, such waves became available and holography was born in the field of light optics. Although efforts were made in electron holography, it has been only quite recently that electron beams have been obtained with coherences higher by one order of magnitude [13.3]. Such an electron beam is emitted from the apex of a tungsten tip where a strong electric field is concentrated (Fig. 13.2). Since there is no need to heat the cathode, the energy spread of the electron beam narrows to about 1/3 of that for a thermionic electron beam. In addition to this feature, the extremely small source size improves the coherence of the field-emission beam.

This field-emission (Appendix A.10) electron beam has led to the practical use of electron holography. The first step in electron holography consists of

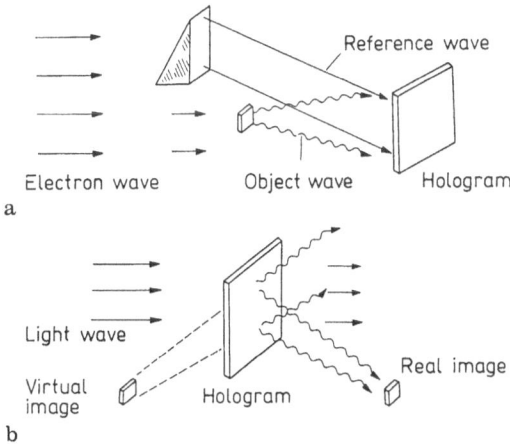

Fig. 13.1a, b. Principle of electron holography

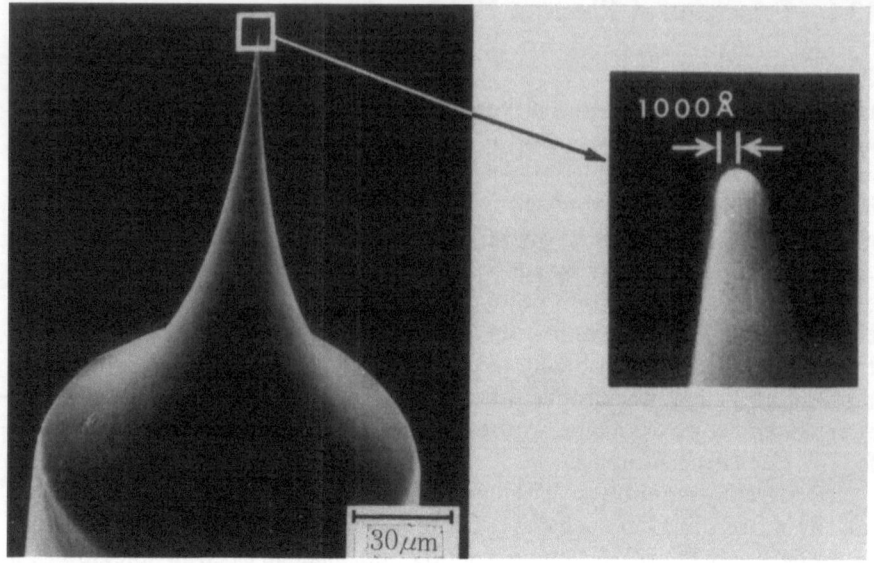

Fig. 13.2. Tungsten tip cathode of a field emission-type electron gun

forming a hologram in an electron microscope, where an electron biprism is installed between the objective and magnifying lenses (Fig. 13.3). This biprism consists of a central fine filament (diameter 0.3 μm) and a ground electrode on either side. The electron beam, which has travelled through an object, passes the filament on one side and a reference electron beam on the other. A positive

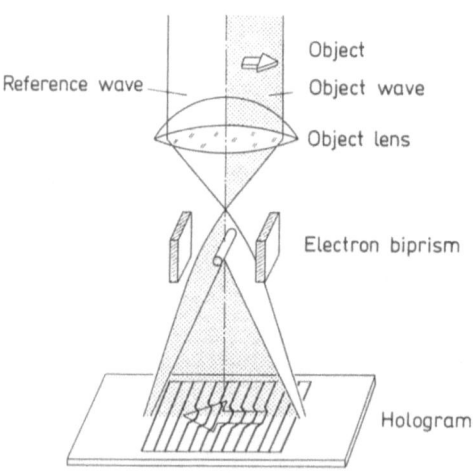

Fig. 13.3. Principle of electron hologram formation

electric potential of 50–200 V applied to the filament makes the two electron beams overlap to form an interference pattern on the lower plane. Finally, the interference pattern is magnified by electron lenses and recorded on film as a hologram.

When the hologram is illuminated with a collimated laser beam in the second step, an image of the original object is reconstructed in one of the two diffracted beams. This image is equivalent to an electron microscopic image.

Next, we explain the method for obtaining an interference micrograph [13.4]. The optical system for interference microscopy (Appendix A.18) is shown in Fig. 13.4. Two collimated laser beams are incident to the hologram, and their directions are oriented differently. One beam forms a reconstructed image in the observation plane, and the other transmits the hologram to overlap the reconstructed image. Other beams are obstructed by the aperture shown in the figure, so that an interference image between the reconstructed image and a plane wave can be observed.

However, there are some cases where the phase distribution is too small to be observed. Here, a unique method peculiar to holography can be utilized. This is called "phase-difference amplified interference microscopy", which makes good use of conjugate images. Assume that the direction of beam A in Fig. 13.4 is adjusted so that a conjugate image is formed in the observation plane and that this image overlaps the reconstructed one. Then, an interference pattern of the two images can be observed. Since the amplitudes of the twin images are complex conjugates of each other (i.e., the phase distributions of the two images are reversed in sign), the resultant interference pattern represents an interference image with a doubly amplified phase distribution.

Examples of interference images are shown in Fig. 13.5. The specimen used here is composed of magnesium-oxide smoke particles formed when magnesium is burned in air. The three-dimensional shape of the particles cannot be viewed from the electron micrograph (a), since only the intensity distribution of the transmitted electron beam is recorded. In the interference micrograph (b), on the other hand, the thickness contour lines overlap the electron micrograph, so the three-dimensional information can be observed. This is because the phase shift of an electron beam during passage through a uniform object is proportional to

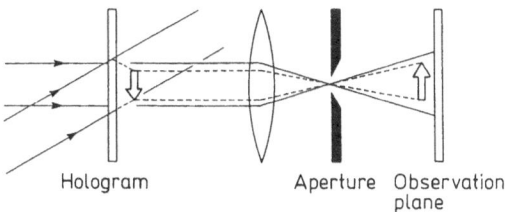

Hologram   Aperture Observation
                                      plane

Fig. 13.4. Optical system for interference microscopy

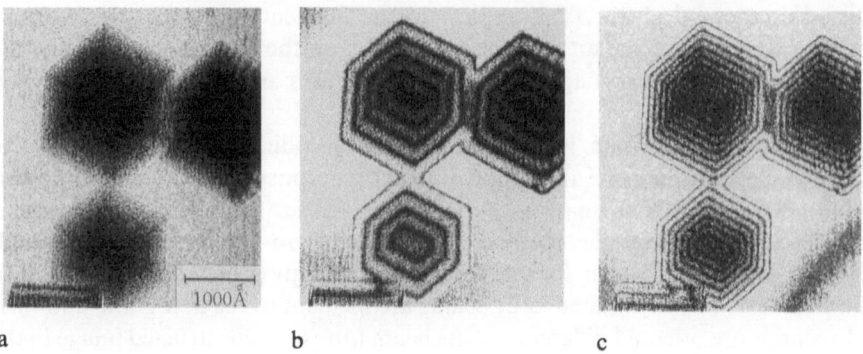

a   b   c

Fig. 13.5a–c. Hologram of a MgO smoke particle. **a** Ordinary electron microscopy; **b** Interference microscopy (× 1); **c** Interference microscopy (× 2)

the thickness of the object. A detailed thickness distribution can be observed in the amplified interference micrograph in Fig. 13.5c.

13.2 Principles of Domain Structure Observation

The simplest case for observing a ferromagnetic thin film by interference electron microscopy is shown in Fig. 13.6. Parallel electrons are incident on a uniformly magnetized film and are deflected by the Lorentz force during passage through the film. Since an interference micrograph represents the contour map

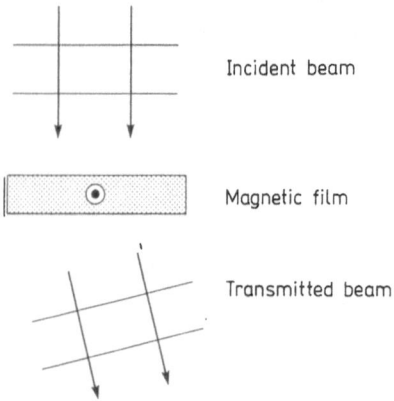

Incident beam

Magnetic film

Transmitted beam

Fig. 13.6. Electron beam transmitted by a magnetic thin film

of the transmitted wavefront, the wavefront of the electron beam is considered next. The wavefront is assumed to be perpendicular to the beam direction. The incident parallel electrons correspond to a plane wave and the transmitted electrons correspond to a tilted plane wave as shown in the figure.

The influence of the magnetic field on an incident electron wave is now discussed. The magnetic field affects the wave by rotating the wavefront about an axis determined by the magnetic lines of force. This fact simplifies the interpretation of interference micrographs. The height of the wavefront is always the same at any point along a magnetic line of force. Consequently, the contour lines of the wavefront indicates the magnetic lines of force.

So far we have established that the contour fringes in an interference micrograph follow the magnetic lines of force. The next question is what determines the density of the fringes. It is easy to calculate the contour lines drawn in units of electron wavelength. The procedure is as follows:[1] The deflection angle of an electron beam during passage through the film is first calculated. This angle is equal to that of the wavefront inclination. Consequently the spacing of adjacent contour lines can be obtained, which have a wavelength difference of a wavefront height between them. If the wavelength is given by de Broglie's relation, then we arrive at the simple conclusion: "Contour fringes in an interference micrograph appear for every constant magnetic flux of h/e".

13.3 Applications to Magnetic Domain Structure Observation

Lorentz microscopy (Appendix A.21) is often employed to achieve highly resolved images of magnetic domain structures in ferromagnetic thin films. In this method, magnetic domain walls can be viewed in black or white contrast by defocusing the electron microscopic image. Defocusing is indispensable for the contrast since the magnetization in the film does not influence the intensity distribution of the in-focus electron micrograph. The only effect seen is a slight deflection of the transmitted electron beam, which has no effect on the image. This is because all electrons starting from an object point are focused onto an image point through an electron lens independent of the angle. The magnetic domain structures cannot be observed until the image is defocused. In this

---

[1] The calculation is as follows. The deflection angle α is given by the film thickness t divided by the radius r of the circular electron trajectory in a uniform magnetic field B: $\alpha = t/r = eBt/mv$. When h/λ is substituted for mv according to de Broglie's relation, then $\alpha = (eBt)(\lambda/h)$. Since the spacing d corresponding to a wavelength difference of the wavefront height is given by λ/α, the relation $B\,dt = h/e$ can be obtained. This means that the magnetic flux contained between two adjacent contour lines $B\,dt$ is given by h/e.

defocused image, however, details of the image cannot be discerned. For example, the outer shapes of fine particles such as those shown in Fig. 13.5, are completely blurred by the large amount of defocusing.

In an interference micrograph, the magnetic lines of force appear as contour fringes on an electron micrograph. An example of this is shown in Fig. 13.7. The specimen is a cobalt smoke particle which is prepared by gas evaporation in an inert gas atmosphere. The three-dimensional shape is a truncated triangular-pyramid as illustrated in Fig. 13.7a. From the figure, the narrow fringes parallel to the three edges show that the thickness increases linearly to 500 Å. The inner region is uniform in thickness, so the circular fringes indicate that the magnetic lines of force are rotating inside the particle.

The wavefront of an electron beam is completely reconstructed during the optical reconstruction stage of the holography. Therefore, a Lorentz micrograph can also be obtained during this stage. An example illustrating the electron, interference, and Lorentz micrographs of a hexagonal cobalt particle is shown in Fig. 13.8. All of the micrographs are optically obtained from a single hologram. It is seen from the Lorentz micrograph that it cannot determine the magnet-ization rotation inside the particle. This is because the image is greatly blurred during the production of the many Fresnel fringes inside and outside the particle. The effect of the magnetic domain structure overlaps this pattern and complicates its interpretation. Also, by comparing these micrographs one can recognize distinctive features of the interference micrograph.

Fine particles have often been employed as electron holography samples, since they can be easily used during experimentation to produce a reference electron wave near the specimen (Fig. 13.3). A ferromagnetic thin film can also be observed if it is cut in half so that the above mentioned reference wave can pass by. An interference micrograph of a permalloy thin film is shown in Fig. 13.9. A schematic diagram of a magnetization configuration mode of a cross-tie domain wall, first predicted by Huber et al. [13.5] is shown in

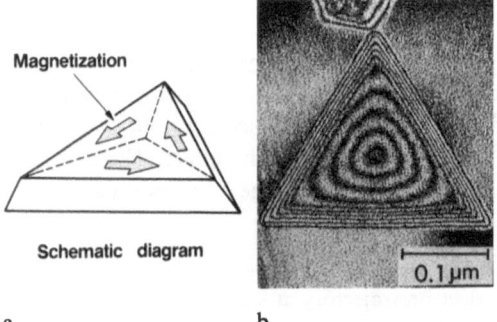

Fig. 13.7a, b. Micrograph of a Co smoke particle. **a** Sample shape; **b** Hologram by interference microscopy

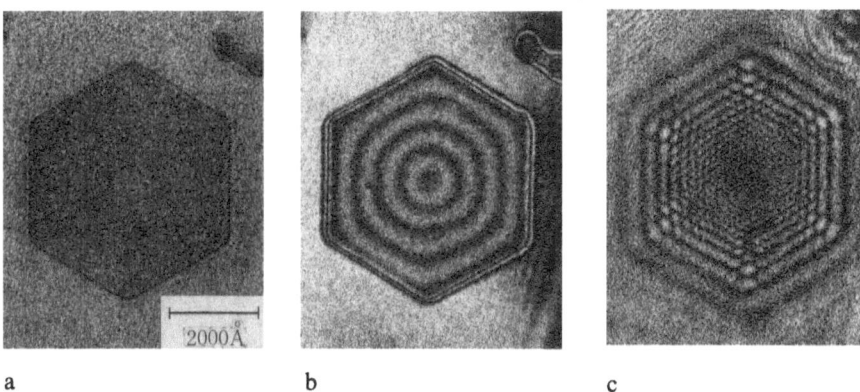

a   b   c

Fig. 13.8a–c. Micrographs of a hexagonal Co particle. **a** Ordinary electron microscopy;
b Interference microscopy (× 1); **c** Lorentz microscopy

Fig. 13.9a. This model has already been confirmed by Lorentz microscopy. However, the detailed distribution of the magnetization can be directly observed in the amplified interference micrograph (2 × phase difference) shown in Fig. 13.9b. Since a unit magnetic flux of $h/2e$ is contained between two adjacent contour lines, the magnetic flux flowing inside the film can be quantitatively observed.

Although the contour fringes are directly interpreted as magnetic lines of force, they arise from the contour lines of the wavefront. The wavefront of the interference micrograph (c) is like a chain of mountains. The centers of the concentric fringes correspond to the mountain tops, and the saddle points lie between them. At these top and saddle points, the wavefront is locally flat and the incident wavefront passes through the film without any tilting. Consequently, the magnetization is found to be perpendicular to the film plane at the mountain tops and saddle points as shown in the predicted model (Fig. 13.9a).

Up to this point, several examples of the magnetic lines of force inside ferromagnetic thin films have been demonstrated to be observed on electron micrographs. By using this method, magnetic fields outside the ferromagnetic samples can also be observed quantitatively. In fact, this is the only quantitative technique for measuring the magnetic flux in the microscopic region. The most suitable sample for such observations is the propagation circuit of a magnetic bubble memory. This circuit consists of tiny horse-shoe magnets. The rotation of the applied magnetic field shifts the location of the magnetic poles, and the magnetic bubbles located below propagate together with the rotation. The magnetic field distribution plays an important role in propagating these bubbles smoothly and correctly. However, it may be possible that the magnetic fields are distorted due to the magnetic domain structure effect in the tiny magnets.

An interference micrograph of the magnetic fields from magnets [13.6] is shown in Fig. 13.10. The specimen is composed of permalloy horse-shoe

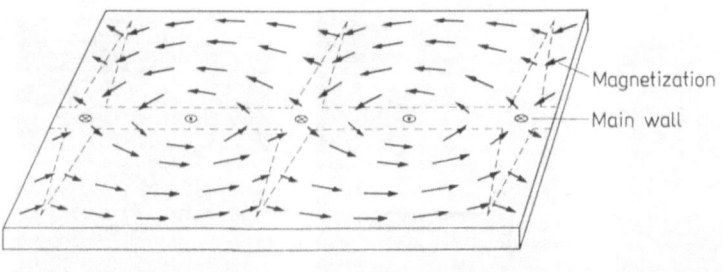

a

b

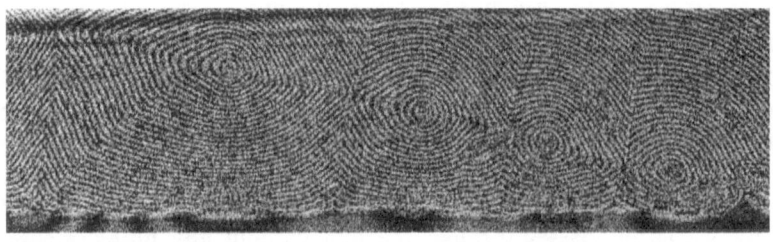

c

Fig. 13.9a–c. Magnetic domain walls in a permalloy thin film. **a** Magnetization distribution; **b** Micrographs by Lorentz microscopy; **c** Micrographs by interference microscopy

magnets 3500 Å thick supported by a thin carbon film. An in-plane magnetic field is applied when electron holograms are set up in the electron microscope. The field direction is indicated by the arrow in the micrograph. The magnetic lines of force originating from one end of a magnet are directed partly to the other end of this magnet and partly to the end of the adjacent magnet.

The next example is an application to high density magnetic recording [13.7]. Magnetic recording density has been increasing rapidly and its unit length is nearing the 1 μm barrier. In order to cope with this situation, a method is required to observe recorded magnetization patterns in detail. Interference electron microscopy may have potential for this purpose.

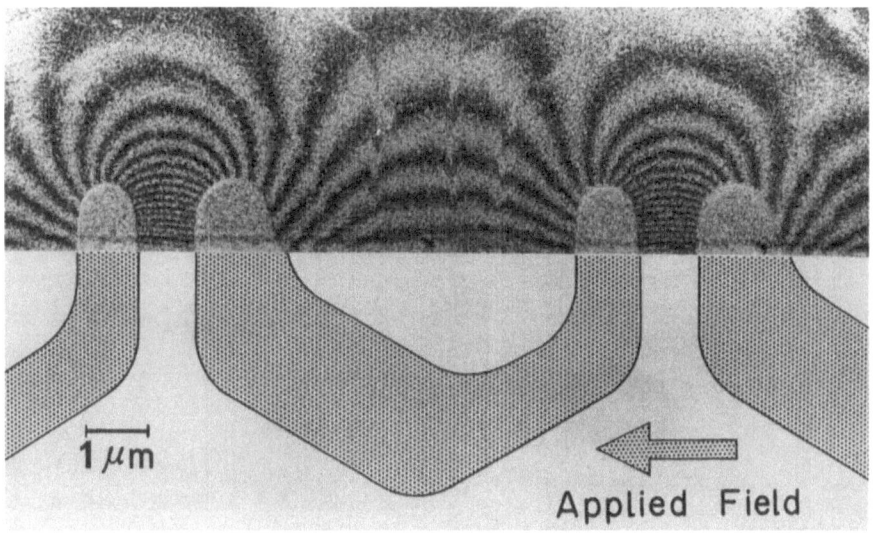

Fig. 13.10. Interference micrograph of the magnetic fields from permalloy horse-shoe magnets in a bubble track

An interference micrograph of a magnetization pattern recorded on a cobalt thin film is shown in Fig. 13.11. A schematic diagram of the recording arrangement is shown in Fig. 13.11a. A magnetic head slides on the film and bit patterns are recorded on it. The recorded film is then peeled off from the glass base and observed as an interference micrograph. The film is observed from above and is cut off in the middle, so the upper part is free space. The arrows indicate the magnetization directions and the magnetic lines of force can actually be observed along these arrows in the recorded regions. In addition, the directions of the magnetization in adjacent regions are opposite. Since the magnetic lines of force have neither sink nor source, vortex-like streams can be observed at the boundary regions where oppositely directed magnetic fields collide head-on with each other. These lines leak outside the film edge, so the leakage flux can be quantitatively measured by counting the number of contour fringes.

Such observations of the detailed magnetic field distribution are useful for the investigation of high density magnetic recording. In fact, the possibility of recording with a 0.15 μm unit length has been confirmed by this method.

So far we have discussed the practical applications of microscopic magnetic field observations. As far as the theory is concerned, interference micrographs cannot be fully explained by only a magnetic field B. Rather, the vector potential A is introduced as the more fundamental physical quantity. In classical mechanics, vector potentials are convenient for mathematically solving electromagnetic problems. In quantum mechanics, on the other hand, vector potentials have a physical significance. This can also be inferred from the fact that the

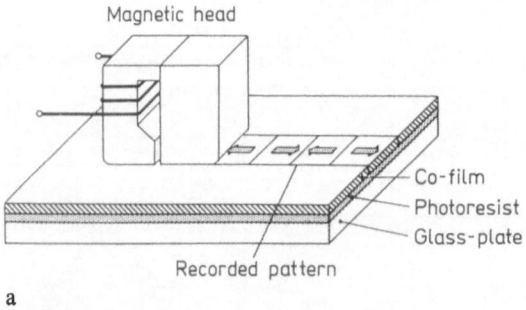

a

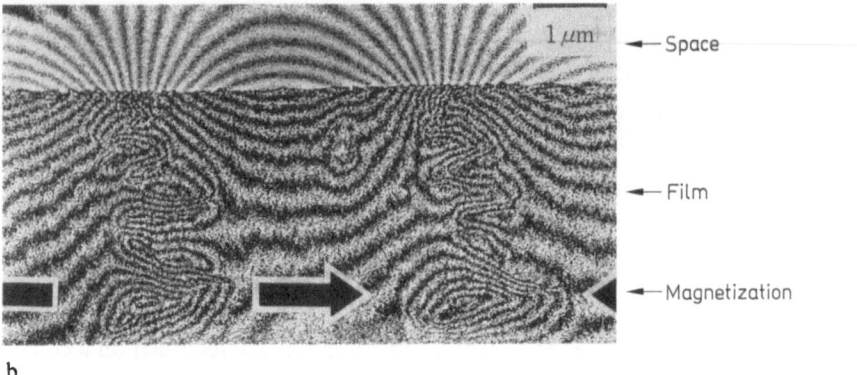

b

Fig. 13.11a, b. Observation of magnetic recordings. **a** Recording device; **b** Micrograph produced by interference microscopy

Schrödinger equation is described by electromagnetic potentials instead of fields. This was not fully recognized until 1959 when the Aharonov–Bohm (AB) effect (Appendix A.1) was proposed [13.8]. In the following we discuss the experimental results [13.9] which indicate the existence of the AB effect.

An interference micrograph of a tiny toroidal ferromagnet is shown in Fig. 13.12. The magnet is made of permalloy, and the magnetic field flows through it without any leakage. It is not possible to determine from the outside whether it is a magnet or not. However, the phase shift can be detected between two electron beams passing through regions inside the hole and outside the magnet, when an electron beam is incident on this magnet and the phase distribution of the transmitted electron beam is measured as an interference micrograph. This is shown in Fig. 13.12b. Although the electron beams pass through the two field-free regions, a physically observable effect is detected. The vector potentials in the field free regions influence the electron beam, and this phenomenon is called the AB effect.

In the 1970s the theory of gauge fields was rediscovered and its significance as a unified theory explaining all physical phenomena was realized. In this

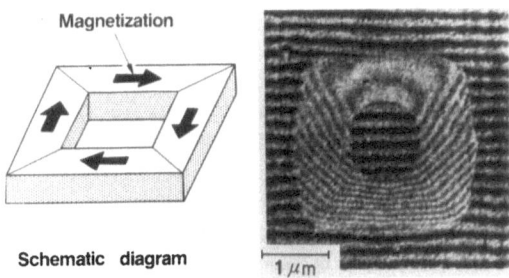

Fig. 13.12a, b. Hologram showing the Aharonov–Bohm effect. **a** Sample of troidal permalloy; **b** Hologram

theory, vector potentials are extended to gauge fields, which are considered to be fundamental physical quantities. The AB effect is direct evidence that electromagnetism follows the gauge principle. The experiment mentioned above was carried out to test the existence of the AB effect.

13.4 Summary

A new electron holography technique has been introduced, which visualizes microscopic magnetic lines of force as an interference micrograph. This method has recently been put to practical use by the development of coherent field-emission electron beams. Although the applications are still limited, this procedure allows the quantitative and direct observation of magnetic lines of force, and so further applications to practical problems can be expected in the near future. The author would like to thank Prof. Sōshin Chikazumi for his encouragement and guidance during this investigation.

References

13.1. D. Gabor: Proc. R. Soc. London A **197**, 454–487 (1949)
13.2. A. Tonomura, T. Matsuda, H. Tanabe, N. Osakabe, J. Endo, A. Fukuhara, K. Shinagawa, H. Fujiwara: Phys. Rev. **B25**, 6799–6804 (1982)
13.3. A. Tonomura, T. Matsuda, J. Endo, H. Todokoro, T. Komoda: J. Electron Microsc. **28**, 1–11 (1979)
13.4. J. Endo, T. Matsuda, A. Tonomura: Jpn. J. Appl. Phys. **18**, 2291–2294 (1979)
13.5. E.E. Huber, D.O. Smith, J.B. Goodenough: J. Appl. Phys. **29**, 294–295 (1958)

13.6. T. Matsuda, A. Tonomura, R. Suzuki, J. Endo, N. Osakabe, H. Umezaki, H. Tanabe, Y. Sugita, H. Fujiwara: J. Appl. Phys. **53**, 5444–5446 (1982)
13.7. N. Osakabe, K. Yoshida, Y. Horiuchi, T. Matsuda, H. Tanabe, T. Okuwaki, J. Endo, H. Fujiwara, A. Tonomura: Appl. Phys. Lett. **42**, 736–748 (1983)
K. Yoshida, T. Okuwaki, N. Osakabe, H. Tanabe, Y. Horiuchi, T. Matsuda, K. Shinagawa, A. Tonomura, H. Fujiwara: IEEE Trans. Mag. MAG-**19**, 1600–1604 (1983)
13.8. S. Olariu, I.I. Popescu: Rev. Mod. Phys. **57**, 339–436 (1985)
13.9. A. Tonomura, T. Matsuda, R. Suzuki, A. Fukuhara, N. Osakabe, H. Umezaki, J. Endo, K. Shinagawa, Y. Sugita, H. Fujiwara: Phys. Rev. Lett. **48**, 1443–1446 (1982)

Appendix: Notes on technical terms

A36. Swap gate and replicate gate
A37. Upper critical field

A.1 Aharonov–Bohm Effect

In 1959, Y. Aharonov and D. Bohm theoretically predicted the strange phenomenon that an electron beam is physically influenced even when it passes through
field-free regions. This is called the Aharonov–Bohm (AB) effect after their
names. Although two kinds of AB effects, magnetic and electrostatic, were
proposed, the magnetic AB effect is frequently discussed and therefore is
explained in the following.

   An electron beam is split into two coherent beams which enclose an infinite
solenoid. If the two beams overlap each other to form an interference pattern,
then a phase shift proportional to the magnetic flux inside the solenoid is
produced. The significance of this effect has recently increased, since it provides
the experimental evidence that a vector potential is a typical example of a gauge
field.

A.2 Antisymmetric exchange interaction
(Dzyaloshinski–Moriya interaction)

This is a type of magnetic interaction acting between two magnetic moments in
crystals lacking inversion symmetry. It changes the sign when the moments are
exchanged, and is represented as $D \cdot (S_i \times S_j)$. The direction of D is determined
by the crystal symmetry and its magnitude by the transfer integral and the
energy difference between the ground and excited states. The origin of the
interaction is the combined effect of the spin-orbit and exchange interactions.
The appearance of weak ferromagnetism in γ-Fe_2O_3 etc. was successfully
explained by this interaction. It plays an important role in the helical magnetic
structure in MnSi, magnetic anisotropy in disordered systems such as spin
glass, etc.

A.3 β-VTR and VHS-VTR

There are two types of consumer video tape recorders, namely β (Beta)-type and
VHS-type. They are helical-scan type VTRs with two heads and use a 1/2″ tape.
Because of the azimuth recording scheme, they have achieved 10 times higher

recording density, resulting in longer recording time, than the conventional EIAJ standard VTRs which used to be standard in consumer VTRs. In addition, the technology of color-under converted signal processing has made recording of pictures better and more stable with a simpler mechanism. β-type and VHS-type recorders have been developed by adapting the PI (phase invert) method and the PS (phase shift) method respectively, in order to reduce cross-talks in color signals recorded on a tape. In principle, these two methods are not different, but as the two VTRs have different tape cassettes and different size cylinders, they are not interchangeable.

A.4   Bloch walls

Between two magnetic domains, there is a region where the spin changes direction gradually from the direction of one of the domains to the other. This region is called a magnetic domain wall. A Bloch wall is a kind of domain wall in which the spin rotates without changing the component of the spin direction normal to the wall, so as not to produce magnetic poles in the wall. The thickness of the wall is determined by the balance between the exchange energy which favours the gradual spin rotation and the anisotropy energy which increases as the spin direction deviates from the easy axes.

A.5   Coherent potential approximation (CPA)

In transition metal alloys, the potential for valence electrons has a random form because of the randomness of the distribution of the two different atoms. In order to calculate the energy and the density of states of electrons, it is convenient to consider an effective periodic potential V_c. V_c depends on the electron energy, and it should be determined self-consistently. It is represented as a complex quantity because Bloch electrons are attenuated by the random potential. The CPA was first introduced by Soven in 1967, and later extensively developed by Kanamori et al. The standard procedure of CPA is to first solve the potential between the two atoms by the Hartree–Fock approximation and then to introduce V_c to obtain the one site Green function and the density of states.

CPA is a good approximation when the potential of the two atoms is not so different. If the difference is large, local effects become significant. For example, in a Ni–Mn alloy, the Mn moments are aligned parallel to the Ni moments for small Mn concentrations, but as the Mn concentration is increased, the moments of the Mn atoms occupying the nearest neighbor sites tend to have opposite directions. The direction of the Mn moment is determined by the number of Mn atoms in the nearest neighbor sites. Such phenomena depending on the environment of each site are called "environment effects".

CPA is considered a powerful means of calculating the electronic states not only in magnetic materials but also in various random systems.

A.6. ΔE effect

When a tensile stress is applied to a ferromagnet, an extra-expansion is caused due to the reorientation of magnetic domains and the rotation of the spontaneous magnetization. This is the inverse effect of magnetostriction and independent of the sign of the saturation magnetostriction. As a result, the Young's modulus E is decreased by ΔE compared to the hypothetical non-magnetic state of the same material. This effect is the ΔE effect. $\Delta E/E$ is larger for larger saturation magnetizations λ_s, as is confirmed for Fe–Ni alloys. Using the ΔE effect, materials having a small temperature coefficient of Young's modulus have been developed. Elinvar (Fe 47%, Ni 36%, Cr 12%) or Vibralloy (Fe 46%, Ni 35%, Mo 16%) are typical examples which have temperature coefficients of 7×10^{-6}, and less than 10^{-6}, respectively.

A.7   Exchange enhancement

In the itinerant electron model, Pauli's paramagnetic susceptibility is represented by $\chi_p = \chi_0/(1 - \alpha)$, where $\alpha = U_0 \rho(E_F)$ (U_0 stands for the Coulomb interaction and $\rho(E_F)$ is the density of states at the Fermi level). For $\alpha < 1$, the susceptibility is enhanced by a facter of $1/(1 - \alpha)$. For $\alpha \to 1$, χ_p diverges, and for $\alpha > 1$, the system undergoes a transition to a ferromagnetic state. This effect plays an important role, for example, in the strong paramagnetism in Pd where α is near 1 or in the weak itinerant ferromagnetism of $ZrZn_2$ where α is slightly larger than 1.

A.8   Exchange stiffness constant and spin wave constant

In a ferromagnet, if the magnetic moments of adjacent neighbors align with finite angles between them, the energy E_{ex} increases because the exchange interaction $- 2J\, S_i \cdot S_j$ acts to align the magnetic moments. E_{ex} can be expressed as $E_{ex} = A((\partial \alpha/\partial x)^2 + (\partial \alpha/\partial y)^2 + (\partial \alpha/\partial z)^2)$ by assuming that the direction $\alpha(r)$ of the magnetic moment varies continuously with distance r. The coefficient A is called the exchange stiffness constant and is expressed as $A = JS^2/a$ in a simple cubic lattice. This A is frequently used in discussions of the shape and motion of magnetic domain walls. The spin wave constant D is defined from the spin wave dispersion relation of a ferromagnet $\hbar\omega_q = Dq^2$. D is expressed as $D = 2JSa^2$ in a simple cubic crystal.

A.9 Faraday rotation, and the magnetic Kerr effect

When linearly polarized light is incident on a substance along the magnetic field H, the polarization plane of the light is rotated as the light proceeds into the substance, and the total rotation angle is expressed by $\theta = V\,dH$, where d is the thickness of the sample. The coefficient V is called the Verdet constant. This effect is called Faraday rotation and is caused by the difference in the refractive index between the left and right circular polarized light which constitute linearly polarized light. The difference in the refractive index arises from the different electron transition for the two different circular polarizations due to the magnetic field. In magnetic substances, the rotation angle depends on the magnetization because the internal field contributes to the rotation, so that the Faraday rotation measurement is used to obtain information of the magnetization. The polarization is also rotated when light is reflected at the surface of a substance in a magnetic field. This is the magnetic Kerr effect.

A.10 Field emission phenomena

Electrons in a solid have to get over the potential barrier at the solid surface in order to emerge outside of it. The methods to emit electrons from solids are mainly classified into two; One is to excite electrons thermally to get over the barrier by raising the temperature of the solid. The other is to make electrons tunnel through the potential by applying an extremely high electric field at the solid surface.

The former thermionic electrons have often been employed as an electron source for electron microscopes et al., while recently the latter, field-emission, electron beam has been developed as a bright beam. The cathode of field-emission electrons consists of a pointed needle with a tip radius of 0.1 μm, since a strong electric field is required at the surface. Although it has to be operated in an ultra high vacuum atmosphere, it has ideal features for a coherent electron beam. This is because the beam has a point source as small as 100 Å, and the electron velocities are uniform due to the room temperature operation.

A.11   First order phase transition

A first order phase transition is the transition in which the order parameter is changed abruptly from a finite value to zero at a certain temperature. This is in contrast to a second order transition where the order parameter (e.g. the spontaneous magnetization in a ferromagnet) vanishes continuously and the state is transformed to another state (e.g. paramagnetic state). In the first order transition, there is a latent heat at the transition temperature and various physical quantities exhibit thermal hysteresis. This is because there are two energy minima in the free energy curve and the transition occurs with a jump of the potential between these two minima. In the case of a first order transition in magnets, an additional factor other than magnetic interaction (e.g. magneto-striction) is necessary to stabilize the ordered state.

A.12   Fluctuation dissipation theory

This theory deals with the dynamical response against an external field with the dissipation of energy, such as current against electric field (conductivity), or motion of particles against external force. It predicts that the response is determined by the fluctuation of the system in the absence of the external field. The most typical example is Einstein's relation for the Brownian motion $\mu = D/kT$. This shows that the mobility μ is obtained from the diffusion constant D which is derived from the fluctuation of particles. In (3.5), which represents the response of the magnetic material against the magnetic field (susceptibility) for neutron scattering, the imaginary part (the dissipative part) is determined by the spin fluctuation $\langle S_q^\alpha S_{-q}^\alpha \rangle$ in the equilibrium state.

A.13   Gauge field theory

Forces in the universe are reduced to four fundamental forces, electromagnetic, weak, strong and gravitational forces. They have been regarded, for a long time, as being of completely different origin. However, it has become possible to describe these forces in a unified manner by the theory of gauge fields. This theory is based on the gauge invariance (symmetry) which is explicit in electro-magnetism. There, vector potentials are extended to gauge fields, which are regarded as the most fundamental physical entity. The Aharonov–Bohm effect is the experimental evidence which indicates both the gauge invariance and the significance of vector potentials. The existence of the AB effect, confirmed with electron holography techniques, removed the problem of leakage field effects which had existed in former experiments.

A.14 Generalized susceptibility

Usually, the magnetic susceptibility is defined by $\chi = M/H$ when the magnetization M is induced in the presence of a magnetic field H. When magnetization $M(r')$ is induced at a point r' in the presence of a magnetic field $H(r)$ which varies as a function of position, the susceptibility is defined by the relation $M(r') = \sum_r \chi(r - r')H(r)$. By a Fourier transformation, a relation $M(Q) = \sum M(r') \exp(iQ \cdot r') = \chi(Q)H(Q)$ holds. $\chi(Q)$ in this equation is called the generalized susceptibility. $\chi(Q = 0)$ is the ordinary susceptibility, which in ferromagnetic substances, diverges at the Curie temperature. For an alternating field which changes sign at neighboring sites, $Q = \pi/a$ (a: atomic distance), and in antiferromagnetic substances, $\chi(\pi/a)$ diverges at the transition temperature. Thus, the magnetism of a substance is determined by the Q dependence of $\chi(Q)$.

A.15 Helimagnetic materials

Helimagnetic materials are a class of ordered magnetic materials whose period of the spin arrangement is different from the crystal periodicity. The spin structure $S_j(r)$ is generally represented by (3.9). Among helimagnets, there are the screw structure, the cycloid structure, transverse spin density waves, longitudinal spin density waves, etc., as described in Sect. 3.3. In neutron scattering, helimagnets commonly exhibit satellites, which are useful for distinguishing them from ferro- or antiferromagnetic substances. In localized spin systems, helimagnets occur when the exchange interaction with the nearest neighbor and that of the next nearest neighbor compete. On the other hand, in itinerant electron systems, they occur when the Fermi surface has a peculiar shape.

A.16 Hubbard model

The electronic states in crystals are generally described by the transfer integral t_{ij} between the two nearest neighbor atomic sites i and j, the Coulomb repulsive energy U for two electrons with opposite spins when they occupy the same site, and the intra-atomic exchange energy J_{lm} between two electrons which occupy orbits l and m in the same atom. The Hamiltonian for the case where the former two energies are dominant is given by (4.4). This model is called the Hubbard model. It is a very useful model for magnetism and electrical conductivity, particularly to derive whether the system becomes a metal or an insulator, or what kind of magnetism will be present in the system.

A.17   Induced ferromagnetism

When sufficiently high magnetic fields are applied to antiferromagnetic or helimagnetic materials, all the spins are directed towards the field direction. Such a state is called the induced ferromagnetic state. In antiferromagnetic substances with a strong anisotropy ($FeCl_2$, for example), the transition from the antiferromagnetic phase to the induced ferromagnetic phase takes place as a first order transition. This is called a metamagnetic transition. At low temperatures, these induced ferromagnetic states are no different from ordinary ferromagnetic states, but as the temperature is increased, the state undergoes a transition to a paramagnetic state without any transition point, so that it is sometimes simply called a paramagnetic state on the phase diagram. The state of MnSi in magnetic fields higher than 6 kG is a typical induced ferromagnetic state.

A.18   Interference microscopy

Although the intensity distribution of a light or electron beam can usually be observed in microscopy, there are no means to directly detect the phase distribution. However, there are some cases where specimens do not give rise to an intensity change, but only to a phase change. Interference microscopy is employed for observing such specimens, and displays the phase contour map of the transmitted (or reflected) beam by overlapping a reference plane wave onto it. In light interference microscopy the refractive index or thickness distribution of a specimen can be observed. In electron interference microscopy, the distributions of magnetic fields can be measured as well, since an electron has an electric charge e. In this case, the contour lines can be interpreted as in-plane magnetic lines of force.

A.19 Itinerant weak ferromagnetism

Itinerant weak ferromagnetic substances are a class of metallic magnetic substances which barely satisfies Stoner's condition for ferromagnetism, having small spontaneous magnetization M_0 and a low Curie temperature T_C. $ZrZn_2$ ($M_0 = 0.12\mu_B$, $T_C = 22$ K), Sc_3In ($M_0 = 0.057\mu_B$, $T_C = 7.5$ K) and Ni_3Al are typical examples. Their temperature dependence of the spontaneous magnetization obeys a T^2 law, and they exhibit a large magnetovolume effect (Invar effect). These properties are explained by the Stoner theory. However, the theory does not explain the fact that the susceptibility follows the Curie–Weiss theory above the Curie temperature. The spin fluctuation theory by Moriya et al. can explain various properties of these substances at finite temperatures including the temperature dependence of the susceptibility.

A.20 Localized spin model and itinerant electron model

The localized spin model is one of the models for understanding the magnetism of matter. In this model, electrons which carry the magnetic moment localize at atomic positions and the magnetic structure is determined by the interactions between the localized magnetic moments (interatomic interaction). Usually, the magnetism of insulators such as oxides is well explained by this model. As the Heisenberg type interatomic interaction, $- JS_i \cdot S_j$, is used in this model, the model is also known as the Heisenberg model. In the case where electrons which carry the magnetic moment are free to move in the crystal, as in the case of transition metals, the magnetic properties are thought to be determined by the band structure, i.e. the motion of electrons. This model is called the itinerant electron or band model. In this model, electron spins are aligned by the intra-atomic interaction which acts only at a single atomic site. The magnetic properties of most of the transition metals and alloys may be understood as an intermediate situation between these two extreme models.

A.21 Lorentz microscopy

Lorentz microscopy is high resolution electron microscopy technique for observing magnetic domain structures in ferromagnetic thin films. The resolving distance in optical microscopes is limited to 0.2 µm, and Lorentz microscopy can be used especially for the purpose of observing detailed magnetic structures finer than the wavelengths of light. An in-focus electron micrograph of a ferromagnetic thin film provides no information about the magnetic domain structure. When the image is defocused in the microscope, black and white contrast appears along magnetic domain walls. The reason for this is as follows: Electrons having passed through a single magnetic domain are slightly deflected in the direction perpendicular to the magnetization there. Consequently, the enhancement or deficiency in the electron intensity can be observed along a boundary between two magnetic domains (domain wall) at the defocused plane.

A.22 Magnetic after effect

When ferromagnetic substances are subjected to a change of magnetic fields, sometimes the response of the magnetization exhibits a delay. Such phenomena are called magnetic after effects, except for the cases where the delay is caused by eddy currents or metallurgical phenomena. A typical example of a magnetic after effect is that of Fe. It is caused by the relaxation of the position of the interstitially located C or N atoms which depends on the spontaneous magnetization. Such an effect is called a diffusion magnetic after effect. Magnetic after effects due to the thermal fluctuation of spins is called a thermal fluctuation magnetic after effect. This effect causes problems in fine particles which have only mono-domains, and in magnetic tapes. The decrease of permeability in high permeability materials is also a kind of magnetic after effect.

A.23 Metamagnetism

It is well known that when high magnetic fields are applied to an antiferromagnetic substances, a spin-flip transition occurs at some critical field H_c, and above H_c, the spin axes are directed normal to the field. When the anisotropy of the system is large so that the system has an Ising character, the spins are not directed perpendicular to the field. In such a case, at a crical field corresponding to H_c, the spin system undergoes a transition to a state where all the spin directions are parallel to the field, and a large jump of the magnetization occurs. Such a transition is called a metamagnetic transition. It was first discovered in $FeCl_2$ by the MIT group. A two-stage metamagnetic transition was found in $CoCl_2 \cdot 2H_2O$ by Haseda et al. and since then a variety of phenomena related to metamagnetism have been found.

A.24   Néel wall

In ferromagnetic thin films, the energy of Bloch walls becomes large because of the magnetostatic energy arising from the magnetic poles appearing on the surfaces. In 1955, Néel considered that a more favorable spin arrangement in the domain wall is such that the spin rotates always in the plane perpendicular to the domain walls (parallel to the film surfaces). This type of domain wall is called a Néel wall. It can be shown from the calculation of the thickness dependence of the domain wall energy, Bloch walls are more stable for thick films, and Néel walls for thin films. For intermediate thickness for which the energies of the two types of walls become comparable, cross tie-like domain walls are observed (Fig. 13.9). There is an attractive interaction between the two Néel walls because of the magnetic poles distributed on the film surfaces, so that sometimes double structures are formed in the case of Neel walls.

A.25   Quasi-dipole interaction

There are two types of interactions acting between two spins S_i and S_j, the exchange interaction $E_{ex} = -2JS_i.S_j$, and the magnetic dipole interaction $E_d = C(2S_{iz}S_{jz} - S_{ix}S_{jx} - S_{iy}S_{jy})$. Usually the former interaction depends only on the mutual spin orientation between S_i and S_j, and not on the spin direction relative to the crystal axis, whereas the latter depends on both. When the spin-orbit interaction energy $E_{LS} = \lambda L \cdot S$ is significant in the atoms, the spin orientation affects the direction of the orbit. This effect gives rise to an anisotropy in the exchange interaction via the change of overlap of the electron wavefunction. The resultant anisotropic exchange energy is represented as $E_{ex}^A = -2[J_1 S_{iz}S_{jz} + J_2(S_{ix}S_{jx} + S_{iy}S_{jy})]$. This form resembles the sum $E_{ex} + E_d$, so that such an interaction is called the quasi-dipole interaction.

A.26   Random phase approximation (RPA)

The RPA is an approximation for treating the problem of electron correlations in many electron systems. In the case of spin dynamics, the fluctuation of spin S_q is assumed to be influenced only by the surrounding spin fluctuations with a wave vector component q and not by any other wave vector component. Thus in RPA, when the magnetization is induced by an external field $H(r)$, the effect of the interaction of the surroundings acting on the q component M_q of the magnetization is expressed as an effective field in a form $H_q^{int} = IM_q$. Combining this field with a relation $M_q = \chi_0(H_q + H_q^{int})$, the generalized susceptibility is obtained as $\chi(q) = M_q/H_q = \chi_0/(1 - I\chi_0)$. This is equivalent to the molecular field approximation in the case of $q = 0$, so that it is also called dynamical molecular field approximation. In contrast to RPA, the spin fluctuation theory is a theory which takes into account the surrounding spin fluctuations more accurately.

A.27 The rigid band model and the Slater–Pauling curve

In the rigid band model, the band structure is assumed to be always constant and only the number of electrons in the band is changed by changing the concentration of the alloy. This model was proposed to explain the Slater–Pauling curve (Fig. 1.1) which implies that the spontaneous magnetization of transition metal alloys can be explained by the number of outer core electrons. The rigid band model is known to be useful if the constituents of the alloy are located near to each other in the periodic table. From recently calculated results on the density of states of electrons in alloys, it has been confirmed that the density of states in different alloys is nearly the same if they have the same crystal structure.

A.28 s–d interaction, and s–f interaction

In transition metals, $3d$ electrons are responsible for the magnetism. The spin-dependent interaction between such $3d$ electrons and s electrons in the conduction or valence bands is generally called the s–d interaction. In rare-earth metals, there is a spin dependent s–f interaction between magnetic $4f$ electrons and s conduction electrons. When transition elements or rare earth elements are doped into metals such as Cu or Au, there is an indirect exchange interaction between the impurity d or f spins via the s–d or s–f interaction with conduction electrons. This indirect interaction is called the RKKY interaction and its energy is given by,

$$E_{sd}(R_{12}) = -C\left[\frac{2k_f R_{12}\cos(2k_f R_{12}) - \sin(2k_f R_{12})}{(2k_f R_{12})^4}\right](S_1 \cdot S_2)$$

As the distance between the two spins is increased, the interaction energy oscillates between a ferromagnetic and an antiferromagnetic phase, and is damped in inverse proportion to the cube of the distance. This interaction determines the magnetic states in alloys and intermetallic compounds, and gives rise to the spin glass state at low temperatures. In addition, in the case of very dilute alloys, the inelastic scattering of conduction electrons by the s–d or s–f interaction gives rise to the Kondo effect.

A.29 Shift register memory

Computer memories for which any address can be accessed directly by leads are called random access memory (RAM). Memories using MOS and bipolar transistors belong to this category. On the other hand, in memories such as magnetic bubble memories, charge coupled device (CCD) memories and magnetic disk memories, recorded information is always circulating within the memory device at a constant speed, and to read information in any arbitrary address, it is necessary to wait for it until it comes to an accessible exit. Writing and reading are done in a serial way at a constant speed. This kind of memory is called a shift register memory or a serial memory.

A.30 Spin density wave (SDW)

In some crystals, the magnetic spin moment on each atomic site $S(r)$ shows a sinusoidal variation like $S(r) = S_0 \cos(qr)$. Such a wave is created by the correlation of itinerant electrons and is called a spin density wave. A typical example is the ordered antiferromagnetic state in metallic Cr. The periodicity of the spin density wave is not necessarily the same as that of the crystal periodicity. When they are the same, it is called commensurate SDW, and when different, an incommensurate SDW. There are also longitudinal SDWs and transverse SDWs depending on the direction of magnetic moment with respect to the wave propagation. In the case of Cr, there is a longitudinal SDW 122 K and a transverse SDW in the temperature range $120 \, \text{K} < T < T_N = 312 \, \text{K}$.

A.31 Spin fluctuation theory

This theory was first developed by Moriya et al. to take account of the interaction among spin fluctuations beyond the random phase approximation (RPA), and it can explain the magnetism of metals at finite temperatures. Particularly in the case of itinerant weak ferromagnetism where only the spin fluctuations at small q are important, the interaction can be represented by a rather simple form, and the generalized susceptibility is given in a form $\chi(q) = \chi_0/(1 - I\chi_0 + \lambda)$, as shown in (3.22). Here, λ is a new term which did not exist within the framework of the RPA. λ is proportional to temperature because it is proportional to thermally induced fluctuations. This explains why the static susceptibility $\chi(0)$ obeys a Curie–Weiss law above T_C, which is not explained by the RPA. Moreover, this theory predicts that the magnetic moment on an atomic site $\langle M_L \rangle$ increases with increasing temperature in itinerant weak ferromagnetic materials (Fig. 3.9). This dependence was confirmed in MnSi etc. Thus this theory presents a unified picture of magnetism incorporating the localized spin model and the itinerant electron model.

A.32   Spin glasses, cluster glasses and mictomagnetism

In magnets which are known as spin glasses, the temperature variation of the magnetization in a weak magnetic field (less than 10 Oe) shows a sharp peak at a certain transition temperature T_g and tends to decrease below this temperature. Below T_g, there is magnetic ordering but no long range order. Time dependent residual magnetization is observed with the field cooling process and, therefore, the ordered state is considered to be a metastable state, as in a glass. These phenomena were firstly found in dilute $Au_{1-x}Fe_x$ alloys. It is still controversial whether or not T_g is a cooperative transition temperature or due to the fact that the relaxation time is simply longer than the observation time. For a dilute alloy, T_g may be the former, because the relaxation time varies by a factor more than 10^3 around T_g. When ferromagnets are diluted with non-magnetic atoms Au in a AuFe alloy) or with atoms which has the effect of destroying the ferromagnetism (Cr in an FeCr alloy), spin glass behavior also appears just above the composition where the ferromagnetism disappears (percolation threshold).

A magnetic cluster behaves as a superparamagnet above T_g and the magnetic moment of the cluster exhibits a disordered magnetic arrangement below T_g. These spin glasses are known as the cluster glasses.

In an alloy with a composition on the ferromagnetic side of the percolation threshold, a cluster glass exists around the ferromagnetic network and exhibits a complex coexistent state. This is known as mictomagnetism.

A.33   Spin waves (magnons)

In ordered magnetic materials, the deviation of the saturation magnetization (perfectly ordered state) propagates as a wave due to the exchange interaction $-2\sum J_{ij} S_i \cdot S_j$. This wave is the spin wave, and the quantized quasi-particle of the spin wave is the magnon. In simple ferromagnetic materials, the dispersion of the spin wave is given by $\hbar\omega_q = 2S_0(J(0) - J(q))$ (3.16), where $J(q) = \sum J_{ij}\exp(iq(r_i - r_j))$ is the Fourier component of J_{ij}. When q is small, it is reduced to $\hbar\omega_q = Dq^2$, where D is the spin wave constant. The magnetization decrease due to the spin wave excitation is proportional to $T^{3/2}$, and the coefficient of this proportionally is proportional to $D^{-3/2}$, so that D can be obtained from the temperature dependence of the magnetization. As the temperature is increased, the interaction between different spin waves becomes important (2 magnon theory), and the dispersion relation $\hbar\omega_q(T)$ is given by (3.17), being temperature dependent. Therefore $D(T)$ also becomes temperature dependent. The dispersion of spin waves can be directly measured by neutron scattering, and some examples of the measurements are described in Sects. 3.5–3.7. Spin waves in antiferromagnetic materials can be investigated by far-infrared optical absorption. There are two kinds of absorption bands, 2 magnon absorption and the magnon side-band. The former is an excitation of two spin waves (2 magnons) with wave vectors k and $-k$ in the two sublattices of the antiferromagnet. The latter is an excitation of an electron to an excited state (exciton excitation) simultaneously with one magnon, and it gives information on the dispersion at the Brillouin zone boundary.

A.34 Stoner theory and Stoner excitations

The theory which was proposed by Stoner in 1938 is the classical theory for explaining the ferromagnetism of a metal at finite temperatures. The theory is based on the assumption that electrons which carry the magnetic moment move around the crystal (band model). The ferromagnetism based on this model is thought to arise from the difference of the number of d-electrons in the minority and majority spin bands. The appearance of ferromagnetism occurs with the condition $ID(E_F) > 1$ (Stoner's condition), where I is the intra-atomic interaction and $D(E_F)$ is the density of states at the Fermi energy. The decrease of spontaneous magnetization at finite temperatures is caused by the excitation of electrons from the majority to the minority spin band. This excitation is called a Stoner excitation. The region of the energy and the momentum space where the Stoner excitation can occur is called the Stoner continuum (Fig. 4.13). The temperature dependence of the spontaneous magnetization due to this excitation is predicted to be proportional to T^2. According to the Stoner theory, the magnetic moment at an atomic position vanishes above the Curie temperature, which is the same behavior as the spontaneous magnetization. In most metal magnets, however, the temperature dependence of the spontaneous magnetization obeys a $T^{3/2}$ law which is predicted from spin wave theory, and the magnetic moment at the atomic positions does not disappear above the Curie temperature (Sect. 4.4).

A.35   Superexchange interactions

In ionic crystals, there exists an exchange interaction between spins S_i and S_j on magnetic ions through other kinds of ions (anions) in between. In the case of MnO, for example, the ground state $Mn^{2+}-O^{-2}-Mn^{2+}$ has a component of the virtually excited state $Mn^{2+}-O^{-}-Mn^{+}$ giving rise to an exchange interaction in a form $-2J(S_i \cdot S_j)$. The sign of J is determined by the number of d electrons in the magnetic ion, the energy splitting among the orbital states due to the crystal field and the wave-function overlap between the magnetic ions and anions. The magnitude of J is determined by the energy difference between the ground and excited states, and the transfer integral between the magnetic ion and the anions. The magnetic structures in many transition metal oxides and halides are explained by this superexchange interaction.

A.36   Swap gates, and replicate gates

A magnetic bubble memory device consists of two major lines for the reading and writing of information and many minor loops for storing information. A swap gate couples minor loops with a major line for writing, whereas a replicate gate couples minor loops with a major line for reading. Each minor loop in a memory is connected to one swap gate and one replicate gate. In the writing process in the memory, first a series of bubbles are generated by a bubble generator on a writing major line, and these bubbles are transferred and stored in minor loops through a swap gate. At the same time, information to be erased which has been stored on minor loops is taken out to the writing major line. In the reading process, on the other hand, a bubble is copied when it is transferred to a replicate gate. One of the bubbles is left on a minor loop and the other is taken out to a major line and led to a bubble detector. Bubbles corresponding to information are always stored on minor loops due to the existence of both swap gates and replicate gates, thus the non-volatility of the system is assured.

A.37   Upper critical field

Superconductors exhibit the Meissner effect, by which they do not allow the penetration of a magnetic flux. When external magnetic fields H are applied, the energy of the superconductor is increased by $1/2\mu_0 H^2$, because of the exclusion of the magnetic flux, and above a critical field H_c, they undergo a transition to the normal state. Such superconductors are called type I superconductors. On the other hand, type II superconductors are those which partly allow penetration of some magnetic flux above a critical field H_{c1} lower than H_c. The difference between type I and type II superconductors arises from the ratio of the penetration depth λ to the coherence length ξ. When $\kappa = \lambda/\xi$ is large, the system becomes a type II superconductor because then the gain of diamagnetic energy by allowing flux penetration can become larger than the loss of condensation energy near the penetration region, and when κ is small, the system becomes a type I superconductors. The boundary is about $\kappa = 1/\sqrt{2}$. In type II superconductors, the diamagnetic susceptibility decreases with increasing field above H_{c1}, from the absolute diamagnetic value μ_0, and finally when it becomes zero at a critical field H_{c2}, there is a transition to the normal state. This critical field H_{c2} is called the upper critical field. Type II superconductors with high H_{c2} are very important for practical use in superconducting magnets. H_{c2} is a decreasing function of temperature, external magnetic field and the current flowing through the sample.

Subject Index

Springer Series in Solid-State Sciences

Editors: M. Cardona   P. Fulde   K. von Klitzing   H.-J. Queisser

Springer Series in Solid-State Sciences

Editors: M. Cardona   P. Fulde   K. von Klitzing   H.-J. Queisser